"十三五"国家重点出版物出版规划项目
材料科学研究与工程技术图书
石墨深加工技术与石墨烯材料系列

石墨烯的性能、制备、表征及器件
GRAPHENE
PROPERTIES, PREPARATION, CHARACTERISATION AND DEVICES

[奥] Viera Skákalová
[新西兰] Alan B. Kaiser   著

王永亮　王继华　译

哈尔滨工业大学出版社
HARBIN INSTITUTE OF TECHNOLOGY PRESS

内容简介

本书共分3部分。第1部分介绍石墨烯的制备，分别讨论了石墨烯在碳化硅中的外延生长、石墨烯薄膜增长的化学气相沉积法、化学法制备石墨烯和电化学剥离制备石墨烯；第2部分是石墨烯的表征，分别介绍用透射电子显微镜法对石墨烯原子结构的研究、石墨烯的扫描隧道显微镜分析、石墨烯的拉曼光谱研究以及低维碳材料的光电子能谱分析；第3部分是石墨烯及石墨烯器件的电子传输性能，重点介绍石墨烯及石墨烯器件的电子传输性能，分别是石墨烯的电子传输、双层石墨烯的电子传输、吸附物对石墨烯电子传输的影响、石墨烯中的单电荷传输以及石墨烯自旋电子学和石墨烯的纳米电机学。

本书可作为生物材料、吸附材料等相关领域研究者的参考资料，也可作为上述专业的教材使用。

图书在版编目(CIP)数据

石墨烯的性能、制备、表征及器件/(奥)维尔拉·斯卡卡洛娃(Viera Skákalová)，(新西兰)艾伦·凯撒(Alan B. Kaiser)著；王永亮，王继华译. —哈尔滨：哈尔滨工业大学出版社,2019.8

ISBN 978－7－5603－7330－0

Ⅰ.①石… Ⅱ.①维… ②艾… ③王… ④王… Ⅲ.①石墨-纳米材料 Ⅳ.①TB383

中国版本图书馆 CIP 数据核字(2018)第 084944 号

材料科学与工程图书工作室

策划编辑	杨 桦　许雅莹　张秀华
责任编辑	范业婷　刘 瑶　庞 雪
封面设计	卞秉利
出版发行	哈尔滨工业大学出版社
社　　址	哈尔滨市南岗区复华四道街 10 号　邮编 150006
传　　真	0451－86414749
网　　址	http://hitpress.hit.edu.cn
印　　刷	哈尔滨市石桥印务有限公司
开　　本	787mm×960mm　1/16　印张 23　字数 412 千字
版　　次	2019 年 8 月第 1 版　2019 年 8 月第 1 次印刷
书　　号	ISBN 978－7－5603－7330－0
定　　价	88.00 元

(如因印装质量问题影响阅读，我社负责调换)

黑版贸审字 08-2019-122 号

Elsevier (Singapore) Pte Ltd.
3 Killiney Road, #08-01 Winsland House I, Singapore 239519
Tel: (65) 6349-0200; Fax: (65) 6733-1817

Graphene:Properties,Preparation,Characterisation and Devices,1 Edition
Viera Skákalová,Alan B. Kaiser
ISBN:9780857095084
Copyright ©2014 Woodhead Publishing Limited. All rights reserved.

This translation of Graphene:Properties,Preparation,Characterisation and Devices,1 Edition by Viera Skákalová and Alan B. Kaiser was undertaken by Harbin Institute of Technology Press and is published by arrangement with Elsevier (Singapore) Pte Ltd.

Graphene:Properties,Preparation,Characterisation and Devices,1 Edition by Viera Skákalová and Alan B. Kaiser 由哈尔滨工业大学出版社有限公司进行翻译,并根据哈尔滨工业大学出版社有限公司与爱思唯尔(新加坡)私人有限公司的协议约定出版。
《石墨烯的性能、制备、表征及器件》(第 1 版)(王永亮 王继华 译)
ISBN: 9787560373300
Copyright ©2019 by Elsevier (Singapore) Pte Ltd.

All rights reserved. No part of this publication may be reproduced or transmitted in any form or by any means, electronic or mechanical, including photocopying, recording, or any information storage and retrieval system, without permission in writing from Elsevier (Singapore) Pte Ltd. Details on how to seek permission, further information about the Elsevier's permissions policies and arrangements with organizations such as the Copyright Clearance Center and the Copyright Licensing Agency, can be found at our website: www.elsevier.com/permissions.

This book and the individual contributions contained in it are protected under copyright by Elsevier (Singapore) Pte Ltd. and Harbin Institute of Technology Press(other than as may be noted herein).

This edition is printed in China by Harbin Institute of Technology Press under special arrangement with Elsevier (Singapore) Pte Ltd. This edition is authorized for sale in the People's Republic of China only, excluding Hong Kong SAR, Macau SAR and Taiwan. Unauthorized export of this edition is a violation of the contract.

本书简体中文版由 Elsevier(Singapore) Pte Ltd. 授权哈尔滨工业大学出版社有限公司在中国人民共和国境内(不包括香港特别行政区、澳门特别行政区以及台湾地区)出版与发行。未经许可之出口,视为违反著作权法,将受民事及刑事法律之制裁。
本书封底贴有 Elsevier 防伪标签,无标签者不得销售。

注意

本书涉及领域的知识和实践标准在不断变化。新的研究和经验拓展我们的理解,因此须对研究方法、专业实践或医疗方法作出调整。从业者和研究人员必须始终依靠自身经验和知识来评估和使用本书中提到的所有信息、方法、化合物或本书中描述的实验。在使用这些信息或方法时,他们应注意自身和他人的安全,包括注意他们负有专业责任的当事人的安全。在法律允许的最大范围内,爱思唯尔、译文的原文作者、原文编辑及原文内容提供者均不对因产品责任、疏忽或其他人身或财产伤害及/或损失承担责任,亦不对由于使用或操作文中提到的方法、产品、说明或思想而导致的人身或财产伤害及/或损失承担责任。

译者前言

石墨烯是一种新型纳米材料，有"新材料之王"之称。随着材料科学的不断进步和发展，人们对石墨烯的关注度日益高涨。对石墨烯的制备、结构及其性能认识和了解的需求也迫在眉睫。由维尔拉·斯卡卡洛娃和艾伦·凯撒著的《石墨烯的性能、制备、表征及器件》一书为我们更好地了解石墨烯提供了方便快捷的途径，具有一定的指导和参考价值。

《石墨烯的性能、制备、表征及器件》一书共分3部分。第1部分介绍石墨烯的制备（第1~4章），分别讨论了石墨烯在碳化硅中的外延生长、石墨烯薄膜增长的化学气相沉积法、化学法制备石墨烯和电化学剥离制备石墨烯；第2部分是石墨烯的表征（第5~8章），分别介绍用透射电子显微镜法对石墨烯原子结构的研究、石墨烯的扫描隧道显微镜分析、石墨烯的拉曼光谱研究以及低维碳材料的光电子能谱分析；第3部分是石墨烯及石墨烯器件的电子传输性能（第9~14章），重点介绍石墨烯及石墨烯器件的电子传输性能，分别是石墨烯的电子传输、双层石墨烯的电子传输、吸附物对石墨烯电子传输的影响、石墨烯中的单电荷传输以及石墨烯自旋电子学和石墨烯的纳米电机学。

本书引用了大量的实验数据和参考文献，从理论和实践角度，对石墨烯的性能、制备、表征以及应用等做了详细全面的论述，并提供了很多有参考价值的实验数据处理的图表和计算公式的推导。每章都有结论，对前面的介绍和论述要点做一总结。全书条理清晰，布局严谨，具有极强的指导性和参考性。

本书的第1部分和第3部分的第9~12章由王继华翻译；第2部分和第3部分的第13章和14章由王永亮翻译。全书由王永亮和王继华统稿。

感谢哈尔滨理工大学韩志东教授对翻译工作的大力支持和帮助。
  由于译者的水平有限,在翻译过程中难免有疏漏之处,敬请指正。

王永亮 王继华

2019 年 1 月

前　　言

  2004 年,石墨烯作为一种新型材料在世界科学舞台上崭露头角,同时也引起了公众和材料界科学家们的关注。50 年前,谁会想到可以获得单层原子结构?

  这个令人兴奋的发现使人联想到 1986 年高温超导体的发现。几百名科学家试图把超导的起始温度 T_c 升至室温。Alex Müller 和 Georg Bednorz 创下最快拿到诺贝尔奖的纪录,即第二年(1987)获得了诺贝尔物理学奖。尽管高 T_c 的超导体应用于各个领域,但遗憾的是,T_c 从来没达到过室温(至少目前是这样)。

  Andrei Geim 和 Kostya Novoselov 用了 6 年时间研究石墨烯而获得诺贝尔物理学奖。其他团队在同一时间进行研究也达到相同目标,这个目标就是从石墨晶体中获得单层碳原子,即现在我们熟知的六边形单层碳原子。哥伦比亚大学的 Philip Kim 团队提出一个具有前瞻性的想法,他们认为将由石墨制成的原子力显微镜(AFM)的针尖紧压在某物质上面,那么石墨留下的某些痕迹可能就是单层石墨。

  后来,这个方法被 Andrei Geim 和 Kostya Novoselov 成功用于分离石墨烯层,近乎达到技术的极限。实际上,用胶带贴在石墨上,再将胶带撕下,就会粘下石墨烯层。有些碳层是一个原子厚度(单层),有些碳层是双层或多层。其实我们用石墨铅笔在纸上写字时,就可以得到单层石墨烯,问题不是形成石墨烯单层,而是如何去辨别它们的存在。

  用于沉积石墨烯的硅衬底上覆盖着 300 nm 的氧化层,这对石墨烯的沉积有利还是不利呢? 后来我们才知道这种硅衬底是最好的。在光学显微镜下,由于二氧化硅层的上下界面对光的有利反射,硅衬底可以有明显衬度。Andrei Geim 和 Kostya Novoselov 及其团队用此方法分辨硅衬底上石墨烯的层数。

为什么这个发现可以获得诺贝尔物理学奖？最重要的是 Andrei Geim 和 Kostya Novoselov 利用其他人的研究经验,进行了大量相关的实验,证明了石墨烯是一种令人称奇的材料。正是这些奠基性实验揭示了以前从未观察到的二维材料中的电子行为,才使他们获得了诺贝尔奖。另外,石墨烯即使在宽泛的使用温度下,仍具有优良的机械强度和电性能,这意味着石墨烯即将用于我们日常生活的很多设备之中。

关于石墨烯,理论总是先于实验。1947 年,加拿大物理学家 Philip Wallace 的论文中计算了 2D 石墨烯层中电子的电子能带结构(即层中电子的不同能态的能量)。2D 石墨烯的计算比 3D 的石墨更容易。Wallace 认为石墨烯的能带结构完全不同于普通金属或半导体,石墨烯可以被称为零态密度费米能级的金属,同样也可以被称为零带隙的半导体。

石墨烯的能带结构研究表明,可以通过场效应晶体管(FET)装置或化学、电化学方法掺杂以改变和控制其能带结构,因此石墨烯可以在电子领域实现多种应用。

我们知道,与石墨烯关系密切的其他形式的纳米碳材料在石墨烯之前就已经被发现了。结构的变化会引起材料电性能的改变,这就是为什么石墨烯家族的每个成员都有自己的特点和独特的工业应用前景。石墨烯最重要的结构变化是富勒烯(fullerenes,通过由独特的五边形碳原子环和六边形碳原子环组合而形成的微小碳球)、圆锥形的纳米角(nanohorns of conical shapes)、垂直衬底生长的纳米墙及 3D 空间的纳米泡沫。

碳纳米管备受关注,如果把石墨烯按照对边碳原子匹配方式卷曲,就形成了碳纳米管。这些中空的碳纳米管可能由直径不到一个纳米的单壁或者多壁(多壁碳纳米管)组成。跟石墨烯一样,碳纳米管也有一系列优良的特性,例如,高的强度、柔韧性、电导率和热导率,这正是纳米级电子复合材料中有待应用的性能。在这方面,碳纳米管是石墨烯最强有力的竞争对手。

前言

　　本书对石墨烯、单层石墨烯及两层和少层石墨烯的主要方面都有深度探讨。在开篇章节中,介绍了制备石墨烯的最新方法。这些方法分别是通过碳化硅晶体的热分解进行石墨烯的外延生长,大面积石墨烯在结晶物质上的化学气相沉积生长,以及用化学和电化学方法从石墨上进行层剥离。每种方法都会得到不同品质、尺寸和数量的石墨烯。为了评定这些参数,需要对合成石墨烯进行表征。

　　第2部分介绍了表征石墨烯原子和电子结构的最强有力的测试技术。对这些独特表征技术的原理和应用也进行了论述。利用透射电子显微镜观察到石墨烯小到原子级别的悬浮结构,为研究晶粒尺寸和结构缺陷以及化学修饰提供了信息。利用扫描隧道显微镜可以看到石墨烯的原位生长或者转移至导体表面的原子分辨率图像。在隧道谱模式下,可以通过改变针尖和衬底之间的偏置电压获得微分电导率,从而得到各种环境条件下石墨烯的电子结构。通过拉曼和光电子能谱,我们会得到由于结构无序、弯曲度和厚度引起电子结构变化的详细谱图。对于这些光谱学方法的重要观点将在第2部分的最后两章介绍。

　　第3部分重点介绍单层和双层石墨烯的电子传输以及应用型电子器件。除了介绍独特新颖的电子传输性能外,还分析了吸附对电子传输的影响。石墨烯压缩区域的量子(单电子)传输以及电子无序对传输性能的影响,将在不同章节进行论述。最后两章介绍了石墨烯自旋电子理论和石墨烯纳米机电系统(NEMS),在这些应用领域,石墨烯被认为是具有发展潜力的材料。

目 录

第1部分 石墨烯的制备 ... 1

第1章 碳化硅(SiC)上石墨烯的外延生长 ... 1
1.1 引言 ... 1
1.2 超高真空下单晶 SiC 的热分解 ... 2
1.3 环境压力下单晶 SiC 的热分解 ... 11
1.4 单晶 SiC 薄膜和多晶 SiC 基板的热分解 ... 14
1.5 插层制备外延石墨烯 ... 15
1.6 结论 ... 17
1.7 参考文献 ... 18

第2章 化学气相沉积(CVD)制备石墨烯 ... 24
2.1 引言 ... 24
2.2 镍衬底上的化学气相沉积 ... 26
2.3 铜表面大尺寸石墨烯 ... 27
2.4 单晶铜上的石墨烯生长 ... 30
2.5 周期性堆叠的多层石墨烯 ... 31
2.6 CVD 石墨烯的同位素标记 ... 33
2.7 结论 ... 36
2.8 参考文献 ... 36

第3章 化学法制备石墨烯 ... 45
3.1 引言 ... 45
3.2 氧化石墨烯的合成 ... 46
3.3 氧化石墨烯(GO)的还原 ... 47
3.4 氧化石墨烯的物理化学结构 ... 48
3.5 氧化石墨烯的电传输 ... 53
3.6 氧化石墨烯与还原的氧化石墨烯的应用 ... 57
3.7 结论 ... 63
3.8 参考文献 ... 63

第4章 电化学剥离制备石墨烯 ... 75
4.1 引言 ... 75

4.2　电化学剥离制备石墨烯相关基本概念 ……………………… 76
　　4.3　石墨烯和石墨烯基材料的应用 ……………………………… 84
　　4.4　结论 …………………………………………………………… 85
　　4.5　参考文献 ……………………………………………………… 85
第2部分　石墨烯的表征 …………………………………………………… 91
　第5章　透射电子显微镜与石墨烯 ……………………………………… 91
　　5.1　引言 …………………………………………………………… 91
　　5.2　石墨烯结构基础 ……………………………………………… 93
　　5.3　石墨烯的电子衍射分析 ……………………………………… 95
　　5.4　石墨烯及其缺陷的像差校正 TEM 和 STEM 分析 ………… 97
　　5.5　石墨烯电子显微分析的启示 ………………………………… 101
　　5.6　结论 …………………………………………………………… 106
　　5.7　参考文献 ……………………………………………………… 107
　第6章　石墨烯的扫描隧道显微镜分析 ………………………………… 113
　　6.1　引言 …………………………………………………………… 113
　　6.2　不同惰性衬底上沉积石墨烯片的形貌、完整性和电子结构
　　　　 ………………………………………………………………… 113
　　6.3　SiC 和金属衬底上外延生长石墨烯的形貌、完整性及电子结构
　　　　 ………………………………………………………………… 119
　　6.4　点缺陷的扫描隧道显微镜（STM）和扫描隧道谱（STS）分析
　　　　 ………………………………………………………………… 131
　　6.5　石墨烯纳米带的 STM/STS 研究 …………………………… 132
　　6.6　结论 …………………………………………………………… 134
　　6.7　参考文献 ……………………………………………………… 134
　第7章　石墨烯的拉曼光谱 ……………………………………………… 141
　　7.1　引言 …………………………………………………………… 141
　　7.2　拉曼散射原理 ………………………………………………… 141
　　7.3　石墨烯内的声子 ……………………………………………… 144
　　7.4　石墨烯的电子结构 …………………………………………… 146
　　7.5　石墨烯拉曼光谱 ……………………………………………… 148
　　7.6　结论 …………………………………………………………… 161
　　7.7　参考文献 ……………………………………………………… 162
　第8章　低维碳材料的光电子能谱 ……………………………………… 166
　　8.1　引言 …………………………………………………………… 166

8.2	光电子发射光谱学	167
8.3	碳 sp^2 杂化体系的电子性质:C 1s 芯能级	170
8.4	化学态的识别:键合环境	173
8.5	价带的电子结构	174
8.6	结论	174
8.7	参考文献	174

第3部分　石墨烯及石墨烯器件的电子传输性能　178

第9章　石墨烯的电子传输:高迁移率　178
- 9.1　引言　178
- 9.2　散射强度的衡量参数　179
- 9.3　石墨烯制备方法　182
- 9.4　石墨烯的散射来源　183
- 9.5　增加载流子迁移率的方法　188
- 9.6　高迁移率石墨烯的物理现象　195
- 9.7　结论　196
- 9.8　参考文献　197

第10章　双层石墨烯的电子传输　205
- 10.1　引言　205
- 10.2　双层石墨烯的历史发展　206
- 10.3　双层石墨烯体系中的传输性能　211
- 10.4　双层石墨烯中传输性能的多体效应　221
- 10.5　结论　232
- 10.6　参考文献　233

第11章　吸附物对石墨烯电子传输的影响　241
- 11.1　引言　241
- 11.2　吸附物与石墨烯的相互作用　242
- 11.3　转移引发的金属和分子吸附　244
- 11.4　吸附对石墨烯场效应晶体管的影响　248
- 11.5　石墨烯中聚合物残留物的去除　253
- 11.6　结论　258
- 11.7　参考文献　260

第12章　石墨烯中的单电荷传输　266
- 12.1　引言　266
- 12.2　单电荷隧穿　267

12.3 石墨烯的电性能 ………………………………………………… 269
12.4 石墨烯中的单电荷遂穿 ………………………………………… 274
12.5 石墨烯的电荷局域化 …………………………………………… 282
12.6 结论 ……………………………………………………………… 288
12.7 参考文献 ………………………………………………………… 288

第13章 石墨烯自旋电子学 …………………………………………… 297
13.1 引言 ……………………………………………………………… 297
13.2 理论基础及重要概念 …………………………………………… 298
13.3 纯自旋电流的试验获得及其性能 ……………………………… 304
13.4 结论及展望 ……………………………………………………… 310
13.5 参考文献 ………………………………………………………… 311

第14章 石墨烯的纳米电机学 ………………………………………… 316
14.1 引言 ……………………………………………………………… 316
14.2 石墨烯与硅 ……………………………………………………… 316
14.3 石墨烯的力学性能 ……………………………………………… 317
14.4 石墨烯微机电系统(MEMS)制备技术 ………………………… 320
14.5 石墨烯纳米谐振器 ……………………………………………… 322
14.6 石墨烯纳米机械传感器 ………………………………………… 329
14.7 结论及发展趋势 ………………………………………………… 330
14.8 参考文献 ………………………………………………………… 331

索引 ……………………………………………………………………… 337

第 1 部分　石墨烯的制备

第 1 章　碳化硅(SiC)上石墨烯的外延生长

本章概述了石墨烯在各种碳化硅(SiC)衬底上的外延生长、生长机理和原子尺寸表征；重点论述了借助单晶 SiC 在超高真空和环境压力下热分解而进行的外延生长石墨烯(EG)，并论述了多晶 SiC 薄膜的热分解过程和制备 EG 的插层方法。

1.1　引　　言

能否实现石墨烯在电子、光电子、化学和生物传感器等方面的技术应用，取决于高品质石墨烯的大规模制备。近几年，许多深入细致的工作都致力于研究获得单层或多层石墨烯的方法，包括用胶带[1-2]对石墨颗粒的微机械剥离，将石墨粉末[3]进行的化学剥离，对氧化石墨烯[4-7]进行的化学或物理还原，碳氢化合物在过渡金属衬底上[8-17]的化学气相沉积(如铜、镍、钌、铱和铂在金属、半导体或绝缘衬底上固态碳源的热分解)，以及在真空或大气压下商业购买 SiC 衬底的热分解等[18-19]。在 SiC 上热生长的外延石墨烯(EG)膜可以与 CMOS 的毫微光刻法兼容实现图案化(pattern)，由此可见未来石墨烯器件将有一个良好的发展前景[20-21]。尤其是高品质器件，如场效应晶体管[22]、光子探测器[23]和化学传感器[24]都已经开始运用 SiC 上的 EG。本章概述了石墨烯在各种碳化硅(SiC)衬底上的外延生长、生长机理和原子尺寸表征；重点论述了借助单晶 SiC 在超高真空和环境压力下热分解而进行的外延生长石墨烯(EG)，并论述了多晶 SiC 薄膜的热分解过程和制备 EG 的插层方法。

1.2 超高真空下单晶 SiC 的热分解

SiC 上结晶石墨层的形成通过超高真空下高温加热来实现,这是 Van Bommel 等在 1975 年发现的[25]。结晶石墨层就是后来人们所说的 EG,但在当时并未引起重视。在报道机械分离法首次分离出独立石墨烯的同时,Berger 和他的同事发现 SiC 上的 EG 有着与独立石墨烯几乎完全相同的性能,并且可以与 CMOS 器件制造的光刻工艺相匹配[18],可以通过控制退火温度和时间在 SiC 上外延生长具有可控原子层的单晶石墨烯。EG 的生长有赖于 SiC 表面的极性(即硅和碳表面),但是对于不同的 SiC 晶型(如 3C、4H 和 6H)来讲,EG 生长几乎是相同的。为了解释这个生长行为,我们先对 SiC 结构进行简要说明。

SiC 包含 1∶1 化学计量比的碳和硅。每个硅(或碳)原子通过共价键与四个最近的碳(或硅)原子以四面体配位(sp^3 杂化)。这些四面体的 Si—C 键以碳和硅交替排列的方式构成六边形双层。Si—C 双层沿着与双层面垂直的方向,可以存在多种堆垛和取向,导致 SiC 有 200 多种构型。其中有两个主要的结构:一个是立方体对称,即面心立方体(fcc);另一个是六面体对称,即密排六方体(hcp)。在这些类型中 3C-SiC,4H-SiC 和 6H-SiC 是最重要的,数字 3(4 或 6)表示每个单元核的双层数目,C(H)表示立方体(六面体)对称。这样,3C-SiC 的的双层面就是(111)晶面,4H-SiC 和 6H-SiC 双层面就是(0001)晶面。图 1.1 所示是沿着与双层面垂直的平面进行的碳化硅的三种典型多形体的堆垛顺序,具体为 3C-SiC(ABCABC…)、4H-SiC(ABACABAC…)和 6H-SiC(ABCACBABCACB…),这是与 3C-SiC 的(110)晶面及 4H-SiC 和 6H-SiC 的(1120)晶面相对应的。不同的堆垛顺序,使每种类型具有独特的物理和电学性能。Si-C 双层的终端(Si 在顶端或 C 在顶端)影响了硅升华和碳偏析过程,从而导致石墨烯的形成存在明显差异。我们先关注 Si-终端或 Si-面的 6H-SiC(0001)来阐述 SiC 上石墨烯的性能,然后再讨论 SiC 中 C-终端或 C-面石墨烯的形成。

EG 的生长可以由 SiC 的热分解实现。高温下,硅原子开始从表面蒸发。偏析的碳原子在表面形成富碳的表面层,包括从界面石墨烯(Interfacial graphene, IG)层,到单层 EG、双层 EG 和少层 EG,该生长包括一系列的表面重构过程。下面以 6H-SiC(0001)的 EG 生长为例,描述基板退火温度与表面重构的关系,所采用的表征手段为原位低能电子衍射(LEED,如

图 1.2(a)~图 1.2(d)所示)和扫描隧道显微镜(STM,如图 1.2(e)~图 1.2(h)所示)[26]。在大约 850 ℃超高真空硅流量下,6H-SiC(0001)退火富硅的 3×3 超结构在 SiC 基板上形成硅吸附层和硅四聚体(tetramers)(图 1.2(a)和图 1.2(e))。

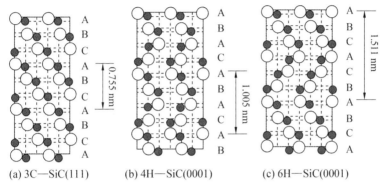

图 1.1 碳化硅的三种典型多形体的堆垛顺序

因此表面硅层的覆盖面为 $\frac{4}{9}$ ML,ML 是指单层[27-28]。在 950 ℃且没有硅流量的情况下,对基板进一步退火,会引起更多硅原子蒸发,导致富硅程度降低的重构(图 1.2(b)和图 1.2(f)),即硅原子存在于 Si-终端晶体的四面体或 T_4 位置(硅覆盖 $\frac{1}{3}$ ML),这就是 $\sqrt{3}\times\sqrt{3}\,R30°$ 重构[29-30]。将基板加热至 1 100 ℃,使硅原子从 SiC 蒸发,伴随着表面碳原子的聚集形成周期约为 1.8 nm 的蜂窝超结构,如图 1.2(c)和图 1.2(g)所示,这就是 $6\sqrt{3}\times6\sqrt{3}\,R30°$ 重构,也简称石墨烯缓冲层或 IG[31-32]。为了统一,今后我们把这个简称为 IG。将 6H-SiC 样品在 1 200~1 400 ℃进行退火,会在 IG 上面形成单晶 EG 层,厚度范围从单层到数层。SiC 上单层 EG 的 LEED 和 STM 图像分别如图 1.2(d)和图 1.2(h)所示。

SiC 上的 EG 形成可以由 C 1s 轨道的 X 射线光电子能谱(XPS)清楚地看到。基于同步加速器的 SiC 的高分辨率 C 1s XPS 光谱与退火温度的函数关系如图 1.3 所示。为了提高表面灵敏度,采用 350 eV 光子能量和 40°的发射角。在富硅的 $\sqrt{3}\times\sqrt{3}\,R30°$ 重构表面,C 1s 光谱只出现了低于费米能量(E_F)在 282.9 eV 体相 SiC 的相关峰。在被 IG 部分覆盖的表面,

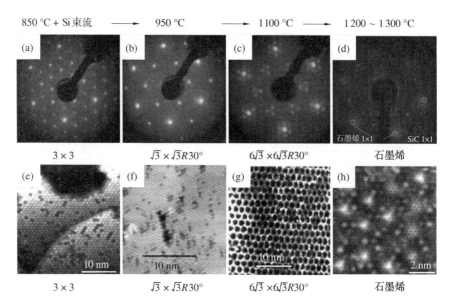

图1.2 退火引起的6H-SiC(0001)表面重构，LEED图((a)~(d))和对应的STM图像((e)~(h))[26]

285.1 eV出现IG相关峰。IG全覆盖表面的C 1s光谱以285.1 eV处的峰为主导，肩峰在283.9 eV。光谱中与石墨烯相关的284.4 eV的C 1s峰，可以在高于1 100 ℃退火的样品中得到，说明表面石墨化。C 1s光谱中该峰可以表明全覆盖EG的表面。

SiC中EG的形成可由拉曼光谱进行分析。6H-SiC(0001)中单层和双层EG、机械剥离单层石墨烯(MCG)、石墨以及6H-SiC(0001)基板的典型拉曼光谱如图1.4所示[33]。在单层或双层EG中，出现了类似于SiC基板的与SiC相关的峰，出现在约1 520 cm^{-1}和约1 730 cm^{-1}处。其他三个与EG有关的峰也可以观察到：缺陷引发的D带在1 368 cm^{-1}处，面内振动G带在1 597 cm^{-1}处，双声子2D带在2 725 cm^{-1}处。图1.4中插图清楚地表明双层EG的2D带比单层EG的2D带更宽(95 cm^{-1}相比于60 cm^{-1})，并发生在更高频率(2 736 cm^{-1}相比于2 715 cm^{-1})，图中用曲线表示趋势。由于SiC(a=0.307 nm)和石墨烯(a=0.246 nm)之间大的晶格失配而引起晶界压力，导致单层EG的G带(1 597 cm^{-1})和2D带(2 715 cm^{-1})与单层MCG(1 580 cm^{-1}的G带和2 673 cm^{-1}的2D带)相比有明显蓝移。

IG层是富碳的，被认为具有与石墨烯相似的原子结构，但IG层没有与石墨烯相同的电子性能。作为对比，通过角分辨光电子能谱(ARPES)给

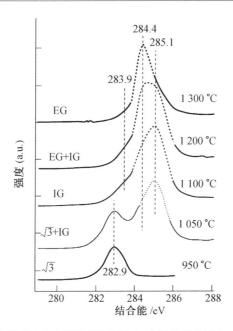

图1.3 不同退火温度SiC表面的基于同步加速器的高分辨率C 1s光谱演变[31]

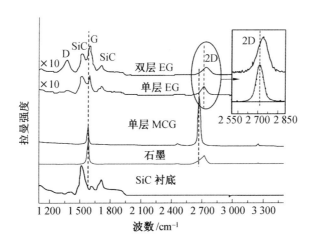

图1.4 SiC上单层EG和双层EG与石墨和单层MCG的典型拉曼光谱对比[33]

出了IG和单层EG的全价带结构,如图1.5(a)和1.5(b)所示[34]。两图均表现出低于E_F的强色散σ带,在8~23 eV,表明IG有与石墨烯相似的C—C距离。从12.5 eV到2.5 eV区间的特征由SiC价带的辐射控制。图1.5(b)清楚地表明在E_F附近的K点的石墨烯线性发散的π键。图1.5

(a)是 IG 的宽化结构。除了低于 E_F 表面与表面相关的 1.8 eV 和 0.6 eV 处无发散外,在 E_F 附近没有占有态,说明在 IG 的碳 p_z 轨道和 SiC(0001)的悬挂键之间存在部分耦合。

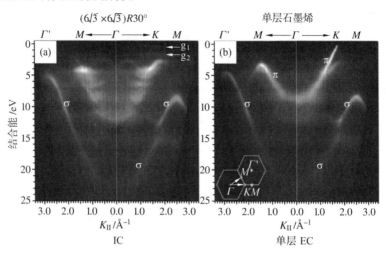

图 1.5　6H-SiC(0001)上 IG 和单层 EG 的 ARPES 图像[34]（1 Å^{-1}=0.1 nm）

SiC 上的单层、双层和多层 EG 在 E_F 周围 K 点的 π 键分散情况差别非常大,导致传输特性不同。图 1.6(a)~图 1.6(d)分别给出了 SiC 上单层到四层的 EG 狄拉克点(E_D)附近 π 键的 ARPES 结果和相应的紧束缚模拟[34]。低于 E_F 的 E_D 位置归因于电荷通过 IG 从 SiC 转移到石墨烯。EG

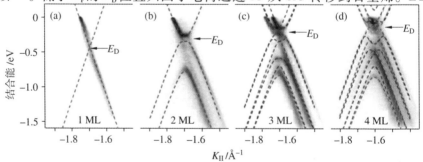

图 1.6　(a)~(d)分别是 6H-SiC(0001)上 1~4 层 E_F 附近的 π 和 π* 带

注:$h\nu$=94 eV,$T \approx 30$ K,其中对应 K 点,$K_{\parallel}$=1.703 Å^{-1},是六角形 Brillouin 区的转角。虚线部分表示计算的紧束缚带结构。π 键数量随着层数而增加,归因于层内分裂[35]

中 E_F 相对于 E_D 点的位置可以通过插层来调整,见 1.5 节所述。图 1.6(a) 是单层 EG 的价带,由于多体的相互作用表现出强烈重正化,即偏离预期的线性色散。双层 EG 的能带显示出 0.15 eV 的带隙,这是由于电荷从基板转移,如图 1.6(b) 所示。这导致了分级的电荷载流子浓度,因此两层有不同的现场库仑电位[36]。EG 三层和四层复杂的带结构分别如图 1.6(c) 和图 1.6(d) 所示。

超高真空下 SiC 上生长的 EG 通常是不均匀的。相同样本上会有多种层数的石墨烯层同时存在。这是因为生长温度对硅解吸附是充分的,但是对于石墨烯薄膜的均匀生长却不充分。低能电子显微镜图像(LEEM)如图 1.7(a) 所示,超高真空下 SiC(0001) 上生长的 EG 是一种单层到四层的混合物。不同对比度区域对应不同层数的 EG 层,分别标为 (1)~(4)。由电子反射率与入射电子束的动能之比可以直接确定石墨烯层的数量,如图 1.7(b) 所示[37-38]。反射率曲线里的局部极小(倾角)的数量代表了石墨烯层的数量,面积明显与 1~4 层区域分布相对应。

此外,石墨烯层的数量也可以由 μ-LEED 决定,如图 1.7(a) 所示。由于低能量电子的穿透深度浅,当石墨烯层的数量增加时,石墨烯点(中心或中间点)周围的六个缓冲层衍射点会逐渐消退。在三层区域内可勉强观察到这六个点,当层数超过三层就看不见了。

制备的石墨烯样品的不均匀性也可通过薄膜应变的不均匀性表示。对于 SiC 上的单层 EG,其厚度无法通过光电子发射光谱表征,应变在不足 300 nm 的范围内发生变化;超过 1 μm 时,这样的应变也可能变为均匀状态。由图 1.7(c) 和图 1.7(d) 可以看到,2D 峰位置的拉曼光谱图与由 AFM 得到的石墨烯薄膜物理形貌图相对应,说明物理形貌的变化可能引起石墨烯薄膜应变的相应改变。

正如 STM 所观察到的,SiC 上的 EG 经历了自下而上的生长模式。STM/STS 是研究 SiC(0001) 上生长 EG 的局部结构和电子性能的有效方法。图 1.8(a) 和图 1.8(b) 所示是大范围形貌及放大图像,表明单层和双层 EG 并存[39]。附加的线形轮廓表明层间高度差异仅有 (0.07±0.01) nm。图 1.8(b) 是把图 1.8(a) 的黑色正方形区域放大的 STM 图像。石墨烯结构在低隧道偏压条件下会更容易分辨,如图 1.8(d) 所示,图中上层的石墨烯从单层区域到双层区域是连续的。图 1.8(b) 中的插图表示单层区域的蜂窝结构和双层区域的三角形晶格,如图中六边形所示。双层或更厚的 EG 包括石墨的 Bernal 或 AB 堆垛,打破了石墨烯六边形晶格的对称,使每个晶胞含有两个不等碳原子。因此,对于双层 EG 而言,STM 揭示了一

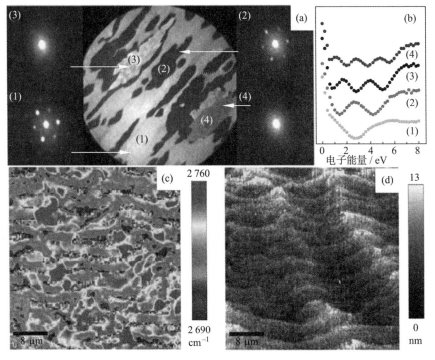

图 1.7 SiC(0001)上 EG 的不均匀性

(a)具有各种层数石墨烯层区域的 LEEM 图像。可视区域为 20 μm,电子能量为 4.2 eV。插图是四个标记区在电场 $E=53.3$ eV 时得到的 μ-LEED 图。(b)四个有代表性区域,(1)~(4)分别对应 1~4 ML 厚的石墨烯的电子反射光谱。(c)2D 峰位置的拉曼光谱。(d)不存在宏观缺陷区域的 AFM 图像。((a)和(b)转自参考文献[37],(c)和(d)转自参考文献[38])

个三重对称形貌。图 1.8(b)右上角插图中蜂窝六角晶格存在两个不等的碳原子。这个从六角蜂窝晶格向三重对称形貌的过渡被用来区分 SiC 上的单层、双层或多层 EG。STS 数据显示了石墨烯 1~4 层的局部电子特性,如图 1.8(c)所示[40]。与 ARPES 的结论一致(图 1.6),随着厚度增加,E_D 向 E_F 偏移。

石墨烯层碳原子的面密度(3.82×10^{15} cm^{-2})是 SiC 双层(1.22×10^{15} cm^{-2})的三倍。说明三个连续的 SiC 双层可以形成一个石墨烯层。图 1.9 提出一个可能的机理。图 1.9(b)中的线形轮廓来自图 1.9(a)中 STM 图像里的直线 AD,表明一个 0.07 nm 高的台阶和一个 0.25 nm 高的台阶。后者缘于 SiC 双层,前者比 SiC 双层高度和石墨层间距(0.335 nm)

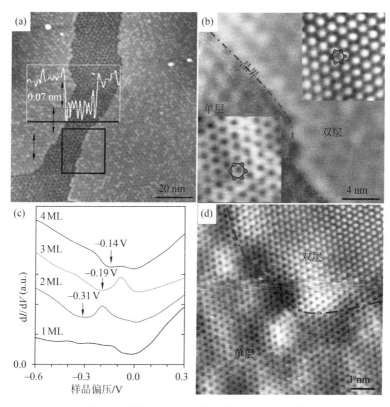

图1.8 单层和双层石墨烯共存的STM图像

(a)100 nm×100 nm,V_T=1.78 V。(b)20 nm×20 nm,V_T=0.5 V,且晶界处石墨烯是连续的。(c)1~4层EG的STS结果。(d)8 nm×8nm,V_T=−0.1 V。(b)中插图分别代表单层和双层EG原子分辨的STM图像;左下方六边形表示单层EG的六边形晶格,右上方六边形表示双层EG两个不等三角子晶格。((a)、(b)和(d)转自参考文献[39],(c)转自参考文献[40])

小很多,但是与它们之间的差异保持了很好的一致性。因此,图1.9(c)中试样说明了单层和双层EG相邻的原子结构。在单层EG的IG下方的SiC双层受热分解,伴随硅从界面的升华和碳的偏析,形成一个新的IG层。这导致起始的IG层转移到新形成的IG层上面,并转变为一个新的EG层,由此实现从单层向双层转变。从由下至上生长的模式看,相邻双层EG顶部层和单层EG源于同一EG层,因此它们会保持连续。这就可以解释观察到的单层和双层EG之间物理连续分界线的出现。由于EG层下面SiC层的分解,使EG层高度降低,可以通过新形成的第二个EG层来弥补(层间距为 0.34 nm),这与测得的 EG 单层和双层之间高度差(如(0.07±0.01)nm)相一致[41]。

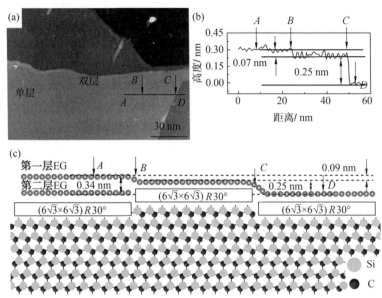

图 1.9 (a)6H-SiC(0001)上外延声子大面积单层和双层石墨烯的 STM 图像(150 nm×100 nm,V_T=1.5 V)。(b)沿着(a)中 AD 线的高度分布;(c)拟建模型[39]

石墨烯也可以在 SiC($000\bar{1}$)上生长,但是完全不同于 SiC(0001)上的生长机理。碳表面的石墨化速度通常比硅表面的快。超高真空生长的 EG 有很多不同取向区域,并形成 3D 结构,这种多层石墨烯膜表现出优良的平整性和独特的层堆叠,并表现为 n 个独立石墨烯单层(参见 1.3 节)。从 4H-SiC($000\bar{1}$)生长的 10 层石墨烯的 LEED 图表明,石墨烯可有许多生长形式:从 SiC[$10\bar{1}0$]的方向旋转 30°($R30$)或±2.2°($R2^{\pm}$)形成的层,三种旋转相在薄膜中相互交织,导致在 $R30$ 和 $R2^{\pm}$ 层之间高密度的堆叠边界缺陷。这种堆叠缺陷具有如图 1.10(a)所示的 $\sqrt{13}\times\sqrt{13}R46.1°$ 晶胞,可以直接在 STM 中观察到,如图 1.10(b)和图 1.10(c)所示[42]。这种堆叠缺陷使邻近石墨烯分离,因此它们的能带结构与分离的石墨烯几乎相同,如多层外延石墨烯(MEG)堆叠的最上层(中性)的 ARPES 所揭示的,如图 1.10(d)和图 1.10(e)所示[43]。多层 EG 的狄拉克锥保持不变但彼此不同。锥体部分的 $K_\perp$ 位移是层与层之间角旋转的结果,如图 1.10(d)和图

1.10(e)所示。

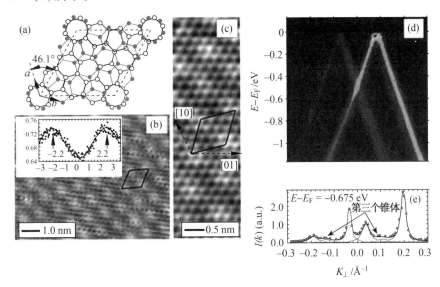

图 1.10 SiC(000$\bar{1}$)上多层 EG 旋转堆垛层错具有的 $\sqrt{13} \times \sqrt{13}\,R46.1°$ 超晶格 (a)拟建模型。(b)较大面积的 STM 图像。(c)放大的 STM 图像。(d)6 K 温度下 11 层 EG 膜的 ARPES 图像,具有三个线性狄拉克锥。(e)$E=E_F-0.675$ eV 中三个锥的动量分布曲线,黑实线是 6 个洛伦兹(细实线)的总和((a)~(c)转自参考文献[42],(d)和(e)转自参考文献[43])

1.3 环境压力下单晶 SiC 的热分解

为了提高 EG 的均匀性,生长条件应向热力学平衡转变,如提高生长温度的同时而不增加硅升华的速度。温度越高,表面碳原子动能越大,转变为石墨烯层时出现的缺陷和晶界就越少。最早使用的可行方法是 Emtsev 等提出的[19],通过向生长环境中引入惰性气体来实现,尤其是氩气。与超高真空下硅原子可以自由从表面逃逸相比,氩分子的出现为硅原子的反弹并返回提供了一定的可能性,因此降低了硅升华的速率。当氩气的压力接近真空时,可以获得具有大尺寸和极少缺陷的石墨烯薄膜[45]。

SiC 上石墨烯生长所需的环境压力(ambient pressure)可由特制的气密炉来实现,这些炉可以是石英管或真空室。在环境压力下,通过对流能更加有效地散热,因此需要用适量的水给炉壁冷却,采用能承受高退火温度的石墨支架。图1.11是氩气下用于制备石墨烯的采用感应加热和水冷却的石英管炉[44]。当SiC样本为晶圆大小(wafer size)时,应考虑采用降低温度梯度的方法。SiC以2~3 ℃/s的速度被缓慢加热或冷却。退火温度为1 500~2 000 ℃,保持15 min。在SiC的硅表面生长时氩气气流以0.9~1 Pa的压力通入到炉内。在碳表面,退火温度保持1 500 ℃,氩气压力升到9 Pa,以抑制表面硅的快速升华[45]。

图1.11 氩气下用于制备石墨烯的采用感应加热和水冷却的石英管炉[44]

超高真空和环境退火制备的EG如图1.12所示。图1.12(a)是氢刻蚀后的SiC(0001)初始表面。可以看到宽度为300~700 nm、高度约为1.5 nm的阶梯。在超高真空制备的样本中(图1.12(b)和图1.2(c)),表面形态发生重大变化,LEEM图像表明石墨烯岛大小约为几百纳米。比较而言,环境压力下的石墨烯生长的形态得到极大改善。原子力显微镜(AMF)图像显示阶梯尺寸从几百纳米增加到几微米,平均台阶的高度为8~15 nm,如图1.12(e)所示。LEEM表明连续的石墨烯层覆盖在由双层或三层石墨烯形成的台阶上。除了晶粒大小的区别,环境气氛制备的石墨烯膜的缺陷也大大减少。图1.12(d)所示是超高真空和氩气条件下制备的样本的拉曼光谱。对于真空中生长的样本来说,G峰、D峰和2D峰分别在1 596 cm^{-1}处、1 356 cm^{-1}处和2 706 cm^{-1}处。对于在氩气里生长的样本,

G峰和2D峰分别在1 592 cm^{-1}处和2 717 cm^{-1}处。D峰却没有出现,2D峰半峰宽在37 cm^{-1}处,比超高真空中生长的样品(54 cm^{-1})窄。拉曼光谱说明氩气中生长的石墨烯比真空中的样品缺陷更少。

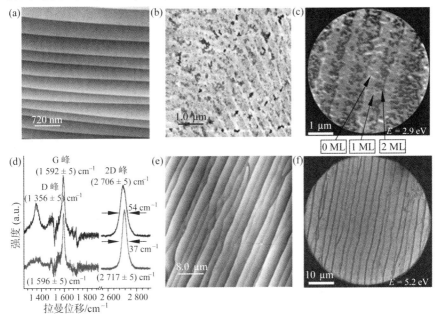

图1.12 超高真空和氩气中生长石墨烯

(a)氢刻蚀后初始表面的AFM图像,梯度为1.5 nm。(b)1 280 ℃超高真空下在6H-SiC(0001)上退火生长的厚度为1ML石墨烯的AFM图像。(c)SiC(0001)上超高真空中生长的厚度为1.2 ML石墨烯膜的LEEM图像。由于不同厚度而表现出不同衬度。浅灰、中灰和深灰分别对应厚度为0 ML、1 ML和2 ML。(d)6H-SiC(0001)上氩气(下)和超高真空(上)中外延生长石墨烯的拉曼光谱。D和G谱线考虑了衬底发射并将其扣除。(e)6H-SiC(0001)上经过氩气退火($p=90$ kPa,$T=$1 650 ℃)得到额定厚度为1.2 ML石墨烯的AFM图像。(f)相当于(d)中样品的LEEM图像,存在50 μm长、至少1 μm宽的大台阶的石墨烯[19]

氩气中制备的石墨烯的传输性能可以通过对载流子迁移率的测试来衡量。表1.1是对氩气中和真空中生长的石墨烯采用霍尔条和van der Pauw法测得的迁移率。氩气中生长的石墨烯的迁移率是真空中生长的石墨烯的2~4倍。在27 K温度下可得到的最大迁移率数值是2 000 cm$^2 \cdot$ V$^{-1} \cdot$ s^{-1}。

表1.1 在 $T=300$ K 和 27 K[19] 温度下,采用霍尔条和 van der Pauw 法测得的超高真空中生长和氩气中生长石墨烯的霍尔迁移率　　　　　　　　　　　　　　$cm^2 \cdot V^{-1} \cdot s^{-1}$

方法	结构	300 K	27 K
氩气	霍尔条	900	1 850
	van der Pauw 法	930	2 000
超高真空	霍尔条	470	—
	van der Pauw 法	550	710

在发现氩气辅助的 SiC 上的石墨烯沉积之后,又进一步提出了采用束流控制硅原子离开表面层实现石墨烯生长的方法。可以用硅烷气体直接控制硅蒸气压[46]或用销钉的方式将样本限制在石墨圈内。后一种被称为约束控制升华(CCD)的方法,可以把硅升华速率控制在 10^{-3} 数量级。

1.4 单晶 SiC 薄膜和多晶 SiC 基板的热分解

使用 SiC 外延生长石墨烯的石墨烯器件在实验室已经得到广泛构建。然而,单晶 SiC 晶圆价格昂贵,且尺寸不超过 100 mm。SiC 基板的高成本和小尺寸阻碍了有效成本下大批量生产石墨烯器件。石墨烯膜生长只需要几个 SiC 原子层的分解,外延 SiC 薄膜和多晶 SiC 基板可以提供更加经济的方式实现 EG 生长。立方 SiC(3C-SiC)薄膜可以在硅晶圆的多个低指数表面生长。这些低指数表面包括 Si(111)、Si(100)和 Si(110)。厚度从 100 nm 到几微米的 3C-SiC 膜可以通过双源(SiH_4/C_3H_8)或单源的($SiCH_6$)化学气相沉积法在 Si 基板上生长。在 1 200~1 300 ℃真空中可实现 3C-SiC 薄膜上石墨烯生长,与超高真空中 EG 在单晶 SiC 上生长类似。在这些薄膜上进行以氩气介导的石墨烯生长则是不可能的,这是由于退火温度超过硅基板的熔点(1 414 ℃)。通过拉曼光谱对超高真空中 SiC 膜上生长的石墨烯进行表征证实了石墨烯的形成,可以看出,石墨烯特征峰(如 G 峰、2D 峰)以及附加的缺陷相关峰(D 峰和(D+G)峰),如图 1.13 所示。SiC(111)和 SiC(110)上石墨烯的强 D 峰和可分辨的(D+G)峰表明石墨烯内部有大量的缺陷[47]。

多晶的 3C-SiC 也可以用于石墨烯生长,从拉曼图谱中可知,多晶的 3C-SiC 能制出与单晶 SiC 石墨烯质量相似的石墨烯[48]。多晶 SiC 市面上可以买到,比单晶 SiC 的价格更低、尺寸更大。多晶 SiC 可以在超高真空

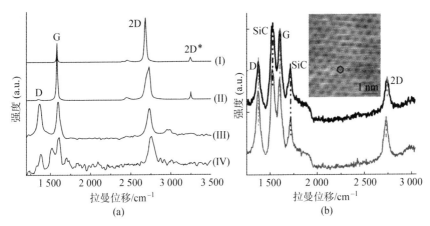

图1.13 3C-SiC薄膜和多晶SiC基板上生长石墨烯的拉曼光谱
(a)不同方法得到的石墨烯光谱:(Ⅰ)SiO_2/Si(110)上剥离SLG,(Ⅱ)SiO_2/Si(100)上剥离SLG,(Ⅲ)3C-SiC(110)/Si(110)上EG和(Ⅳ)6H-SiC(0001)块晶上的EG。D光谱是从原始数据中扣除6H-SiC光谱后得到的。在扣除后的光谱中,扣除不完会引起出现SiC的成分。(b)样品中两点的拉曼光谱给出了石墨烯相关的D带、G带和2D带。插图是多晶SiC基板上石墨烯晶格的STM图像((a)转自参考文献[47],(b)转自参考文献[48])

或氩气中分解,在氩气中分解的方法还没有相关的报道。超高真空制备的石墨烯层的拉曼光谱如图1.13(b)所示,在1 524 cm^{-1}和1 716 cm^{-1}处出现了石墨烯相关峰(G峰、2D峰和D峰)和SiC相关峰。光谱中未发现明显的(D+G)峰,表明石墨烯的质量与单晶基板上的EG类似[48]。多晶基板上的石墨烯STM图像如图1.13(b)中插图所示。蜂窝结构的石墨烯晶格也清楚地显示了单晶石墨烯的生长情况。

1.5 插层制备外延石墨烯

石墨烯层通过范德瓦耳斯力与基板存在较弱的相互作用,有利于分子或原子进入或插入石墨烯和基板连接处,该现象在SiC和金属基板都可以看到。例如,镍(111)上的EG可以插入其他金属原子,如铁、金或铝,并通过XPS、ARPES和STM等手段证明[49-50]。插入的原子作为缓冲层,能够消除石墨烯与衬底间的相互作用,并在K点恢复石墨烯的线性色散。插层可能会影响石墨烯的化学反应活性,例如,铂插入钌(0001)上的石墨烯会降低在氧气中氧化反应活性[51]。

在 SiC 基板上,插层往往伴随着副反应。原子掺入 IG 和 SiC 之间,在界面处会发生原子与硅悬挂键反应,释放 IG 并形成一个附加石墨烯层。不论顶部石墨烯层是否存在,插层反应都会发生。这样就有机会把 IG 层可逆性地变为石墨烯并消除 IG 的 n 型掺杂对 EG 的影响。很多插层原子可以与硅原子反应,包括氢气、氧气、氟、金、铁、锂、锗,甚至硅自身也会发生[52-59]。然而也有例外,铯和铷可能因为具有很大的原子半径而不能插层到外延石墨烯中[60]。

插层法通常分两步:第一步,原子在石墨烯表层沉积(几个原子层);第二步,使表层退火,促使吸附的原子通过缺陷或晶界扩散并插层到界面,与硅原子反应。退火温度因所采用的原子不同而不同。对于锗元素,退火温度为 920 ℃[55],而对于锂元素,虽然 330 ℃ 的退火有助于锂原子均匀分散,但在室温就可以发生插层反应。表 1.2 总结了各种原子插层的退火温度。对于气体分子来说,可以通过原子源(氢气)[52]、高压退火(氧气,1 Pa,250 ℃)[53]或分子分解($C_{60}F_{48}$,氟)实现插入[54]。

ARPES 证实通过氢插层会使 IG 转变为 SiC 上的单层 EG,如图 1.14(a) ~ 图 1.14(e)所示[52]。

表 1.2　石墨烯表面固体原子插层的退火温度

元素	Au[59]	Fe	Li[57]	Ge[55]	Si[56]
温度	727 ℃	600 ℃	—	720 ~ 920 ℃	800 ℃

K 点的线性色散只有在氢插入后才会出现,900 ℃ 以上消失,也就是在此温度下氢会发生解吸附。同样氢插入可以将 SiC 上的单层 EG 转变成双层 EG,如图 1.14(f) ~ 图 1.14(j)所示。值得注意的是,插入后的 E_F 与狄拉克点相吻合,表明石墨烯固有的 n 型掺杂消失了。其他元素的插层可能引起附加 n 型掺杂(Li)、p 型掺杂(F)或是依靠退火温度(Ge)的双掺杂。

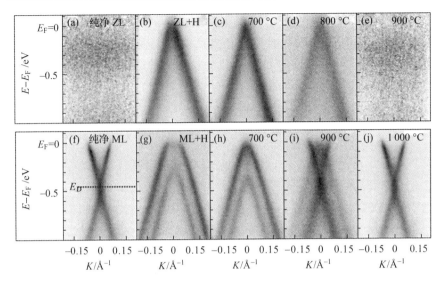

图 1.14 在垂直于石墨烯布里渊区的 Γ 方向采用 ARPES 测试获得的 π 带的色散
(a)SiC(0001)上生长中的零层石墨烯。(b)氢处理和(c)~(e)不同退火处理。(f)生长中的单层(ML)。(g)氢处理后和(h)~(j)不同退火处理[52]

1.6 结　论

本章论述了在超高真空或环境压力下,通过热分解实现单晶 SiC 晶圆、多晶 SiC 和 SiC 薄膜上 EG 的生长及原子尺度表征。可以在单晶 SiC 基片上得到具有可控层数量和超过 1 μm 晶粒的高品质石墨烯膜。此外,SiC 上 EG 生长与目前硅基的微制造光刻工艺相匹配,使 EG 有望实现石墨烯基电子器件的大规模生产。由于 SiC 上 EG 具有原子级别的平整度、化学惰性和结构简单的表面,可以用作高品质拓扑绝缘体 Bi_2Se_3 或超导体 FeSe 的优质衬底,或有机覆盖的石墨烯基异质结。然而存在几个与 SiC 上生长 EG 相关的技术问题需要妥善解决,以便进一步扩大石墨烯基器件的商业化应用,包括:

(1) SiC 基板成本高,尤其是单晶 SiC 基板,需要寻找可以生长高品质 EG 膜的低成本替代物。

(2) 目前技术生长 EG 的温度很高,通常高于 1 200 ℃,而且需要巨大的能量输入,因此低温下如何实现 SiC 上高品质 EG 的生长对于进一步发展 EG 基器件至关重要。

(3) EG 的电子特性受到其与底板 SiC 耦合的严重影响；使 EG 基器件相比于微机械剥离石墨烯，性能优势有所降低。因此，提高使 EG 与基板 SiC 电子解耦的切实可行的方法，对于 EG 基器件的制造一体化是非常必要的。

(4) SiC 上的 EG 因为有更小的晶粒尺寸，所以缺陷比膨胀石墨烯更多，因此要完善 EG 生长工艺有必要解决晶粒尺寸问题。

1.7 参考文献

[1] NOVOSELOV K S, JIANG D, SCHEDIN F, et al. Two-dimensional atomic crystals[J]. Proceedings of the National Academy of Sciences of the United States of America, 2005, 102: 10451.

[2] NOVOSELOV K S, GEIM A K, MOROZOV S V, et al. Electric field effect in atomically thin carbon films[J]. Science, 2004, 306: 666.

[3] LI X L, ZHANG G Y, BAI X D, et al. Highly conducting graphene sheets and Langmuir-Blodgett films[J]. Nature Nanotechnology, 2008, 3: 538.

[4] STANKOVICH S, DIKIN D A, PINER R D, et al. Synthesis of graphene-based nanosheets via chemical reduction of exfoliated graphite oxide[J]. Carbon, 2007, 45: 1558.

[5] GILJE S, HAN S, WANG M, et al. A chemical route to graphene for device applications[J]. Nano Letters, 2007, 7: 3394.

[6] LI D, MULLER M B, GILJE S, et al. Processable aqueous dispersions of graphene nanosheets[J]. Nature Nanotechnology, 2008, 3: 101.

[7] PARK S, RUOFF R S. Chemical methods for the production of graphenes[J]. Nature Nanotechnology, 2009, 4: 217.

[8] LAND T A, MICHELY T, BEHM R J, et al. STM investigation of single layer graphite structures produced on Pt(111) by hydrocarbon decomposition[J]. Surface Science, 1992, 264: 261.

[9] NAGASHIMA A, NUKA K, ITOH H, et al. Electronic states of monolayer graphite formed on TiC(111) surface[J]. Surface Science, 1993, 291: 93.

[10] BAE S, KIM H, LEE Y, et al. Roll-to-roll production of 30-inch graphene films for transparent electrodes[J]. Nature Nanotechnology, 2010, 5: 574.

[11] LI X S, CAI W W, COLOMBO L, et al. Evolution of graphene growth on Ni and Cu by carbon isotope labeling[J]. Nano Letters, 2009, 9: 4268.

[12] MARCHINI S, GÜNTHER S, WINTTERLIN J. Scanning tunneling microscopy of graphene on Ru(0001)[J]. Physical Review B, 2007, 76: 075429.

[13] ELENA L, NORMAN C B, PETER J F, et al. Evidence for graphene growth by C cluster attachment[J]. New Journal of Physics, 2008, 10: 093026.

[14] SUTTER P W, FLEGE J-I, SUTTER E A. Epitaxial graphene on ruthenium[J]. Nature Materials, 2008, 7: 406.

[15] YU Q, LIAN J, SIRIPONGLERT S, et al. Graphene segregated on Ni surfaces and transferred to insulators[J]. Applied Physics Letters, 2008, 93: 113103.

[16] REINA A, JIA X, HO J, et al. Large area, few-layer graphene films on arbitrary substrates by chemical vapor deposition[J]. Nano Letters, 2008, 9: 30.

[17] CORAUX J, N'DIAYE A T, ENGLER M, et al. Growth of graphene on Ir(111)[J]. New Journal of Physics, 2009, 11: 023006.

[18] BERGER C, SONG Z M, LI T B, et al. Ultrathin epitaxial graphite: 2D electron gas properties and a route toward graphene-based nanoelectronics[J]. The Journal of Chemical Physics B, 2004, 108: 19912.

[19] EMTSEV K V, BOSTWICK A, HORN K, et al. Towards wafer-size graphene layers by atmospheric pressure graphitization of silicon carbide[J]. Nature Materials, 2009, 8: 203.

[20] BOSTWICK A, MCCHESNEY J, OHTA T, et al. Experimental studies of the electronic structure of graphene[J]. Progress in Surface Science, 2009, 84: 380.

[21] FIRST P N, DE HEER W A, SEYLLER T, et al. Epitaxial graphenes on silicon carbide[J]. MRS Bulletin, 2010, 35: 296.

[22] MOON J S, CURTIS D, BUI S, et al. Top-gated epitaxial graphene FETs on Si-face SiC wafers with a peak transconductance of 600 mS/mm[J]. IEEE Electron Device Letters, 2010, 31: 260.

[23] SINGH R S, NALLA V, CHEN W, et al. Laser patterning of epitaxial graphene for Schottky junction photodetectors[J]. ACS Nano, 2011, 5:

5969.

[24] NOMANI M W K, SHISHIR R, QAZI M, et al. Highly sensitive and selective detection of NO_2 using epitaxial graphene on 6H–SiC[J]. Sensors Actuators B,2010,150:301.

[25] VAN BOMMEL A J,CROMBEEN J E,VAN T A. LEED and auger electron observations of the SiC(0001) surface[J]. Surface Science,1975, 48:463.

[26] CHEN W,WEE A T S. Self-assembly on silicon carbide nanomesh templates[J]. Journal of Physics D:Applied Physics,2007,40:6287.

[27] STARKE U,SCHARDT J,BERNHARDT J,et al. Novel reconstruction mechanism for dangling-bond minimization: combined method surface structure determination of SiC(111)-(3×3) [J]. Physics Review Letters,1998,80:758.

[28] REUTER K, BERNHARDT J, WEDLER H, et al. Holographic image reconstruction from electron diffraction intensities of ordered superstructures[J]. Physics Review Letters,1997,79:4818.

[29] LI L,TSONG I S T. Atomic structures of 6H–SiC (0001) and (000-1) surfaces[J]. Surface Science,1996,351:141.

[30] RAMACHANDRAN V,FEENSTRA R M. Scanning tunneling spectroscopy of Mott-Hubbard states on the 6H–SiC(0001) $\sqrt{3}\times\sqrt{3}$ surface[J]. Physics Review Letters,1999,82:1000.

[31] CHEN W,XU H,LIU L,et al. Atomic structure of the 6H–SiC(0001) nanomesh[J]. Surface Science,2005,596:176.

[32] KIM S,IHM J,CHOI H J,et al. Origin of anomalous electronic structures of epitaxial graphene on silicon carbide [J]. Physics Review Letters, 2008,100:176802.

[33] NI Z H,CHEN W,FAN X F,et al. Raman spectroscopy of epitaxial graphene on a SiC substrate[J]. Physical Review B,2008,77:115416.

[34] EMTSEV K V,SPECK F,SEYLLER T,et al. Interaction,growth and ordering of epitaxial graphene on SiC{0001} surfaces:a comparative photoelectron spectroscopy study[J]. Physical Review B,2008,77:155303.

[35] OHTA T,BOSTWICK A,MCCHESNEY J L,et al. Interlayer interaction and electronic screening in multilayer graphene investigated with angle-

resolved photoemission spectroscopy[J]. Physics Review Letters,2007, 98:206802.
[36] OHTA T,BOSTWICK A,SEYLLER T,et al. Controlling the electronic structure of bilayer graphene[J]. Science,2006,313:951.
[37] VIROJANADARA C,YAKIMOVA R,ZAKHAROV A A,et al. Large homogeneous mono-/bi-layer graphene on 6H−SiC(0001) and buffer layer elimination[J]. Journal of Physics D: Applied Physics,2010,43: 374010.
[38] ROBINSON J A,PULS C P,STALEY N E,et al. Raman topography and strain uniformity of large-area epitaxial graphene [J]. Nano Letters, 2009,9:964.
[39] HUANG H,CHEN W,CHEN S,et al. Bottom-up growth of epitaxial graphene on 6H−SiC(0001)[J]. ACS Nano,2008,2:2513.
[40] LAUFFER P,EMTSEV K V,GRAUPNER R,et al. Atomic and electronic structure of few-layer graphene on SiC(0001) studied with scanning tunneling microscopy and spectroscopy[J]. Physical Review B,2008, 77:155426.
[41] POON S W,CHEN W,WEE A T S,et al. Growth dynamics and kinetics of monolayer and multilayer graphene on a 6H−SiC(0001) substrate [J]. Physical Chemistry Chemical Physics,2010,12:13522.
[42] HASS J,VARCHON F,MILLÁN-OTOYA J E,et al. Why multilayer graphene on 4H−SiC(0001) behaves like a single sheet of graphene[J]. Physics Review Letters,2008,100:125504.
[43] SPRINKLE M,SIEGEL D,HU Y,et al. First direct observation of a nearly ideal graphene band structure[J]. Physics Review Letters,2009,103: 226803.
[44] DE HEER W A,BERGER C,RUAN M,et al. Large area and structured epitaxial graphene produced by confinement controlled sublimation of silicon carbide[J]. Proceedings of the National Academy of Sciences of the United States of America,2011,108:16900.
[45] NORIMATSU W,TAKADA J,KUSUNOKI M. Formation mechanism of graphene layers on SiC(0001) in a high-pressure argon atmosphere[J]. Physical Review B,2011,84:035424.
[46] TROMP R M,HANNON J B. Thermodynamics and kinetics of graphene

growth on SiC(0001)[J]. Physics Review Letters,2009,102:106104.
[47] SUEMITSU M,FUKIDOME H. Epitaxial graphene on silicon substrates [J]. Journal of Physics D: Applied Physics,2010,43:374012.
[48] HUANG H,WONG S L,TIN C C,et al. Epitaxial growth and characterization of graphene on free-standing polycrystalline 3C-SiC[J]. Journal of Applied Physics,2011,110:014308.
[49] DEDKOV Y S,FONIN M,RUDIGER U,et al. Graphene-protected iron layer on Ni(111)[J]. Applied Physics Letters,2008,93:022509.
[50] GIERZ I,SUZUKI T,WEITZ R T,et al. Electronic decoupling of an epitaxial graphene monolayer by gold intercalation[J]. Physical Review B, 2010,81:235408.
[51] JIN L,FU Q,MU R T,et al. Pb intercalation underneath a graphene layer on Ru(0001) and its effect on graphene oxidation[J]. Physical Chemistry Chemical Physics,2011,13:16655.
[52] RIEDL C,COLETTI C,IWASAKI T,et al. Quasi-free-standing epitaxial graphene on SiC obtained by hydrogen intercalation[J]. Physics Review Letters,2009,103:246804.
[53] OIDA S,MCFEELY F R,HANNON J B,et al. Decoupling graphene from SiC(0001) via oxidation[J]. Physical Review B,2010,82:041411R.
[54] WONG S L,HUANG H,WANG Y Z,et al. Quasi-free-standing epitaxial graphene on SiC (0001) by fluorine intercalation from a molecular source[J]. ACS NANO,2011,5:7662.
[55] EMTSEV K V,ZAKHAROV A A,COLETTI C,et al. Ambipolar doping in quasifree epitaxial graphene on SiC(0001) controlled by Ge intercalation[J]. Physical Review B,2011,84:125423.
[56] XIA C,WATCHARINYANON S,ZAKHAROV A A,et al. Si intercalation/deintercalation of graphene on 6H-SiC(0001)[J]. Physical Review B,2012,85:045418.
[57] VIROJANADARA C,WATCHARINYANON S,ZAKHAROV A A,et al. Epitaxial graphene on 6H-SiC and Li intercalation[J]. Physical Review B,2010,82:205402.
[58] WALTER A L,JEON K J,BOSTWICK A,et al. Highly p-doped epitaxial graphene obtained by fluorine intercalation[J]. Applied Physics Letters, 2011,98:184102.

[59] PREMLAL B, CRANNEY M, VONAU F, et al. Surface intercalation of gold underneath a graphene monolayer on SiC(0001) studied by scanning tunneling microscopy and spectroscopy[J]. Applied Physics Letters,2009,94:263115.

[60] WATCHARINYANON S, VIROJANADARA C, JOHANSSON L I. Rb and Cs deposition on epitaxial graphene grown on 6H-SiC(0001)[J]. Surface Science,2011,605:1918.

第2章 化学气相沉积(CVD)制备石墨烯

本章讨论了在镍和铜基板上用化学气相沉积法(CVD)制备石墨烯和最新成果;并对大面积单层的形成、单晶上的生长以及可控形成有序多层进行详细论述;利用同位素标记方法对CVD石墨烯进行深入研究。

2.1 引　　言

用于电子器件的石墨烯的生产主要依赖于化学气相沉积(CVD)[1]。本章重点是讨论石墨烯形成的机理和控制生长过程。尽管对CVD法制备石墨烯的研究已经做了大量工作,但仍有很多问题有待解决。铜和镍由于具有成本低、可刻蚀及晶粒尺寸大等特点,被广泛用作催化剂[1-4]。

虽然其他金属也已被成功用于催化CVD石墨烯生长,如Pt[5]、Co[6]、Ir[7-8]或Ru[9-10],但本章重点介绍铜和镍,因为它们有望实现石墨烯在光学和电子方面的大规模应用。本书并不介绍2011年之前的几个重要研究,重点介绍2012年和2013年的研究。我们希望读者能注意Mattevi等[11]或Ago等[12]发表的这两篇文献,它们涵盖了CVD制备石墨烯的其他方面研究。

通常,在石墨烯化学气相沉积过程中,气态的前驱体注入反应釜中,随着温度升高,与催化剂开始反应,在催化剂表面形成石墨烯。前驱体通常是小分子的碳氢化合物,如甲烷或乙烯,但也会用到具有挥发性的低相对分子质量的乙醇。生长的温度范围从几百摄氏度到催化剂金属的熔点。Li[13]等提出根据催化剂不同,石墨烯生长的两个基本机理。对于多晶的镍,前驱体在表面分解,碳溶解在金属中。当基板冷却时,碳在镍中的溶解度降低,石墨烯首先偏析,然后在镍表面生长[13-14]。因此,控制冷却条件对于制备单层石墨烯(1LG)至关重要[3]。图2.1(a)是一个少层石墨烯在镍上生长的例子。

然而,铜做催化剂,碳就不会溶解在金属中,因为碳在铜中的溶解度即使在很高的温度下也可以忽略不计。高温下碳原子会在表面直接形成石墨烯,而不用精确控制金属基板的冷却速度。铜基板上的CVD法被认为是表面介导和自限过程(self-limiting)[13]。一旦形成单层,进入催化剂 Cu 表面的路径就被封锁,蔓延过程将停止。因此,通过铜催化 CVD 法只能形成单层石墨烯,如图 2.1(b)所示。而在很多情况下,会具有双层或多层的小区域生成,如图 2.1(c)所示[15-16]。多层区域的形成原因和形成过程还不清楚。这个多层区域可能会影响大型石墨烯器件的制造,因为多层区域会打破石墨烯膜的均匀性。所以如何限制多层的出现(图 2.1(d)),是对铜催化 CVD 法进行石墨烯生长提出的一个重要要求[16]。

本章我们先论述镍上石墨烯的生长,然后详细介绍铜上石墨烯的生

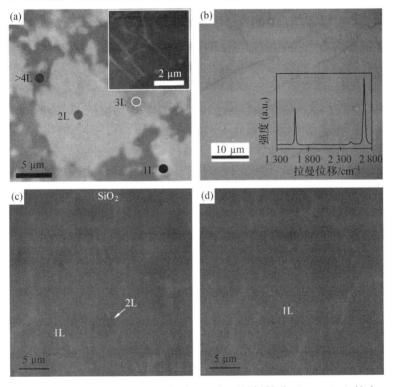

图 2.1 (a)镍片和(b)~(d)铜片上生长石墨烯转移至 Si/SiO₂ 上的光学显微照片

(a)由于光学对比度,可观察到层数变化,插图给出了起皱边缘的 AFM 图像[2]。(b)铜片上形成连续单层,存在褶皱,插图是转移单层的典型拉曼光谱[21]。(c)铜片上短时 CVD 生长给出了石墨烯晶粒的形状,且大概在晶粒中心处存在小尺寸多层。(d)与(c)中条件相同下生长的样本,生长后氢刻蚀[22]

长,讨论大面积石墨烯制备、单晶上生长以及可控有序的形成的最新进展,如可控形成 AB 堆叠的石墨烯双层(AB-2LG)。最后,介绍研究 CVD 石墨烯的先进手段——同位素标记,并列举实例。

2.2 镍衬底上的化学气相沉积

在用于石墨烯合成的衬底中,镍是与石墨烯具有最小晶格失配的元素之一。Ni(111)面晶格常数是 0.249 nm,而高度有序的热解石墨烯 HOPG (0001)为 0.246 nm[17]。因此,镍是与外延生长的结构均匀的石墨烯相匹配的最有发展前景的催化剂之一。Reina 等[3]指出在多晶镍的表面精确控制碳沉积量,可以在大气压力下化学气相沉积过程(AP-CVD)中生长一层或两层石墨烯。另外,在石墨烯成长过程中既控制甲烷浓度又控制基板冷却速度,能够显著提高厚度的均匀性,结果是整个薄膜的 87% 都是一层或两层区域。这个区域延伸覆盖了下面多晶镍薄膜的大量晶界。通过降低冷却速率,可减少多层石墨烯或石墨(大于 2 层)的面密度[18]。

镍基板上生长的石墨烯具有不同晶体形貌,例如单晶和多晶,这可能涉及不同反应机理[19]。通过扫描隧道显微镜(STM)可以看出镍(111)单晶上的单层石墨烯和相同条件下镍膜上生长的多层石墨烯[19-20]。一般认为,镍(111)上石墨烯的生长可以严格以镍(111)晶格作为模板进行,这是因为 Ni-C 强烈的相互作用,而获得单层石墨烯。此外镍(111)有原子尺度光滑的表面且无晶界[20]。多晶镍薄膜上形成的多层石墨烯片,通常与伯纳尔堆叠存在偏离,并在碳层中存在轻微旋转。考虑到不同基板的特点,多晶镍膜上的晶界被当作双层甚至多层石墨烯的生长前沿。多晶镍上 CVD 石墨烯中含多层石墨烯的比例更高,因为镍的晶界可以作为多层石墨烯生长的成核点[19,20]。

高温 CVD 生长多层石墨烯通过碳从块体材料的偏析形成,而低温下(低于 600 ℃)则不同,石墨烯可以按照自限单层生长过程进行生长[23]。550 ℃是最佳生长温度。高于这个温度,碳扩散进入块体材料限制表面生长率,低于 500 ℃时,会存在石墨烯与表面碳相的竞争而阻碍石墨烯的形成[23]。

同样,在快速热 CVD 生长中采用冷墙反应器,可以在镍基板上生成高品质的石墨烯膜[24]。生长时间小于 10 s 的情况下生成石墨烯膜,说明石墨烯的直接生长机理起着更重要的作用,而不是沉积机理。较低的氢气束流有利于获得高品质石墨烯膜。薄膜的特性表明镍基板上的石墨烯膜的结构和电学特性与铜基板上CVD制备的膜类似[24]。

除了气体前驱体合成石墨烯方法之外,也有关于其他大量碳源的报道,包括聚甲基丙烯酸甲酯(PMMA)、SU8-2002光阻材料、苯以及其他碳材料[25,26]。在氩气和氢气的混合气体中,退火温度为1 000 ℃,运用SU8-2002光阻材料可在镍片上合成高品质的石墨烯[25]。然而,从扫描电子显微镜(SEM)图像可以看到镍片上生成了不同层数的石墨烯,这是由于下层镍的晶粒取向不同造成碳偏析速度的不同。另一个石墨烯前驱体的例子是乙醇。由于镍的表面粗糙度造成石墨烯层数的变化,石墨烯晶粒的大小取决于下层镍晶粒的大小[26]。

据报道,在真空退火条件下运用铜-镍合金,可以偏析获得晶圆大小的石墨烯,并且其层数可控[27]。提高镍在铜-镍合金中的比例可以得到更厚、更均匀的石墨烯。碳在铜和镍中不同溶解度的协同作用被认为是生成高品质均匀少层石墨烯的偏析机理[28-31]。

镍催化剂也可以用于在绝缘基板上制得石墨烯。通过退火可以得到具有单个或极少晶粒的"镍点",可在"镍点"上生长单层石墨烯[32]。

2.3 铜表面大尺寸石墨烯

近两年,制备与机械剥离石墨烯大小相近的单晶石墨烯的CVD方法得到飞速发展,而且该方法在保证结晶尺寸和质量方面一直占据优势。表2.1总结了近来有关于铜基板的研究文献,在铜基板上石墨烯的生长在铺满整个催化表面之前就被冻结。通常情况下,生长几百甚至几千微米晶粒大小的石墨烯,需要几个基本要点。作为前提条件,成核点的数量要在开始生长之前保持最小值。这是相当困难的,因为成核点可能会发生在杂质、表面边缘、台阶和晶界(在箔片上,图2.2)[33-34]。适当的表面预处理,如电学抛光会有助于表面平整,减少成核点数量[35-36]。值得注意的是,采用单晶催化表层,可以减少晶界的数量,尤其对于Cu(111)。

表 2.1　铜催化制备石墨烯的晶粒大小、形状及层数

压力	温度/℃	晶粒形状	层数	大小/μm	文献
6.666 Pa	1 035	六边形/枝晶	1	500	[38]
环境压力	1 000	六边形	1	20	[39]
环境压力	1 045	方形	1	400	[40]
26.664 Pa	1 000	六叶花状	1	100	[41]
环境压力	1 080	六边形	2	15~100	[42]
环境压力	1 070	六边形	1	150	[43]
100 Pa	1 050	六叶形	1	20	[28]
环境压力	1 000	六边形	1	100	[36]
低压	1 000	四叶形	1~2	50~100	[44]
环境压力	1 090	六边形	1	200	[51]
14.399 kPa	1 077	六边形	1	2 300	[45]

很明显,成核和进一步生长依赖于所有条件和所用参数的相互关系,如催化剂表面、碳前驱体、温度、总压力以及气体前驱体和氢气的压力。要获得低成核密度,宜于使用低浓度的前驱体,在低总压下更容易实现。另外,在达到大气压条件下由于催化剂的低挥发性,更容易获得连续生长。准确控制总压力,有利于为大面积石墨烯生长找到最佳条件,如 Yan 等发表的有关晶粒超过 2 mm 的报道[45]。

图 2.3 描述了石墨烯在铜基板上生长的不同晶粒的形状。简单来分,具有笔直边缘的晶粒包括六边形或正方形,是在较高压力下获得的,如图 2.3(a)和图 2.3(e)所示;而在低压下生长的形状则具有花形,或 4~6 个叶状和树枝状突起,如图 2.3(b)和图 2.3(d)所示。图 2.3(c)中也可以观察到其他形状,甲苯作为碳前驱体在 600 ℃下可得细长的长方形[37]。

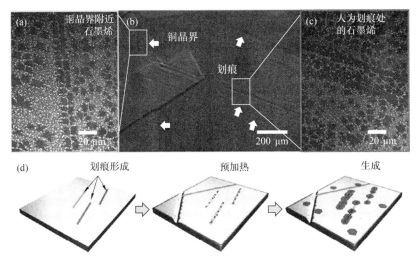

图 2.2 (a)~(c)在 5 s 时铜晶粒边界附近和划痕处附近初步形成石墨烯的不同放大率的 SEM 图像。(d)划痕形貌和石墨烯生长的示意图[33]

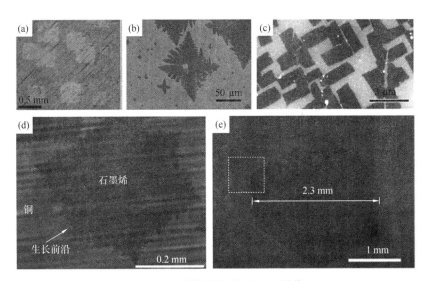

图 2.3 石墨烯域晶粒的 SEM 图像

(a)AP-CVD 所得方形晶粒[40]。(b)在 26.664 Pa 下生长的四叶形晶粒[44]。(c)LP-CVD 法用甲苯做原料所得的细长矩形[37]。(d)LP-CVD 法所得的有突起的六边形[38]。(e)14.399 kPa 下生长的六边形区域[45]

2.4 单晶铜上的石墨烯生长

具有远程结晶序的表面上石墨烯的外延生长尤其受到人们的关注,如铜单晶,这是由于在这种情况下影响石墨烯生长的一个重要参数被固定:铜箔以结晶面的形式表现出高度不均匀性[46-48],甚至择优晶粒取向可以通过金属箔片类型、生产者、预处理或退火温度控制[49-52]。另一方面,基板的结晶性好像只在低压下对石墨烯生长有影响[46-50],而在高压下,可以跨越多个铜晶面大面积连续生长石墨烯[53]。铜晶面上的石墨烯取向不敏感可能是因为 Cu—C 之间相互作用比较弱[54-55]。然而一些关于晶格控制的表面上石墨烯生长的实验数据并不是这样[5,12,48,56-65],即使在大气压下[48,56]也是。

首次 Cu(111)单晶上石墨烯合成,以乙烯作为前驱体气体,在 8 μm 汞柱、1 000 ℃[5]温度条件下进行。STM 表现出两个不同的 Moiré 条纹,这两个条纹表明石墨烯晶格(a=0.246 nm)和下层 Cu(111)晶格(0.256 nm)存在两种不同旋转排列方式。当两个晶格按 0°旋转对齐时,表现出约 6.6 nm 的周期,却很少表现出 7°角误差(约 2 nm 周期性)。然而这些样品表现出高强度的拉曼 D 峰,表明石墨烯的质量较差(见文献[5]中图 4(b))[5],这让我们联想到低温下甲烷生长石墨烯的例子。2011 年 Nie 等研究了生长温度对在 Cu(111)单晶上形成石墨烯的镶嵌度(mosaicity)的影响。发现当温度从 690 ℃升高到 950 ℃时,铜和石墨烯之间的旋转角降低到大约±1.5°。即使在更高温度下仍存在镶嵌度,这部分源于经过台阶或分枝生长的石墨烯层所引起的薄片内边界旋转[34]。生长温度(AP-CVD 条件下)、拉曼 D 峰强度和铜-石墨烯旋转无序性之间的相互关系可通过低能量电子衍射/电子显微镜(LEED/LEEM)对蓝宝石上异质外延 Cu(111)测试得到[56]。900 ℃下形成一些与 Cu(111)存在 0°和 30°旋转的小的石墨烯晶粒,在拉曼光谱中可以看到强烈的 D 带;1 000 ℃下只能看到一种旋转排列(0°)并没有拉曼 D 带[48,56]。Tao 等比较了在低温无氢环境下 Cu(111)表面生长的高品质石墨烯膜和相同条件下 Cu 片上生长的石墨烯膜[66]。

对于 Cu(100)情形则不同。具有 0.255 nm 原子间距的 Cu(100)的正方形晶格给石墨烯生长提供了有别于 Cu(111)的能量分布,从而影响石墨烯生长。一些研究[59,62,63,65]直接报道了这两种不同类型(100)和(111)的铜单晶表面差异,而 Rasool 等[60]运用 STM 从原子尺寸重点研究了 Cu(100),发现石墨烯沿着 Cu(100)晶格的各个方向生长,由此形成了据有大量晶界的多晶薄膜。他们还发现薄膜连续生长并越过台阶、转角和螺纹错位,这就表明基板原子可能对生长过程起了至关重要的作用[60],这与先前的研究一致[52,67]。

在异质外延 Cu(111)和 Cu(100)上生长的薄膜分别沉积在 MgO(111)和 MgO(100)上,结果表明,晶粒的结构、大小和取向都会受到铜晶面的影响,并且 Cu(111)非常适于控制石墨烯取向生长[59]。与 Rasool 等[60]所研究的不同,异质外延 Cu(100)上的石墨烯相对于下方的 Cu[011]晶格表现出两个不同的取向(0°±2°)和(30°±2°)[59],让人想起由 Wofford 等[50]发现的与正方形 Cu(100)基板相关的四叶石墨烯晶粒。在超高真空,900 ℃温度下,由乙烯在 Cu(111)和 Cu(100)上生长的石墨烯增长的 LEEM 图谱表明,前者均匀取向生长,而后者中具有两种取向(此时为 90°旋转)以及存在无取向的小区域[63]。STM 实验以及拉曼光谱和分子动力学模拟研究表明,在 Cu(111)和 Cu(100)上生长石墨烯的物理吸附应变具有很大差异。最后,采用角分辨光电子能谱(ARPES)结合 LEEM/LEED 对两种基板上的石墨烯电子结构进行了详细论述,表明了两种情况下存在方位角无序性[62]。然而生长温度在 850～900 ℃,可能会降低之前所说的取向有序[59]。对于 Cu(111)和 Cu(100)上石墨烯的 0.3 eV 的狄拉克交叉(Dirac crossing)和 250 meV 的带隙是相似的,但对于 Cu(100),样品暴露到空气中会使氧气插层,从而增加狄拉克交叉的偏移量和带隙。

Zhang 等[68]解释了实验结果和铜与石墨烯间弱相互作用之间的矛盾,计算表明,相对于已证实的石墨烯表面和铜之间弱相互作用,石墨烯边缘和铜表面存在较强相互作用。最后得出结论,石墨烯边缘-催化剂相互作用是 CVD 生长石墨烯取向的主要因素[68]。

2.5 周期性堆叠的多层石墨烯

周期性堆叠的多层石墨烯的制备,尤其是双层或三层的,是目前面临的挑战之一。多层石墨烯的优点是比单层石墨烯更适于制备场效应晶体管(FET)[69-75]。所带来的对称破缺引起能隙的打开,这也是石墨烯作为 FETs 用于逻辑电路的先决条件。

尽管从镍基板上 CVD 方法成功得到大面积的双层或三层石墨烯[3],

但层与层之间往往存在旋转错位现象。对于铜基板也会有类似情况发生[1],由于自限生长,多层区域面积会更小。它们通常还会发生方向偏离,然而,在有些情况下会以 AB 堆积双层为主。这种偶尔出现的情况,激发了人们探索 Bernal(AB)甚至是斜方六面体(ABC)堆叠多层的研究热情。

首先,要找到一个便利方法研究这些小到只有几微米的多层片区的统计学特性。带有光学显微镜的拉曼光谱可能是完成这个任务的便利工具,但该方法用于准确描述旋转无序性时,能力非常有限,有必要运用更多的激发波长[22,76,77]。此外,辨别三层石墨烯的扭转角也超出了拉曼光谱的能力范畴。通过 AFM/STM 对 Moiré 结构的观察可以提供一些扭转角的信息,但如果应用于更多层数的样本,不但耗费时间,同时也很难做到对连续薄膜的层数进行测试。

在这方面电子显微镜具有超凡的能力,它不仅能够给出单层石墨烯的结构[78,79],而且能分析层数和多层样本的夹层信息。要完成这个任务,不一定要使用最先进的校正高分辨透射电镜(TEM),一个普通的具有暗场成像条件且可以测试所选区域的电子衍射(SAED)的透射电镜即可[80,81]。Brown 等[80]运用该技术证实大量低压力下 CVD(LP-CVD)生长的双层实际上是 Bernal 堆叠。此外,无取向差的 2LG 区域的扭转角小于 4°,81% 的扭转的 2LG 至少与一个 AB-2LG 直接相连[80]。该现象可以通过层间电位来解释,即导致扭转 2LG 生长中趋于 AB 堆叠的部分。Nie 等[95]用 LEEM/LEED 研究了 AP-CVD 生长样品中几个 2LG 岛的旋转无序性。所有这些都不存在取向,但是他们所研究对象的数量比 Brown 等[80]的要少。

铜上 CVD 生长的 2LG 的第二层(或第三层等)出现另一个不同于堆叠次序的问题,即位置问题,也就是它生长于单层的上面还是下面的问题。Nie 等[44]研究证实了"从下方生长",也就是第二层直接与铜接触,然而拉曼光谱运用同位素标记法测量结果恰好相反,第二层是在顶部的[22]。

虽然有关确切的反应机理一直都存在争论,但在过去一年里,几个小组采用不同方法,成功得到取向多层的石墨烯大面积生长[29,42,82-84]。一种方法是在生长过程中使用更高压力和更高的甲烷浓度,这一做法推翻了自限效应,获得了大量的多层覆盖[15]。第一步,用 70 cm^3/min 的甲烷在 59.99 Pa[25]和 1 000 ℃ 下得到无取向差的 2LG。大尺寸的 AB-2LG 的首次制备是由 Yan 等[84]提出的层层外延。第一层在 10 $cm^3 \cdot min^{-1}$/5 $cm^3 \cdot min^{-1}$ 的甲烷/氢气,46.66 Pa 和 1 000 ℃ 下开始生长。第二步,分离过程,第一层(铜表面)再次被加热到 1 000 ℃,并向炉下方移动。把一个新铜条放置于初始位置,这次加热到 1 040 ℃,新铜条作为第二层外延生长的催化剂来源。第二步条件为 35 cm^3·min^{-1}/2 cm^3·min^{-1} 的甲烷/氢气和93.32 Pa[84]。最后所获得的 AB-2LG 覆盖并不完全,而是面积的67%,晶粒大小达到 50 μm。第二种方法是两步法,这次是一次完成,是由 Bi 等[42]提出的。他们用 5 cm^3·min^{-1}/300 cm^3·min^{-1} 的甲烷/氢气先在 1 080 ℃、大气压下进行 CVD 40 min,第二阶段在稍低温度下继续几分钟。第二阶段理想温度是 1 040 ℃,会使 AB-2LG 石墨烯晶粒尺寸增加到 100 μm,覆盖率达到 97%[42]。铜的温度越接近熔点,碳在铜里的溶解度就越高(1 080 ℃下原子数分数为 0.04%,1 000 ℃下为 0.001%),这有助于给双层"喂料"以便其进一步生长[42]。然而,在 1 000 ℃没有氢气的环境压力下,CVD 制备的多层薄膜生长厚度取决于甲烷束流(大于10 mL/min 获得厚膜,1 mL/min 获得 SLG)[82]。氢气的有无是决定薄膜取向有(无)序的主要因素[82]。对甲烷和氢气的分压力(以及体系的总压力)影响的系统研究可获得生长 1~4 层的最佳条件,进而获得有序的多层[83]。1 000 ℃下甲烷/氢气的流速为 10 cm^3·min^{-1}/300 cm^3·min^{-1},系统总压力从773.26 Pa 调整到 98.66 kPa,也就是说,压力变化范围从低压增加到接近大气压。压力增加导致层数增加,AB-2LG 在总压力12.44 kPa 下形成,对应甲烷479.95 Pa 的分压。早期生长阶段的"冻结"表明,种子层的厚度(层数)在初始阶段就已经确定,所有垂直划分的层同时生长。

可控获得 AB-2LG 的方法不同之处在于铜-镍合金基板的使用[29]。对于"90-10"合金(质量分数:铜88%,镍 9.9%),1 000 ℃下碳溶解度是0.01%,生长反应机理与 Bi 等研究的机理相似,基于 1 000 ℃下冷却时基板上的碳沉淀比标准铜催化的沉淀显著。在 1 050 ℃下用 2 cm^3/min 氢气和 3 cm^3/min 甲烷(甲烷的分压力是 6 Pa)获得覆盖面积大于96%的均匀 AB-2LG[29]。

2.6 CVD 石墨烯的同位素标记

碳有两个稳定的同位素^{12}C 和^{13}C。同位素^{12}C 的量最大,但是同位素^{13}C瞬间聚集也在自然界中存在。合成石墨烯的碳来源往往是甲烷,因为^{13}C甲烷在市面上可以买到,因此利用该丰富的前驱体完全有可能合成^{13}C石墨烯。此外,^{12}C 甲烷和^{13}C 甲烷也可以任意比混合并获得石墨烯。

拉曼光谱是描述石墨烯特性最便捷的工具。它可以区分单层、双层和多层石墨烯[86],它对掺杂和应力也非常灵敏[86-88]。含有^{13}C 材料的拉曼带源于该同位素质量的增加,因此^{13}C 石墨烯和^{12}C 石墨烯可以在拉曼光谱上很容易区分开。

同位素标记可用于记录生长过程[13,22,56]。例如,可以用来研究铜和镍上生长机理的区别[13],以确定石墨烯制备过程中石墨烯层或额外层(add-layer)的生长速度[22],或者可以追踪铜-镍薄片上的 AB-堆叠 2LG 的形成过程[29]。

同位素标记另一个有价值的应用是用来研究由大量单层石墨烯片的转移而形成的多层石墨烯样品[16,89]。每层可能包含不同量的碳同位素,因此单独层可由拉曼光谱追踪记录[16,89]。图 2.4 是 3LG 的简单例子,即包含纯^{13}C 石墨烯层(下)、^{12}C 和^{13}C 的 1∶1 混合物(上)及纯^{12}C 石墨烯层(中)。单独的 1LG 显示出了不同的拉曼峰。同位素标记的石墨烯样本的拉曼带与^{12}C 石墨烯相比没有发生明显宽化,因此,同位素标记对于拉曼实验是均匀的。把这三个单层的石墨烯合在一起可以得到 3LG。3LG 样品的结果表现出了每个单层的 G 和 G′模式。各个单层拉曼峰的位置随着 1LG 拉曼峰的位置改变而改变。每个独立层的拉曼频率偏移都不同,差异与掺杂[90]和压力[86]有关,也不同于基板上的 1LG。因此,同位素标记多层石墨烯是研究基板和环境对石墨烯影响的理想材料[76]。

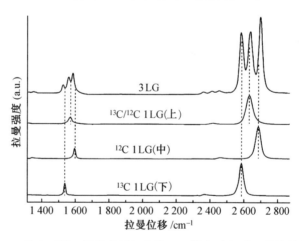

图 2.4 具有不同 C 同位素层组成的 3LG 样本(图上部)的拉曼光谱
注:单独层的拉曼光谱见图下方。垂直虚线为了突出独立层和 3LG 样本之间的拉曼位移中的差异,激发能为 2.33 eV

Bernard 等[22]对^{12}C 甲烷和^{13}C 甲烷上生长的石墨烯的拉曼光谱进行了详细研究,发现了许多与拉曼模式相对应的独特声子作用。利用大量可辨别的拉曼峰,Bernard 等还绘制出石墨烯的平面声子带结构图[92]。碳的同位素标记也可以用于电化学充电过程中多层石墨烯的电子结构变化的研究。图 2.5 所示是 3LG 的原位拉曼光谱电化学测试图。3LG 的掺杂可通过电化学过程实现,并获得了稳定电位下的拉曼光谱。同位素标记让读者看到了掺杂对于单独石墨烯层拉曼光谱的影响。与 1LG[90] 和 2LG[16] 情况一样,3LG 的拉曼光谱也受掺杂水平影响。3LG 中单独层的行为特点取决于它们相对于基板的位置。结果表明了石墨烯环境对于电化学充电过程中拉曼光谱变化的影响程度,认为中间层可表现出最小扰动下石墨烯的行为。

同位素标记可用于研究加工过程对单独石墨烯层的影响。众所周知,由于石墨烯和 SiO_2/Si 基板的热膨胀系数相反,热处理会引起石墨烯层的应力变化。同位素标记 2LG 的拉曼光谱表明层顶部和底部的拉曼带的频率会因温度变化而不同[92]。这也表明基板膨胀产生的应力只对底部直接与 SiO_2/Si 基板接触的石墨烯层产生强烈影响。

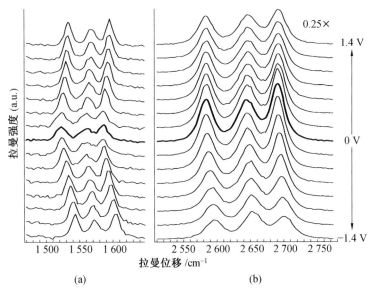

图 2.5　^{13}C(底层)、$^{12}C/^{13}C$(中层)和 ^{12}C(顶层)的 3LG 样本的(a)G 带和(b)G′带的原位拉曼光谱电化学测试图
注:激光激发能为 2.33 eV,电极电位为 −1.4~1.4 V(0.2 V 梯度,从底部到顶部)。中间的粗线光谱对应于 0 V Ag/Ag$^+$

同位素标记另一个应用是对多层石墨烯缺陷形成机理的研究[93-94]。

要准确测定采用^{12}C或^{13}C制得的单层石墨烯的碳原子原子位移截面,需要在一系列加速电压下采用HR-TEM研究完成[94]。^{13}C石墨烯的撞击截面在给定加速电压下明显较低,两个同位素作用差异有助于对辐射破坏机理的理解[94]。另一个研究是,用100 keV氩离子照射同位素标记的原始2LG样本,并进行拉曼光谱分析[93]。基于D/G和D/G′强度比与缺陷变化有关,预期两种石墨烯的缺陷数量随着辐射通量的增加而增加,但由于底层缺陷密度比顶层的更低,层缺陷聚集的速率明显不同。基于二元碰撞模型(binary collision model)和分子动力学模拟势能分析,后期观察与预期的结果正好相反,两层产生的缺陷数目相同。结论表明,当辐射方法被用于石墨烯和其他2D系统性能调整时,二维材料的退火缺陷可能相当重要。

同位素标记也可用于研究更复杂的石墨烯基杂化膜材料。由于这些材料的特殊性能使其成为研究重点。因此,同位素标记是CVD过程的一个重要手段和特征。

2.7 结 论

高品质石墨烯的大规模制备是让这种奇妙材料成功用于目前可预见的广阔领域的重要挑战之一。这方面,基于CVD的方法最具有发展前景,该方法的优势不仅体现在石墨烯生产上,也体现在日益增加的其他2D材料的生长制备方面,如六方氮化硼、多种金属硫化物和其他物质。本章简要介绍了2011~2013年报道的有关CVD石墨烯的一些成就和遇到的问题;同时也提及会面临的很多新问题。需解决的首要问题就是要弄清石墨烯生长过程本身,以及生长与周围环境条件和催化剂表面状态的关系。一旦完全解决这些问题,二维材料的CVD方法的发展将是不可限量的,甚至可以实现理想纳米带的制备以及一步法合成新一代"全2D"电路。

2.8 参考文献

[1] LI X S,CAI W W,AN J H,et al. Large-area synthesis of high-quality and uniform graphene films on copper foils[J]. Science,2009,324:1312.

[2] KIM K S,ZHAO Y,JIANG H,et al. Large-scale pattern growth of graphene films for stretchable transparent electrodes[J]. Nature,2009,457:706.

[3] REINA A,JIA X,HO J,et al. Large area,few-layer graphene films on ar-

bitrary substrates by chemical vapor deposition[J]. Nano Letters, 2009, 9:30.

[4] YU Q, LIAN J, SIRIPONGLERT S, et al. Graphene segregated on Ni surfaces and transferred to insulators[J]. Applied Physics Letters, 2008, 93: 113103.

[5] GAO L, GUEST J R, GUISINGER N P. Epitaxial graphene on Cu(111) [J]. Nano Letters, 2010, 10:3512.

[6] AGO H, ITO Y, MIZUTA N, et al. Epitaxial chemical vapor deposition growth of single-layer graphene over cobalt film crystallized on sapphire [J]. ACS Nano, 2010, 4:7407.

[7] VO-VAN C, KIMOUCHE A, RESERBAT-PLANTEY A, et al. Epitaxial graphene prepared by chemical vapor deposition on single crystal thin iridium films on sapphire[J]. Applied Physics Letters, 2011, 98:181903.

[8] HATTAB H, N'DIAYE A T, WALL D, et al. Interplay of wrinkles, strain, and lattice parameter in graphene on iridium[J]. Nano Letters, 2011, 12: 678.

[9] SUTTER P W, ALBRECHT P M, SUTTER E A. Graphene growth on epitaxial Ru thin films on sapphire[J]. Applied Physics Letters, 2010, 97: 213101.

[10] YOSHII S, NOZAWA K, TOYODA K, et al. Suppression of inhomogeneous segregation in graphene growth on epitaxial metal films[J]. Nano Letters, 2011, 11:2628.

[11] MATTEVI C, KIM H, CHHOWALLA M. A review of chemical vapour deposition of graphene on copper[J]. Journal of Materials Chemistry, 2011, 21:3324.

[12] AGO H, OGAWA Y, TSUJI M, et al. Catalytic growth of graphene: toward large-area single-crystalline graphene [J]. Journal of Physics Chemistry Letters, 2012, 3:2228.

[13] LI X, CAI W, COLOMBO L, et al. Evolution of graphene growth on Ni and Cu by carbon isotope labeling[J]. Nano Letters, 2009, 9:4268.

[14] SHELTON J C, PATIL H R, BLAKELY J M. Equilibrium segregation of carbon to a Ni(111) surface: a surface phase transition[J]. Surface Science, 1974, 43:493.

[15] BHAVIRIPUDI S, JIA X, DRESSELHAUS M S, et al. Role of kinetic

factors in chemical vapor deposition synthesis of uniform large area graphene using copper catalyst[J]. Nano Letters,2010,10:4128.

[16] KALBAC M,FARHAT H,KONG J,et al. Raman spectroscopy and in situ Raman spectroelectrochemistry of bilayer $^{12}C/^{13}C$ graphene[J]. Nano Letters,2011,11:1957.

[17] FUJITA D,YOSHIHARA K. Surface precipitation process of epitaxially grown graphite(0001) layers on carbon-doped Ni(111) surface[J]. Journal of Vaccum Science & Technology A,1994,12:2134.

[18] REINA A,THIELE S,JIA X,et al. Growth of large-area single layer and bilayer graphene by controlled carbon precipitation on polycrystalline Ni surfaces[J]. Nano Research,2009,2:509.

[19] ZHANG Y,GAO T,XIE S,et al. Different growth behaviors of ambient pressure chemical vapor deposition graphene on Ni(111) and Ni films: a scanning tunneling microscopy study[J]. Nano Research,2012,5:402.

[20] ZHANG Y,GOMEZ L,ISHIKAWA F N,et al. Comparison of graphene growth on single-crystalline and polycrystalline Ni by chemical vapor deposition[J]. Journal of Physics Chemistry Letters,2010,1:3101.

[21] LI X,ZHU Y,CAI W,et al. Transfer of large-area graphene films for highperformance transparent conductive electrodes[J]. Nano Letters,2009,9:4359.

[22] KALBAC M,FRANK O,KAVAN L. The control of graphene double-layer formation in copper-catalyzed chemical vapor deposition[J]. Carbon,2012,50:3682.

[23] ADDOU R,DAHAL A,SUTTER P,et al. Monolayer graphene growth on Ni(111) by low temperature chemical vapor deposition[J]. Applied Physics Letters,2012,100:021601-3.

[24] HUANG L,CHANG Q H,GUO G L,et al. Synthesis of high-quality graphene films on nickel foils by rapid thermal chemical vapor deposition [J]. Carbon,2012,50:551.

[25] LEE H,LEE S,HONG J,et al. Graphene converted from the photoresist material on polycrystalline nickel substrate[J]. Japanese Journal of Applied Physics,2012,51:06FD17.

[26] MIYASAKA Y,MATSUYAMA A,NAKAMURA A,et al. Graphene seg-

regation on Ni/SiO$_2$/Si substrates by alcohol CVD method[J]. Physica Status Solidi C,2011,8:577.

[27] LIU X,FU L,LIU N,et al. Segregation growth of graphene on Cu−Ni alloy for precise layer control[J]. The Journal of Physical Chemistry C, 2011,115:11976.

[28] LIU L,ZHOU H,CHENG R,et al. High-yield chemical vapor deposition growth of high-quality large-area AB-stacked bilayer graphene[J]. ACS Nano,2012,6:8241.

[29] WU Y,CHOU H,JI H,et al. Growth mechanism and controlled synthesis of AB-stacked bilayer graphene on Cu−Ni alloy foils[J]. ACS Nano, 2012,6:7731.

[30] CHEN S,CAI W,PINER R D,et al. Synthesis and characterization of large-area graphene and graphite films on commercial Cu−Ni alloy foils [J]. Nano Letters,2011,11:3519.

[31] LIU N,FU L,DAI B,et al. Universal segregation growth approach to wafer-size graphene from non-noble metals[J]. Nano Letters,2010,11: 297.

[32] WANG Y J,MIAO C Q,HUANG B C,et al. Scalable synthesis of graphene on patterned Ni and transfer[J]. IEEE Transaction on Electron Devices,2010,57:3472.

[33] HAN G H,GÜNEŞ F,BAE J J,et al. Influence of copper morphology in forming nucleation seeds for graphene growth[J]. Nano Letters,2011, 11:4144.

[34] NIE S,WOFFORD J M,BARTELT N C,et al. Origin of the mosaicity in graphene grown on Cu(111)[J]. Physics Review B,2011,84:155425.

[35] LUO Z,KIM S,KAWAMOTO N,et al. Growth mechanism of hexagonal-shape graphene flakes with zigzag edges[J]. ACS Nano,2011,5:9154.

[36] LUO Z,LU Y,SINGER D W,et al. Effect of substrate roughness and feedstock concentration on growth of wafer-scale graphene at atmospheric pressure[J]. Chemistry of Materials,2011,23:1441.

[37] ZHANG B,LEE W H,PINER R,et al. Low-temperature chemical vapor deposition growth of graphene from toluene on electropolished copper foils[J]. ACS Nano,2012,6:2471.

[38] LI X,MAGNUSON C W,VENUGOPAL A,et al. Large-area graphene

single crystals grown by low-pressure chemical vapor deposition of methane on copper[J]. The Journal of American Chemical Society, 2011, 133:2816.

[39] VLASSIOUK I, REGMI M, FULVIO P, et al. Role of hydrogen in chemical vapor deposition growth of large single-crystal graphene[J]. ACS Nano, 2011, 5:6069.

[40] WANG H, WANG G, BAO P, et al. Controllable synthesis of submillimeter single-crystal monolayer graphene domains on copper foils by suppressing nucleation[J]. The Journal of American Chemical Society, 2012, 134:3627.

[41] ZHANG Y, ZHANG L, KIM P, et al. Vapor trapping growth of single-crystalline graphene flowers: synthesis, morphology, and electronic properties[J]. Nano Letters, 2012, 12:2810.

[42] BI H, HUANG F, ZHAO W, et al. The production of large bilayer hexagonal graphene domains by a two-step growth process of segregation and surface-catalytic chemical vapor deposition[J]. Carbon, 2012, 50:2703.

[43] LIU L, ZHOU H, CHENG R, et al. A systematic study of atmospheric pressure chemical vapor deposition growth of large-area monolayer graphene[J]. Journal of Materials Chemistry, 2012, 22:1498.

[44] NI G X, ZHENG Y, BAE S, et al. Quasi-periodic nanoripples in graphene grown by chemical vapor deposition and its impact on charge transport[J]. ACS Nano, 2012, 6:1158.

[45] YAN Z, LIN J, PENG Z, et al. Toward the synthesis of wafer-scale single-crystal graphene on copper foils[J]. ACS Nano, 2012, 6:9110.

[46] WOOD J D, SCHMUCKER S W, LYONS A S, et al. Effects of polycrystalline Cu substrate on graphene growth by chemical vapor deposition [J]. Nano Letters, 2011, 11:4547.

[47] KIDAMBI P R, DUCATI C, DLUBAK B, et al. The parameter space of graphene chemical vapor deposition on polycrystalline Cu[J]. The Journal of Physical Chemistry C, 2012, 116:22492.

[48] OROFEO C M, HIBINO H, KAWAHARA K, et al. Influence of Cu metal on the domain structure and carrier mobility in single-layer graphene [J]. Carbon, 2012, 50:2189.

[49] CHO J, GAO L, TIAN J, et al. Atomic-scale investigation of graphene

grown on Cu foil and the effects of thermal annealing[J]. ACS Nano, 2011,5:3607.

[50] WOFFORD J M, NIE S, MCCARTY K F, et al. Graphene islands on Cu foils: the interplay between shape, orientation, and defects [J]. Nano Letters,2010,10:4890.

[51] WU Y A, FAN Y, SPELLER S, et al. Large single crystals of graphene on melted copper using chemical vapor deposition[J]. ACS Nano,2012,6: 5010.

[52] TIAN J, CAO H, WU W, et al. Graphene induced surface reconstruction of Cu[J]. Nano Letters,2012,12:3893.

[53] YU Q, JAUREGUI L A, WU W, et al. Control and characterization of individual grains and grain boundaries in graphene grown by chemical vapour deposition[J]. Nature Materials,2011,10:443.

[54] KHOMYAKOV P A, GIOVANNETTI G, RUSU P C, et al. First-principles study of the interaction and charge transfer between graphene and metals[J]. Physics Review B,2009,79:195425.

[55] MI X, MEUNIER V, KORATKAR N, et al. Facet-insensitive graphene growth on copper[J]. Physics Review B,2012,85:155436.

[56] HU B, AGO H, ITO Y, et al. Epitaxial growth of large-area single-layer graphene over Cu(111)/sapphire by atmospheric pressure CVD[J]. Carbon,2012,50:57.

[57] ISHIHARA M, KOGA Y, KIM J, et al. Direct evidence of advantage of Cu(111) for graphene synthesis by using Raman mapping and electron backscatter diffraction[J]. Materials Letters,2011,65:2864.

[58] MILLER D L, KELLER M W, SHAW J M, et al. Epitaxial(111) films of Cu, Ni, and Cu_xNi_y on α-Al_2O_3(0001) for graphene growth by chemical vapor deposition[J]. Journal of Applied Physics,2012,112:064317.

[59] OGAWA Y, HU B, OROFEO C M, et al. Domain structure and boundary in single-layer graphene grown on Cu(111) and Cu(100) films[J]. Journal of Physics Chemistry Letters,2011,3:219.

[60] RASOOL H I, SONG E B, MECKLENBURG M, et al. Atomic-scale characterization of graphene grown on Cu(100) single crystals [J]. The Journal of American Chemical Society,2011,133:12536.

[61] TAO L, LEE J, HOLT M, et al. Uniform wafer-scale chemical vapor dep-

osition of graphene on evaporated Cu(111) film with quality comparable to exfoliated monolayer[J]. The Journal of Physical Chemistry C,2012, 116:24068.
[62] WALTER A L,NIE S,BOSTWICK A,et al. Electronic structure of graphene on single-crystal copper substrates[J]. Physics Review B,2011, 84:195443.
[63] ZHAO L,RIM K T,ZHOU H,et al. Influence of copper crystal surface on the CVD growth of large area monolayer graphene[J]. Solid State Communications,2011,151:509.
[64] REDDY K M,GLEDHILL A D,CHEN C-H,et al. High quality, transferrable graphene grown on single crystal Cu(111) thin films on basal-plane sapphire[J]. Applied Physics Letters,2011,98:113117.
[65] HE R,ZHAO L,PETRONE N,et al. Large physisorption strain in chemical vapor deposition of graphene on copper substrates[J]. Nano Letters, 2012,12:2408.
[66] TAO L,LEE J,CHOU H,et al. Synthesis of high quality monolayer graphene at reduced temperature on hydrogen-enriched evaporated copper (111) films[J]. ACS Nano,2012,6:2319.
[67] RASOOL H I,SONG E B,ALLEN M J,et al. Continuity of graphene on polycrystalline copper[J]. Nano Letters,2010,11:251.
[68] ZHANG X,XU Z,HUI L,et al. How the orientation of graphene is determined during chemical vapor deposition growth[J]. Journal of Physics Chemistry Letters,2012,3:2822.
[69] BAO W,JING L,VELASCO J,et al. Stacking-dependent band gap and quantum transport in trilayer graphene[J]. Nature Physics,2011,7: 948.
[70] CASTRO E V,NOVOSELOV K S,MOROZOV S V,et al. Biased bilayer graphene:Semiconductor with a gap tunable by the electric field effect [J]. Physics Review Letters,2007,99:216802.
[71] CRACIUN M F,RUSSO S,YAMAMOTO M,et al. Trilayer graphene is a semimetal with a gate-tunable band overlap[J]. Nature Nanotechnology, 2009,4:383.
[72] LUI C H,LI Z,MAK K F,et al. Observation of an electrically tunable band gap in trilayer graphene[J]. Nature Physics,2011,7:944.

[73] OHTA T, BOSTWICK A, SEYLLER T, et al. Controlling the electronic structure of bilayer graphene[J]. Science, 2006, 313:951.

[74] ZHANG Y B, TANG T T, GIRIT C, et al. Direct observation of a widely tunable bandgap in bilayer graphene[J]. Nature, 2009, 459:820.

[75] ZHANG W, LIN C-T, LIU K-K, et al. Opening an electrical band gap of bilayer graphene with molecular doping[J]. ACS Nano, 2011, 5:7517.

[76] KALBAC M, FRANK O, KONG J, et al. Large variations of the Raman signal in the spectra of twisted bilayer graphene on a BN substrate[J]. Journal of Physics Chemistry Letters, 2012, 3:796.

[77] KIM K, COH S, TAN L Z, et al. Raman spectroscopy study of rotated double-layer graphene: misorientation-angle dependence of electronic structure[J]. Physics Review Letters, 2012, 108:246103.

[78] AN J, VOELKL E, SUK J W, et al. Domain (grain) boundaries and evidence of twinlike structures in chemically vapor deposited grown graphene[J]. ACS Nano, 2011, 5:2433.

[79] HUANG P Y, RUIZ-VARGAS C S, VAN DER ZANDE A M, et al. Grains and grain boundaries in single-layer graphene atomic patchwork quilts[J]. Nature, 2011, 469:389.

[80] BROWN L, HOVDEN R, HUANG P, et al. Twinning and twisting of trilayer and bilayer graphene[J]. Nano Letters, 2012, 12:1609.

[81] PING J, FUHRER M S. Layer number and stacking sequence imaging of few-layer graphene by transmission electron microscopy[J]. Nano Letters, 2012, 12:4635.

[82] DIAZ-PINTO C, DE D, HADJIEV V G, et al. AB-stacked multilayer graphene synthesized via chemical vapor deposition: a characterization by hot carrier transport[J]. ACS Nano, 2012, 6:1142.

[83] SUN Z, RAJI A-R O, ZHU Y, et al. Large-area Bernal-stacked bilayer, trilayer, and tetralayer graphene[J]. ACS Nano, 2012, 6:9790.

[84] YAN K, PENG H, ZHOU Y, et al. Formation of bilayer bernal graphene: layer-by-layer epitaxy via chemical vapor deposition[J]. Nano Letters, 2011, 11:1106.

[85] FERRARI A C, MEYER J C, SCARDACI V, et al. Raman spectrum of graphene and graphene layers[J]. Physics Review Letters, 2006, 97:187401.

[86] FRANK O, TSOUKLERI G, PARTHENIOS J, et al. Compression behavior of single-layer graphenes[J]. ACS Nano,2010,4:3131.

[87] FRANK O, TSOUKLERI G, RIAZ I, et al. Development of a universal stress sensor for graphene and carbon fi bres[J]. Nature Communications,2011,2:255.

[88] MOHIUDDIN T M G, LOMBARDO A, NAIR R R, et al. Uniaxial strain in graphene by Raman spectroscopy: G peak splitting, Grueneisen parameters, and sample orientation[J]. Physics Review B, 2009, 79: 205433.

[89] KALBAC M, KONG J, DRESSELHAUS M S. Raman spectroscopy as a tool to address individual graphene layers in few-layer graphene[J]. The Journal of Physical Chemistry C,2012,116:19046.

[90] KALBAC M, REINA-CECCO A, FARHAT H, et al. The Influence of strong electron and hole doping on the Raman intensity of chemical vapor-deposition graphene[J]. ACS Nano,2010,4:6055.

[91] BERNARD S, WHITEWAY E, YU V, et al. Probing the experimental phonon dispersion of graphene using ^{12}C and ^{13}C isotopes[J]. Physics Review B,2012,86:085409.

[92] KALBAC M, FRANK O, KAVAN L. Effects of heat treatment on Raman spectra of two-layer $^{12}C/^{13}C$ graphene[J]. Chemistry-A European Journal,2012,18:13877.

[93] KALBAC M, LEHTINEN O, KRASHENINNIKOV A V, et al. Ionirradiation-induced defects in isotopically-labeled two layered graphene: enhanced in-situ annealing of the damage[J]. Advanced Materials,2012, 25:1004.

[94] MEYER J C, EDER F, KURASCH S, et al. Accurate measurement of electron beam induced displacement cross sections for single-layer graphene[J]. Physics Review Letters,2012,108:196102.

[95] NIE S, WU W, XING S, et al. Growth from below: bilayer graphene on copper by chemical vapor deposition[J]. New Journal of Physics,2012, 14:093028.

第3章　化学法制备石墨烯

本章论述了通过石墨的化学功能化合成石墨烯,包括探究悬浮液中生成的氧化石墨烯的原子结构。讨论了氧化石墨烯的基本性能和一系列应用方面的最新进展。

3.1 引　　言

由于对石墨烯的兴趣与日俱增,全世界的科学家们热衷于研究各种制备石墨烯的方法。制备多层石墨烯的开拓性方法是对天然石墨的机械剥离。尽管这种方法可以得到高品质的多层,但终因耗时多、手工操作而被冷落。一种可能解决这些问题的方法是基于溶液技术分离石墨层而得到石墨烯的悬浮液。已经出现了许多相关方法,并遵循相同的基本原则,就是利用插层或对单独层进行功能性改性,进而减弱石墨层间的范德瓦耳斯力来实现液态剥离。这一方法为大量生产石墨烯提供可能性及多样性,能很好地实现化学功能化,因此具有广泛的应用发展前景。

石墨层间化合物或膨胀石墨是获得单层石墨烯片的胶体分散液的初始材料。理论上讲,这种方法适用于高品质单层石墨烯的制备。有机溶液中石墨烯层可发生胶体分散,例如,在有机溶剂 N-甲基吡咯烷酮(NMP)中对石墨粉进行超声处理,但是横向尺寸只有几百个纳米,且产率很低[1]。利用这种石墨烯制成的很薄的膜进行电阻率测量,在光学透明度达到 42% 的片上电导率为 $6\,500\,S \cdot m^{-1}$。此外,即使薄膜经过 400 ℃ 的干燥,剩余的 NMP 也不会完全消除,据估计也占质量的 7%。文献[2]报道了一种温和溶解途径,采用中性石墨,不用任何超声方法,就可以获得大尺寸的石墨烯片。这个方法可以通过搅拌 NMP 中的三元钾盐 $K(THF)_xC_{24}$(一种石墨插层化合物)来实现。聚合物包覆的石墨烯衍生物通过以下过程获得,将购买的膨胀石墨进行热处理,并置于溶解有聚(间-对苯乙炔-共 2,5-二辛酯-p-对苯撑乙炔)(PmPV)的二氯甲烷中,进行超声处理[3]。另一个方法(图 3.1)用发烟硫酸插层,四丁基氢氧化物(TBA)膨胀,在 N,N-对二甲基甲酰胺(DMF)中超声,所获得石墨烯 90% 为修饰的单层,最终获得 1,2-二硬酯酰-sn-甘油(snglycer)-3-磷酸乙醇胺-N-甲氧基(聚乙二醇)-

5000(DSPE-mPEG)包覆的石墨烯片的悬浮液[4]。

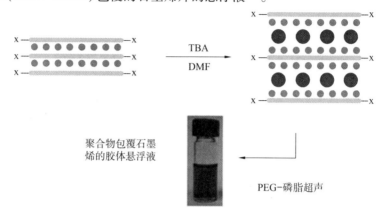

图3.1 采用硫酸(发烟硫酸)(小圆点)插层,随后是TBA(大圆点)插层后石墨进行剥离的示意图。经过磷脂聚合物中超声,插层材料转变为分离石墨烯层构成的黑色胶体悬浮液[4]

已经出现了大量采用液体中剥离而得到稳定的石墨烯胶状悬浮液的方法[5-7],其中最有发展前景、成本低、可大规模生产,并得到广泛研究的合成方法是对氧化的石墨层进行还原,该方法可在多种基板上实现可控密度的沉积。氧化石墨法制备石墨烯的主要优点是可以与现有的薄膜电子技术自上而下的方法兼容实现合成、加工并集成。此外,化学法制备石墨烯在工业规模生产方面具有潜力[8]。Stankovich[9-10]提出了化学合成氧化石墨的技术,将其剥离为单独的氧化石墨烯(graphene oxide(GO))并进行还原。以下章节对该方法会有详细介绍。

3.2 氧化石墨烯的合成

尽管GO工艺对于石墨烯而言相对新颖,但它的历史可追溯到19世纪初期。在此之后,氧化石墨主要由Brodie[12]、Staudenmaier[13]和Hummers[14]提出的方法制备。

最早的报道是英国化学家B. C. Brodie发表的,他通过研究片状石墨的反应活性来探究石墨的结构。把氯酸钾($KClO_3$)加入到石墨和发烟硝酸(HNO_3)的混浆中,得到由碳、氢和氧组成的产物,导致石墨片总质量增加。多次氧化处理导致氧含量进一步增加,经过4次反应后达到一个极限。组成比例为61.04%碳、1.85%氢和37.11%氧(质量分数)。此外,

Brodie 可以将材料分散到水或碱性溶液中,但不能在酸性溶液中。加热到 220 ℃ 使碳的质量分数增加到 80.13%。尽管 Brodie 没能准确确定石墨的相对分子质量,但是他发现了制备氧化石墨的方法。

德国化学家 Staudenmaier 改进了 Brodie 的方法,第一步,按照反应过程将 $KClO_3$ 分为多部分添加;第二步,添加浓硫酸来增加混合物的酸性。大约 60 年后,Hummers 和 Offeman 发现了另一种氧化方法,让石墨与高锰酸钾和浓硫酸的混合物进行反应,达到相似的氧化水平。2014 年以前虽然又出现了很多改进的方法得到氧化石墨,但本质一直未改变。

Hummers 用高锰酸钾($KMnO_4$)和硫酸(H_2SO_4)混合。虽然高锰酸是公认的氧化剂,但在石墨的氧化过程中真正的活性组分是 Mn_2O_7,它以 $KMnO_4$ 和 H_2SO_4 反应形成的棕红色油的形式存在(反应式(3.1))。Mn_2O_7 比 MnO_4^- 的活性大很多,当加热到高于 55 ℃ 或与有机化合物接触时会发生爆炸反应[15,16]。Trömel 和 Russ 发现,Mn_2O_7 能够有选择地氧化脂肪族不饱和双键,而不会氧化芳香族中的双键,这对于石墨结构和氧化过程中的反应路径具有重要的意义[17]。

$$\left.\begin{array}{l} KMnO_4+3H_2SO_4 \longrightarrow K^++MnO_3^++H_2O^++3HSO_4^- \\ MnO_3^++MnO_4^- \longrightarrow Mn_2O_7 \end{array}\right\} \quad (3.1)$$

通过这种方法可对市面上各种原料氧化制备氧化石墨,其中,片状石墨最常用,它是由天然存在的矿石进行提纯除去杂质获得的。由于片状石墨的复杂性以及结构中存在的固有缺陷,因此确定石墨氧化的机理也非常具有挑战性。

这种方法制得的氧化石墨以棕色黏稠泥浆的形式存在,它不仅包括氧化石墨,还包括非氧化的石墨化颗粒和反应副产物的残余物。可以通过离心、沉淀和透析的方法,除去氧化过程中的盐和离子,得到纯 GO 悬浮[18-21]。通过进一步离心,去除未氧化的石墨颗粒和较重的氧化石墨片层,得到单层 GO 的悬浮液。单分散的 GO 片(根据侧面尺寸分离)悬浮液可以通过梯度离心得到[22-23]。

3.3 氧化石墨烯(GO)的还原

虽然 GO 自身有绝缘性,它的碳结构骨架也可以通过热退火或化学还原剂进行还原,得到还原的氧化石墨烯(reduced graphene oxide (RGO))。早期实现该过程的研究主要采用水合肼[10]。但它既有剧毒又有潜在的爆炸性,因此,人们又尝试了许多可取代肼的工艺方法。有研究称硼氢化钠($NaBH_4$)作为 GO 的还原剂比肼更有效[24]。尽管 $NaBH_4$ 会缓慢水解,但这个过程在动力学上进行得很缓慢,新配制的溶液足以作为 GO 的还原剂使用。可用于 GO 还原的还原剂还有对苯二酚[25]、气态氢[26]、维生素 C(抗坏血酸)[27]和强碱性溶液[28]。

更简单但同样有效的还原 GO 的方法是采用温和的氢等离子体,可以达到类似于其他还原方法的效果[29]。通过在真空或充入氩气下,进行热退火,可以进一步提高还原效果。对该方法的研究发现,温度是很重要的影响因素[30]。一直以来,大量研究专注寻找能替代化学方法的途径,例如电化学还原[31-33]、光催化还原[34]及摄影闪光灯还原[35]。

RGO 中的缺陷可以通过高温下(800 ℃)引入碳源(如乙烯)而得到弥补,这类似于单壁碳纳米管(SWNTs)的 CVD 生长条件。这种碳的引入使单独 RGO 层的电阻下降到约 28.6 $k\Omega \cdot sq^{-1}$(电导率为 350 $S \cdot cm^{-1}$)[36]。

3.4 氧化石墨烯的物理化学结构

为了确定氧化石墨的化学结构,研究人员已经做了大量研究,也提出了几种模型(图3.2)。早期有关结构的解释为,氧通过环氧的形式(C_2O)与六边形平面上的碳原子相作用(Hofmann)[37]。这种情况下石墨层的平整性被很大限度地保留了下来,但是在后来的模型中受到了挑战:椅型环己烷构成弯曲的碳层,通过和轴向氢氧根结合以及 1,3 位的醚键结合使碳饱和(Ruess)[38]。这个模型首次提出了 GO 中氢的存在,后来又采用 C═C 双键、酮和烯醇进行修正[39]。此外,材料的酸性表明面内烯醇(—OH)和羧酸基团存在。这种模型后来被 Scholz 和 Boehm[40]采用立体化学进行修正。另一个模型则提出了一个类似于多氟烃的结构[41]。

近期模型集中在非化学计量的无定形模型,而不是晶格模型。例如 Lerf 等提出[43],基于 NMR 研究,把 GO 描述成未氧化苯环及含有(C═C)、1,2-二醇醚和羟基的扭曲六边形随机分布。人们对 Scholz 等提出的氧化形成表面功能团的模型的发展进行了研究,包括几个立体化学改性[42]。Lerf 等用固态核磁共振光谱(NMR)对材料进行研究。^{13}C 标记的氧化石墨的 NMR 研究表明石墨的 sp^2 键合的碳网络受到强烈破坏,碳网络的很大一部分与羟基或环氧化物发生了键合,认为在氧化石墨的各层边缘存在微量的羧基或羰基[44]。

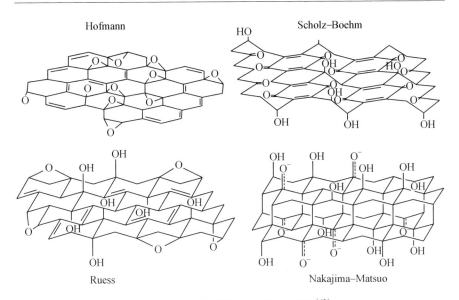

图 3.2 氧化石墨烯的早期结构模型[42]

基于著名的 Lerf 和 Klinowski 模型,出现了 Dékány 模型,认为其存在由环己基连接的规则波纹状醌类结构(图 3.3)。Dékány 模型由两个不同的区域组成:含有叔基醇和 1,3-醚的反联环己基,以及醌类波形网络。该模型认为 GO 中没有羧酸。进一步的氧化反应通过 1,2-醚的形成破坏了醌类烯烃以及用于合成的初始氧化条件下存留的任何芳香族化合物。也有假设认为醌类结构导致了石墨烯的刚性和晶界,也会导致 TEM 图像中常见的宏观褶皱[42]。

近几年,由于拉曼光谱具有非破坏性、快速、高分辨率的特点,并能提供大量的结构和电子信息,因此拉曼光谱成为研究碳纳米结构的重要工具[45]。高度有序的石墨只显示几个明显的拉曼活性带,也就是在 1 600 cm^{-1} 下观察到的石墨晶格的 C-C 伸缩振动(G 带)和约 1 355 cm^{-1} 的由石墨边缘引起的弱无序带(D 带)。图 3.4 是单层 GO 和 RGO 的典型拉曼光谱,激发波长为 532 nm。由于氧化过程中出现石墨晶格的无序性,导致石墨向氧化石墨的转变时 G 和 D 带加宽;还观察到 G 带向更高频率偏移的过程[45]。这种向高频的偏移可能由于隔离双带在高于石墨的频率下发生了共振。GO 的还原把 G 带恢复到几乎与石墨烯相同的位置,说明石墨的晶格出现显著的恢复[46]。机械剥离石墨烯光谱的显著特点是,具有 30 cm^{-1} 的半峰宽的尖锐 2D 带,不同于 GO 半峰宽约 200 cm^{-1}。此外,缺陷引发的 D+D′峰在 2 950 cm^{-1} 处[47]。整个拉曼峰的强度在还原后降低,表

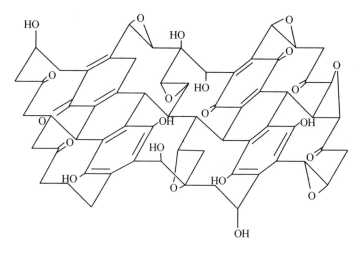

图 3.3 由 Dékány 等提出的 GO 结构[42]

明在还原过程中存在碳损失[48]。

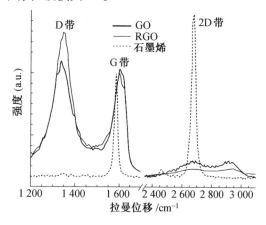

图 3.4 SiO_2/Si 基板上单层 GO、RGO 和机械剥离石墨烯的拉曼光谱[30]

GO 的还原由于去除了氧和碳原子而引起结构的变化[49-50]。拉曼 D 峰和 G 峰的面积比可以用来衡量 sp^2 簇组在 sp^2 和 sp^3 网络中的大小。Tuinstra-Koenig 的关系[51]给出了 D/G 强度比和石墨化样本的晶粒大小之间的关系,并提出该强度比与平均晶粒大小成反比。基于这个关系,可以计算出 GO 中石墨化晶粒的平均大小,数值为 2.5~6 nm[29,49]。文献[9,10,49,52-55]关于还原过程对 GO 的 sp^2 区域大小的影响与 D/G 比例变化的解释一直是模棱两可[9,10,49,52-55]。通常出现显著的 D 峰信号,说明还原样本中有明显的无序性。然而 Tuinstra-Koenig 关系在高于临界缺陷密度时是无效的。对于高无序性材料，如果 sp^2 区域尺寸小于 3 nm，那么 D/G 比例就会随着芳香环数量的增加而增加，这与 Tuinstra-Koenig 关系相背离[45]。

为了辨别 GO/RGO 中 sp^2 区域的确切排布，一些研究采用了显微技术进行直接成像的方法。通过扫描隧道显微镜(STM)探究 GO 的表面情况发现，其结构中包含高度缺陷区域，可能由于近乎完整的区域包围的氧的出现，如图 3.5(a)、图 3.5(b)所示[29,46,57]。STM 的傅里叶变换表明长程结晶有序[46]。

为了弄清 GO/RGO 的原子结构，许多研究运用了透射电子显微镜(TEM)[58-59]。GO 单层对电子束的固有透明度与无定形碳支撑膜相比，不仅可以利用 TEM 获得其晶格图像，而且 GO 也可以用作 TEM 的支撑膜[60]。Gomez-Navarro 和 Meyer 等在碳微栅上制备单层 RGO[59]。衍射分析表明单层仅表现出一个六边形的形貌，如图 3.5(d)所示，这表明片层中存在长程六角有序[61]。而少层区域也具有多重的六边形形貌，如图 3.5(c)所示，表明多层以乱层形式堆叠，不同于石墨和机械剥离少层石墨烯中的 AB Bernal 堆叠所具有的弱相互作用[61]。上述结果考虑到 GO 平面上功能基团减弱了堆叠层间的相互作用[58-60]。

Gomez-Navarro 和 Meyer 等得到单层石墨烯的像差校正高分辨率图像，并由此可以看到 RGO 确切的原子结构。图 3.5(f)是将图 3.5(e)TEM 图像中的不同区域用颜色标识。这些结果清楚地揭示了石墨烯片层的大部分区域存在石墨烯的六边形晶格(浅灰色)。可分辨的结晶良好的区域尺寸为 3~6 nm，数据表明它们的覆盖面积为 60%。而大范围区域却由碳质吸附物所覆盖，也包括陷入的重原子(深灰色)[62-63]。另一个特征为，在干净的区域内观察到大量的拓扑缺陷，与完美的机械剥离石墨烯形成鲜明对比。进一步研究孤立的拓扑缺陷(五边形-七边形组合，黑色)和连续的(聚集)拓扑缺陷发现，其以准非晶单层碳的形式(交叉阴影)出现在这些区域里。连续的拓扑缺陷覆盖了表面的 5%，直径达 1~2 nm。尽管出现大量的缺陷，但仍保持着长程取向有序[58-59]。

TEM 研究进一步表明，与 GO 样本结构相比，除去官能团[59]后 RGO 结构仍然保留拓扑缺陷，GO 显示出更强的被吸附物覆盖的能力且表面上没有连续缺陷的特征[64]。由于 GO 的合成过程，大部分的氧都是以共价键合官能团的形式存在。这导致碳原子大部分是 sp^3 杂化(约 60%)，还原使

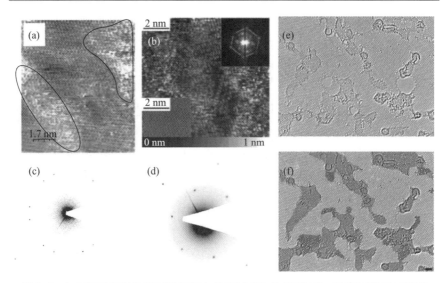

图 3.5 (a)高度有序热解石墨基板上的单层 GO 的 STM 图。轮廓区域表示填充了氧官能团[29]。(b)HOPG 基板上的单层 RGO 的 STM 图像及其傅里叶变换(右上角)左下角是相同条件下的 HOPG 表面 STM 图[46]。(c)双层区域的电子衍射图,显示出片层取向不匹配的堆叠。(d)单层衍射图,单层 RGO 膜的原子分辨率的像差校正的 TEM 图。(e)原始图。(f)调整之后的图,突出不同特征((c)、(d)、(e)和(f)来源于参考文献[56])

这部分降到 20%[49]。单层 GO 的原子力(AFM)厚度约为 1 nm,比理想样品的厚度大很多,这是由于碳原子层上方或下方存在官能团。这一结论被 Pacilé 等进一步证实[64],他们采用近边 X 射线和精密结构(NEXAFS)发现,多层 GO 彼此间的相互作用被大大减弱。这一研究充分利用了偏振对测试的依赖性,通过观察氧 K 边缘光谱,发现羧基、环氧化物和羟基都与苯环相连,羧基可能存在于薄膜的边缘部分[64]。

关于 GO 固有官能团排布的报道一直存在争议性。Pandey 等[65]运用高分辨率的 STM 图像,发现氧原子在指向环氧基团的矩形晶格中规则排布。经密度泛函理论(DFT)计算证实,这一发现极具价值[66]。采用 NEXAFS 的类似研究表明,由于光的偏振使氧边缘谱的精密结构表现出非常尖锐鲜明的特点,说明了氧原子局部的有序排列[64]。然而从图 3.5(c)和图 3.5(d)所示的电子衍射图中看不到衍射斑点,不同于石墨晶格,这一现象表明任何含氧官能团都不具备结构规整性[60]。

正如大量报道所证实的,GO 的结构及其还原形式的确是一个有趣的研究课题。通过这些报道可以得出,GO/RGO 的物理结构是由连续和独立的拓扑缺陷区域包围的石墨化对称区域组成的。还可以得出由于石墨的氧化过程,GO 中的主要官能团包括羟基、环氧基和羰基,它们的量在还原后明显减少。

3.5 氧化石墨烯的电传输

由前面章节可知,氧化石墨可以制成连续胶体悬浮液,所得片层具有绝缘性。然而采用光刻限定电极对 RGO 片层进行双探针传输测试,结果表明还原过程导致电导率提高约三个数量级(图 3.6(a))。室温测量电导率值可达到 $0.1 \sim 3$ S·cm^{-1},与先前的研究一致[29]。这个范围值低于天然石墨烯约三个数量级[67]。

双探针结构的测量电阻被认为是石墨烯层和触点共同作用的结果。对机械剥离石墨烯而言,该方法带来一个问题,即金属触点具有浸润性[68-69]。对 RGO 器件来说,情况则有所不同,触电对测试过程的贡献很小,电压降主要沿着电流隧道产生。如图 3.6(b)所示,具有不同隧道长度的装置在 100 mV 漏源极偏压下进行双探针电阻的测量。电阻随横坐标呈线性增加并经过原点,说明电阻仅源于 RGO 隧道。这点通过光电流显微镜测试也得到进一步证实,研究表明相比于微机械剥离样品,RGO 片层[70]和金属接触点存在很少或没有相互作用[68]。

首次报道的 RGO 场效应晶体管采用了空穴和电子迁移率为 2 cm^{-2}·V·s和0.2 cm^{-2}·V·s 的单层 RGO[29,36,55,71-74]。为制造单 RGO 器件,独立的 GO 片层被沉积到带预先图案化的基板上,用光学显微镜或 AFM 定位,按照标准平版印刷过程实现金属电极接触。室温真空下进行的测量主要表现出 p 型行为。而器件在惰性环境下测量表现出石墨烯的双极场效应(ambipolar field effect)特征。Jung 等通过研究 FET 性能发现了 GO 对空气的高敏感性[30]。当样品暴露于空气中时,如图 3.6(c)所示,在栅极电压扫描过程中 FETs 表现出高滞后,该滞后通过图 3.6(d)中真空测量结果获得。

对于机械剥离石墨烯和单壁碳纳米管(SWNTs)也存在类似的偏移,可以通过氧和(或)水吸附引起的掺杂来进行解释[75]。缺陷位置处和碳纳米管悬挂键的水分子吸附已经得到证实[76]。通过实验[77]和理论研究[75]证明石墨烯表面吸附的水分可作为电子受体。有报道称石墨烯表面水分吸附伴随着空穴注入。当吸收分子(水)的电子亲和势大于基板(石墨烯)的功函数时,有望获得 p 型掺杂。只有当吸收物未占据的电子态能量低于基

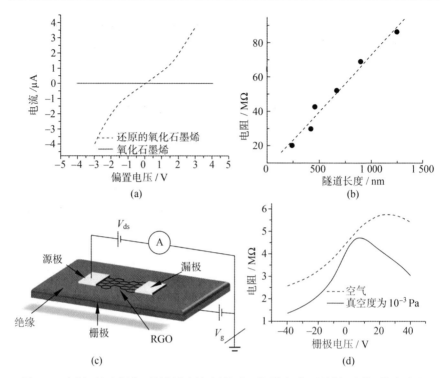

图 3.6 室温下(a)氧化石墨烯(连续直线)和还原的氧化石墨烯(虚线)的电流和电压曲线图。经过 5 s 氢等离子处理后,绝缘 GO 转变成导电的还原态。(b)室温电阻(100 mV 下测量)随着隧道长度增加而线性增加,说明 GO 片层的固有电阻占主导,接触电阻微乎其微。(c)RGO 场效应晶体管的构造图。(d)室温下空气和真空中测得 RGO 片层电阻与栅极电压的关系

注:$V_g = V_G$

板的最高占据态能量时,上述情况才可能发生。理论计算表明该体系符合上述条件[75]。

根据前面章节的叙述,RGO 包含由氧化过程产生的缺陷区域。因此,RGO 就是由高导电区域和无序区穿插组成的(图 3.5)。Mott 描述了在极低温度下,当没有足够热能使电荷从价带跃迁到导带时,电荷传导以局域态之间跳跃的方式进行[78]。能量低时,载流子跳跃距离增加,这就是的变程跃迁。按照 Mott 理论,该体系的电导率表示为

$$\sigma = A\exp\left(-\frac{T_0}{T^{\frac{1}{d+1}}}\right) \quad (3.2)$$

式中,A 是一个常量;T_0 是临界温度;d 是体系的维数。二维体系中,$\sigma \approx$

$T^{-1/3}$。

值得注意的是，在 Mott 的变程跳跃(VRH)中，局域态中的库仑相互作用可以忽略，载流子跳跃也不需要能量。因为 RGO 具有很多限制电荷载流子的缺陷，所以温度依赖性遵循具有金属的温度无关项的修正变程跳跃模型。低温下这种作用可以被识别(图3.7)，并与单壁碳纳米管组成的网络结构相似[79]。如图3.7(a)所示，Kaiser 等[80]认为，这一行为可通过无序势垒区中的 2D VRH 和一个常数项(例如隧穿高导电区之间薄势垒)共同组成的均匀导体模型进行描述。对于这种模型，温度与电导率 σ 的关系为

$$\sigma = \sigma_1 \exp\left(-\frac{B}{T^{1/3}}\right) + \sigma_2 \tag{3.3}$$

其中，第一项是一般 2D VRH 的电导率表达式；第二项表示没有热激发的纯场驱电导；跳跃参数 B 取决于费米能级附近的态密度 $N(E_F)$ 和所涉及的电子波函数的局域长度(localisation length) L_1。对于 2D VRH 来说，B 值由下式确定：

$$B = \left[\frac{3}{k_B N(E_F) L_1^2}\right]^{1/3} \tag{3.4}$$

式中，k_B 是玻耳兹曼常数。

双终端电流的温度依赖性符合图3.7所示的模型。施加负 V_g 会出现使局域长度增加的较低 B 值，如图3.7(b)所示。跳跃参数随着 V_g 的变化而变化，表明栅极偏置改变了在电荷中性点达到最大值的跳跃条件[73,80,81]。

对不同还原程度的 RGO 电传输的研究表明，随着还原程度增加，GO 经历了绝缘-半导体-半金属的过渡[73]。Eda 等报道了 10～50 meV 的价带尾态和导带之间的明显间隙，进一步还原后间隙接近 0。此外还发现，GO 的还原并没有导致局域态的减少。实际上，L_1 基本保持恒定不变，而 $N(E_F)$ 是增加的(也可由拉曼光谱计算相干长度减少得不明显得到验证[29])。因此还原造成的电导增加是局域态数量增加的结果而不是电荷载流子离域增加的结果。这种中间能隙态的形成与石墨烯局域无序的诱导有关[82]。上述内容很好的地解释了当 GO 轻微还原时出现 RGO 中的电传输，并在整个温度范围出现线性曲线，而对于还原程度好的 GO 装置在高于 240 K 温度下，则表现出从 Mott VRH($T^{-1/3}$)到 Arrhenius 型(T^{-1})跳跃的转换，如图3.7(c)所示。

RGO 薄膜电传输的实验研究有利于对不同还原程度 RGO 的传输性能

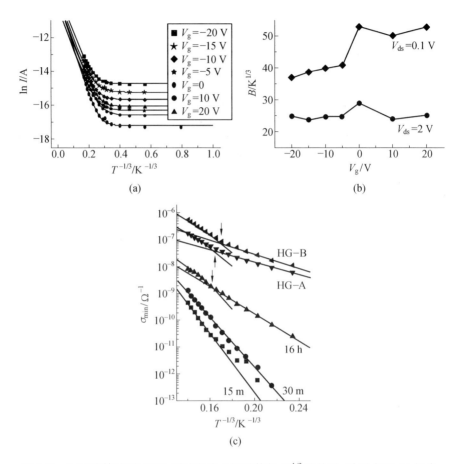

图 3.7 （a）不同栅极电压下,测试电流 I 的对数与 $T^{-1/3}$ 的关系,采用式(3.3)拟合表明 2D 变程跳跃与温度无关项平行[80]。（b）不同偏压下,跳跃参数 B 对应(a)中拟合曲线的跳跃参数 B 与栅极电压 V_g 的关系[80]。（c）RGO 最小电导率 σ_{min} 与 $T^{-1/3}$ 的函数关系图。拟合线性关系表明为 VRH 传输。对于 16 h、HG-A 和 HG-B(HG-A 和 HG-B 是高度还原的 GO,采用无水肼 150 ℃ 退火),箭头指出了发生热激发传输的偏离点[73]

有更深层次的理解。在未还原的片层中,sp^2 杂化的态聚集区域被含氧官能团的区域分割,增加了 GO 的绝缘性。还原过程恢复了 GO 里 sp^2 碳的含量(尽管如之前描述的,sp^2 比率增加并不代表石墨化碳的比例增加),引起 sp^2 聚集区域之间的传输势垒降低。能量势垒的降低促进了 sp^2 位置之间的跳跃或隧穿。还原程度较高,传输由 sp^2 聚集区域间的渗流决定。

3.6 氧化石墨烯与还原的氧化石墨烯的应用

3.6.1 透明导体

RGO 是原子级厚度的碳原子薄片。这一特性使厚度小于 30 mm 的 RGO 在可见范围和近红外区域具有半透明性,而厚膜是不透明的。GO/RGO 膜的透光率和电导率可以通过调整膜的厚度和还原程度而得到调整[83]。因此,RGO 可以广泛用于透明导体,可以在一些器件中代替氧化铟锡(ITO)[84],如有机太阳能电池[85-86]和有机发光二极管[87]。RGO 在更为常见的日用品中也具有潜在的宽阔市场,如抗静电涂层或遮光涂层[88]和低热辐射窗[89]。

理想的石墨烯片电阻为 6 kΩ·sq^{-1},可见-红外范围内具有恒为 98% 的光学透明性[90],尽管比 ITO 还高一个数量级(约 100 Ω·sq^{-1}),但通过掺杂石墨烯,这个值还能提高[91-92]。起初 RGO-硅复合膜被认为可用于透明导体[93],但由于电荷载流子在绝缘基体中渗漏作用不显著,其性能表现平平。从那以后,出现了一些有关 RGO 形成薄膜的研究[1,4,49,52,83,94-96](表 3.1)[103]。

表 3.1 RGO 和剥离石墨烯在波长 550 nm 下的电阻率(或电导率)和透明度[103]

材料	电阻率或电导率	透明度	文献
剥离石墨烯	5 kΩ·sq^{-1}	90	[97]
RGO	1.8 kΩ·sq^{-1}	70	[98]
	1 kΩ·sq^{-1}	80	[83]
	5 kΩ·sq^{-1}	80	[86]
	70 kΩ·sq^{-1}	65	[85]
	11 kΩ·sq^{-1}	96	[99]
	1 425 S·cm^{-1}	70	[100]
RGO-硅	0.45 S·cm^{-1}	94	[93]
RGO-CNT	240 Ω·sq^{-1}	85	[101]
	151 kΩ·sq^{-1}	93	[102]

RGO 薄膜作为透明导体应用的障碍是大部分还原过程需要相当高的温度或强化学试剂。为了解决这一问题,许多团队研究了室温下还原和掺杂 GO 的路线[24,28,31,53,85]。另一个方法是使用大横向尺寸的 GO 片(平均

大于 25 μm)以减小层间连接处的影响。约 50 μm 的横向片层尺寸的空穴迁移率是 365 cm^2·V^{-1}·s^{-1},电子迁移率是 281 cm^2·V^{-1}·s^{-1},表明 π 键可以在还原后得到充分恢复,以满足 RGO 薄膜中的有效载流子传输。

3.6.2 光伏应用

RGO 透明导体的直接应用是用于光伏器件的透明电极[83,85,94,96,104]。在光伏器件中代替 ITO 的重要准则是材料的功函数[81]。石墨烯的功函数是 4.42 eV[105],而 ITO 的功函数是 4.4~4.5 eV[106]。这个巧合使 RGO 能够应用于标准光伏电池中而不改变电池的结构。Wang 等[94]研究了使用 RGO 透明电极的染料敏化太阳能电池的制备和表征(图 3.8)。这种太阳能电池工作机制为通过吸收太阳光使电子激发进入二氧化钛的导带并传输到 RGO 电极。激发空穴顺着类似路径通过空穴输送层,在阴极被收集。太阳能电池结构如图 3.8(a)所示。由于 RGO 比 FTO 具有更高的片层电阻和更低的透过率,因此 RGO 具有更低的短路电流。能量转换效率也因此相对适中。除了好的电子和光学特性,RGO 还具有优良的电化学特性,这使它即使作为对电极材料也是有力的竞争者。最近关于 RGO 用作太阳能电池的对电极的研究,也获得良好效果,如图 3.8(d)所示[108],甚至超过铂-氟掺杂氧化锡(FTO)的对电极[107]。

有机光伏器件(OPVs)采用导电的有机聚合物或少量有机分子用于光吸收和电荷传输。对用 RGO 作为导电透明电极的 OPVs 的研究说明 RGO 性能与染料敏化对电极相似[83,85,109-111]。异质结太阳能电池是 OPVs 的一类,OPVs 的电子供体和受体材料混合在一起成为混合物。这种混合物中相的分离就会形成结节,使电荷载流子到达对电极的扩散距离降低,从而有利于电荷收集和提高效率。RGO 具有大的比表面积和相对高的载流子迁移率,这使它用作 OPV 中的电子受体和传输材料。采用类似于合成 RGO-增塑淀粉(PS)复合材料的方法,可将 RGO 引入聚 3-己基噻吩(P3HT)或聚 3-辛基噻吩(P3OT)。Liu 等[112]报道了将 RGO 加入 P3HT 制得器件,这种光电池系统的能量转化效率约为 1.4%,比 SWNT/P3HT 的组合更好。[113]由于简单低廉的合成方法以及易于与器件匹配,使 RGO 基的本体异质结器件前景广阔,但 GO 在聚合物中的有效分散仍需进一步提高。此外,防止基板降解(避免强化学试剂和高温)的还原过程仍需要研究。

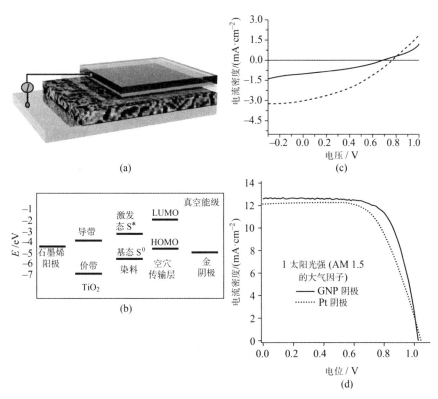

图3.8 (a)采用RGO做窗口电极的染料敏化太阳能电池(DSSC)的示意图。(b)RGO/TiO$_2$/染料/spiro-OMeTAD/金层组成的DSSC能级图。(c)模拟太阳光照下RGO(连续直线)和FTO(断点直线)电流与电压关系[94]。(d)铂氟掺杂氧化锡(FTO)对电极的DSSC电流和电压关系图(虚线);金纳米(GNP)颗粒对电极的连续DSSC(连续直线)[107]

3.6.3 锂离子电池

日益增加的能量需求促进了对寻找新能源的研究,也促进了存储装置技术的提高。电能可以存储在由几个电化学单元构成的电池中,这几个单元相互连接提供所需的电压和电容。每个单元由一个正极和一个负极构成,两极由电解质溶液隔开。近年来,由于锂电池(lithium ion batteries (LIBs))具有优异的充放电特性、独特的电容和耐久性,引起人们极大关注。锂电池占便携式电池全球销售量的63%[114]。

LIB的工作原理为:充电时锂离子从正极向负极流动,而放电时向相反方向流动。锂离子电池用可插锂的化合物作为电极材料。至今已经研究出多种电极材料。已经用于 LIB 的过渡金属氧化物(如 SnO_2、Co_3O_4、Fe_3O_4、TiO_2 和 Mn_3O_4)得到比石墨更高的比容量。尽管这些材料具有比石墨高的比容量,但由于电导率低使其应用受到限制。此外,电化学稳定性对于长时间频繁充放电非常重要,所以这些材料的不稳定性限制了它们的使用[115]。提高比容量的一种可行方法是引入导电材料,如碳纳米管[116-118]。这有利于改善电池的柔性,有望应用于可弯曲电子产品方面。

GO 和 RGO 是目前构建纸状(paper-like)器件的单元。柔韧的、高强度和导电的石墨烯纸可单独作为锂电池中的阳极,有望获得柔韧轻质能源。石墨烯纸可通过真空过滤石墨烯分散液而形成,石墨烯分散液可采用肼还原 GO 得到[119]。石墨烯纸也是一个理想的电池体系,可通过肼蒸汽还原预制的湿 GO 得到。这种未退火纸张表现出类似于聚合物-石墨烯阳极的优良特性和电化学活性[120],充放电流为 50 $mA·g^{-1}$,无黏结剂阳极的可逆比容量是 84 $mA·h·g^{-1}$。

将石墨烯基材料引入过渡金属氧化物后,高放电速率下会提高比容量,同时高充放电次数下也能提高电化学稳定性。石墨烯的高比表面积有利于锂离子的插入,很多研究正试图提高 LIBs 的高循环性(表 3.2)。

表 3.2 石墨烯锂离子电池材料及性能[115]

材料	比容量/ ($mA·h·g^{-1}$)	电流密度/ ($mA·g^{-1}$)	充放电次数	文献
GO 包覆 Co_3O_4	1 000 ~ 1 100	74	约 130	[121]
GO 负载 Co_3O_4	935	50	约 30	[122]
RGO-Mn_3O_4	730 ~ 780	400	约 50	[123]
Mn_3O_4	115 ~ 300	40	约 10	[123]
Co_3O_4	900	—	—	[124]
SnO_2-GO	625	10	约 10	[125]
RGO 覆盖的 Fe_3O_4	1 026	35	约 30	[126]
TiO_2-RGO	160	—	约 100(1 C 充放电)	[127]
GO 纸-Si	2 200	—	约 50	[128]
石墨烯纳米片	290	50	约 20	[129]
RGO	794 ~ 1 054	—	—	[130]

3.6.4 传感器

石墨烯二维材料的独特性能是高的比表面积。石墨烯的这一特性使其应用于极其敏感的装置,主要取决于由于表面的分子吸附而使电学特性发生的变化情况。由于 RGO 对化学和生物的环境变化都极为敏感[131-134],使 RGO 在电化学和生物传感器方面具有很好的发展前景。这些变化不仅反映出 RGO 制备的 FET 的电导率变化,还有电容的变化和掺杂影响。例如,制造灵活的 RGO 化学传感器,采用将聚对苯二甲酸乙二酯(PET)喷墨印刷到薄膜装饰 RGO 片层上,可实现蒸气敏感度为 10^{-9} 的 NO_2 和 Cl_2 的可逆检测。

氢检测对于安全和其他与氢有关的问题都很重要。例如,氢传感器可以检测氢能源汽车和燃料站的泄漏,在气体还未达到爆炸性危险浓度之前就可以被检测到。Sundaram 等[131]指出 RGO 对于氢分子的不敏感性可以通过 Pd 纳米粒子功能化得到缓解(图 3.9(a)和图 3.9(b))。已出现了许多基于氢与 RGO 的电学相互作用制备的氢传感器[135-136]。

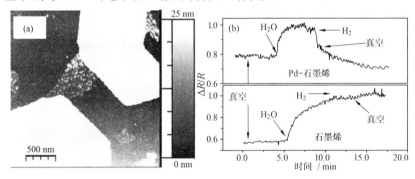

图 3.9 (a)在石墨烯上电化学沉积两个电极的原子力显微镜形貌,Pd 相对于 Pt 为-0.85 V。(b)裸石墨烯层(下)和 Pd 功能化的石墨烯层(上)室温下不同气体压力下的动力学相对应[131]。

床旁诊断是另一个亟待解决成本问题的领域,据估计它有 10 亿美元的市场。无须经过病理实验室手段,它可以对病人做出相对快速的响应。为此就需要基于 RGO 的生物传感器对一系列生物物种进行检测,例如细菌和核酸[137],如图 3.10(a)和图 3.10(b)所示。其他研究还包括针对血糖的检测。Shan 等[139]采用 RGO 和 Kang 等[140]采用热处理的 GO,都实现了葡萄糖氧化酶的直接电化学检测[141](这些生物传感器通过将酶与电极相连,并测量产生的电荷,追踪经过酶的电子数量来检测葡萄糖水平)。

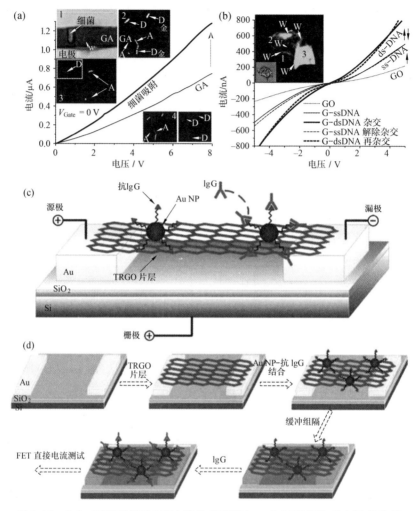

图 3.10 (a) p 型石墨烯胺(GA)器件表面增加一个细菌细胞时电导率的增加(插图 1)。对 GA 上细菌采用共聚焦显微镜测试,证实了静电沉积后的大部分细菌还是活的(插图 3; A = 活的, D = 死的)。对 GA-Au-细菌器件(插图 2 和插图 4)电学测试,立刻进行活/死检测,表明 GA 上二氧化硅上面的细菌细胞是活的,而沉积在 GA 上金电极上面细菌在电学测试后是死的(插图 4(右侧))。(b) DNA 晶体管:ss-DNA 固定于 GO 会提高器件的电导率。在 G-DNA 上的 DNA 杂交(hybridisation)和解除杂交(dehybridisation)会引起电导率完全可逆的增长和恢复。插图显示出了 G-dsDNA 片层的皱褶和折叠[137]。(c) 热还原氧化石墨烯(TRGO) FET 的示意图。抗 IgG 通过 Au 纳米颗粒锚定在 TRGO 上,并成为识别 IgG 的官能团。蛋白质结合(IgG 与抗 IgG)的检测可以通过 FET 的电流直接测试完成。(d) TRGO FET 生物传感器制造过程示意图。TRGO 片层先分散在电极上,然后通过非共价连接的方式与 Au NP-抗体结合[138]

无标记核酸传感器对特定DNA序列或与人类疾病有关的基因突变的检测都非常重要。这些用碳纳米管的电化学DNA传感器,适用于极少量样品的微装置检测(少到10^{-18}L)[142]。用RGO可以制得能有效分离来自不同基底的四个信号,并显示出特异性和敏感性的DNA传感器[143]。RGO FETs也能够无标记检测激素茶酚胺分子和它们的活细胞动态分泌物[144]。

具有高灵敏度和选择性的蛋白质检测器对于快速、低廉地检查疾病是至关重要的,为此,出现了Au NP-抗体结合物修饰GO的FET生物传感器。研究给出了基于GO的生物传感器,用免疫球蛋白G检测轮状病毒病原体。残余缺陷和含氧官能团为石墨烯的生物功能化提供了位点,使探测不仅具有相当高的敏感度,而且还有很好的选择性。

3.7 结　　论

将氧化石墨进行液相剥离是迄今为止制备石墨烯便捷、经济和宏量的方法。简单、多样化的特点使石墨烯可通过化学方法获得某种特殊性能,从而可实现特定目的的应用。基于GO和RGO材料的主要应用瓶颈,即物理、化学和电子结构,已经通过各种高分辨率显微镜和光谱技术得到解决。RGO透明导体虽然前景很乐观,但还需要进一步研究发展以便于与现有技术竞争。然而,低廉的生产成本使它们对市场具有很大的吸引力。其电学和光学特性的优化有望引起石墨烯的首次大规模应用。

3.8　参考文献

[1] HERNANDEZ Y. High-yield production of graphene by liquid-phase exfoliation of graphite[J]. Nature Nanotechnology,2008,3:563.

[2] VALLÉ S C. Solutions of negatively charged graphene sheets and ribbons [J]. Journal of the American Chemistry Society,2008,130:15802.

[3] LI X,WANG X,ZHANG L,et al. Chemically derived, ultrasmooth graphene nanoribbon semiconductors[J]. Science,2008,319:1229.

[4] LI X L. Highly conducting graphene sheets and Langmuir-Blodgett films [J]. Nature Nanotechnology,2008,3:538.

[5] LIU Z,ROBINSON J T,SUN X,et al. PEGylated nanographene oxide for delivery of water-insoluble cancer drugs[J]. Journal of the American Chemistry Society,2008,130:10876.

[6] LIU N. One-step ionic-liquid-assisted electrochemical synthesis of ionicliquid-functionalized graphene sheets directly from graphite[J]. Advanced Functional Materials,2008,18:1518.

[7] HAO R,QIAN W,ZHANG L,et al. Aqueous dispersions of TCNQ-anion-stabilized graphene sheets[J]. Chemical Communications,2008:6576.

[8] SEGAL M. Selling graphene by the ton[J]. Nature Nanotechnology,2009,4:612.

[9] STANKOVICH S. Graphene-based composite materials[J]. Nature,2006,442:282.

[10] STANKOVICH S. Synthesis of graphene-based nanosheets via chemical reduction of exfoliated graphite oxide[J]. Carbon,2007,45:1558.

[11] HIURA H,EBBESEN T W,FUJITA J,et al. Role of sp^3 defect structures in graphite and carbon nanotubes[J]. Nature,1994,367:148.

[12] BRODIE B C. On the atomic weight of graphite[J]. Philosophical Transactions of the Royal Society of London,1859,149:249.

[13] STAUDENMAIER L. Verfahren zur darstellung der graphitsaure[J]. Berichte der Deutschen Chemischen Gesellschaft,1898,31:1481.

[14] HUMMERS W S,OFFEMAN R E. Preparation of graphitic oxide[J]. Journal of the American Chemistry Society,1958,80:1339.

[15] SIMON A,DRONSKOWSKI R,KREBS B,et al. The crystal structure of Mn_2O_7[J]. Angewandte Chemie International Edition in English,1987,26:139.

[16] KOCH K R. Oxidation by Mn_2O_7: an impressive demonstration of the powerful oxidizing property of dimanganeseheptoxide[J]. Journal of Chemical Education,1982,59:973.

[17] TRÖMEL M,RUSS M. Dimanganheptoxid zur selektiven oxidation organischer substrate[J]. Angewandte Chemie,1987,99:1037.

[18] EDA G,FANCHINI G,CHHOWALLA M. Large-area ultrathin films of reduced graphene oxide as a transparent and flexible electronic material[J]. Nature Nanotechnology,2008,3:270.

[19] HIRATA M,GOTOU T,HORIUCHI S,et al. Thin-film particles of graphite oxide 1: High-yield synthesis and flexibility of the particles[J]. Carbon,2004,42:2929.

[20] COTE L J,KIM F,HUANG J X. Langmuir-blodgett assembly of graphite

oxide single layers[J]. Journal of the American Chemistry Society, 2009,131:1043.

[21] LI D,MULLER M B,GILJE S,et al. Processable aqueous dispersions of graphene nanosheets[J]. Nature Nanotechnology,2008,3:101.

[22] SUN X M. Nano-graphene oxide for cellular imaging and drug delivery [J]. Nano Reserch,2008,1:203.

[23] SUN X,LUO D,LIU J,et al. Monodisperse chemically modified graphene obtained by density gradient ultracentrifugal rate separation[J]. ACS Nano,2010,4:3381.

[24] SHIN H-J. Efficient reduction of graphite oxide by sodium borohydride and its effect on electrical conductance[J]. Advanced Functional Materials,2009,19:1987.

[25] WANG G. Facile synthesis and characterization of graphene nanosheets [J]. The Journal of Chemical Physics C,2008,112:8192.

[26] WU Z-S,et al. Synthesis of high-quality graphene with a pre-determined number of layers[J]. Carbon,2009,47:493.

[27] FERNÁNDEZ-MERINO M J. Vitamin C is an ideal substitute for hydrazine in the reduction of graphene oxide suspensions[J]. The Journal of Chemical Physics C,2010,114:6426.

[28] FAN X. Deoxygenation of exfoliated graphite oxide under alkaline conditions: a green route to graphene preparation[J]. Advanced Materials, 2008,20:4490.

[29] GOMEZ-NAVARRO C. Electronic transport properties of individual chemically reduced graphene oxide sheets[J]. Nano Letters,2007,7: 3499.

[30] JUNG I,DIKIN D A,PINER R D,et al. Tunable electrical conductivity of individual graphene oxide sheets reduced at low temperatures[J]. Nano Letters,2008,8:4283.

[31] ZHOU M. Controlled synthesis of large-area and patterned electrochemically reduced graphene oxide films[J]. Chemistry-A European Journal, 2009,15:6116.

[32] RAMESHA G K,SAMPATH S. Electrochemical reduction of oriented graphene oxide films:an in situ Raman spectroelectrochemical study[J]. The Journal of Chemical Physics C,2009,113:7985.

[33] WANG Z J, ZHOU X Z, ZHANG J, et al. Direct electrochemical reduction of single-layer graphene oxide and subsequent functionalization with glucose oxidase[J]. The Journal of Chemical Physics C, 2009, 113: 14071.

[34] WILLIAMS G, SEGER B, KAMAT P V. TiO_2-graphene nanocomposites UV-assisted photocatalytic reduction of graphene oxide[J]. ACS Nano, 2008, 2: 1487.

[35] COTE L J, CRUZ-SILVA R, HUANG J X. Flash reduction and patterning of graphite oxide and its polymer composite[J]. Journal of the American Chemistry Society, 2009, 131: 11027.

[36] LÓPEZ V. Chemical vapor deposition repair of graphene oxide: a route to highly-conductive graphene monolayers[J]. Advanced Materials, 2009, 21: 4683.

[37] HOFMANN U, HOLST R. Über die Säurenatur und die Methylierung von Graphitoxyd[J]. Berichte der deutschen chemischen Gesellschaft, 1939, 72: 754.

[38] RUESS G. Über das graphitoxyhydroxyd (graphitoxyd)[J]. Monatshefte für Chemie/Chemical Monthly, 1947, 76: 381.

[39] CLAUSE A, PLASS R, BOEHM H P, et al. Untersuchungen zur struktur des graphitoxyds[J]. Zeitchrift für anorganische und allgemeine Chemie, 1957, 291: 205.

[40] BOEHM H-P, SCHOLZ W. Untersuchungen am graphitoxyd, IV vergleich der darstellungsverfahren für graphitoxyd[J]. Justus Liebigs Annalen der Chemie, 1966, 691: 1.

[41] NAKAJIMA T, MATSUO Y. Formation process and structure of graphite oxide[J]. Carbon, 1994, 32: 469.

[42] SZABO T, et al. Evolution of surface functional groups in a series of progressively oxidized graphite oxides[J]. Chemistry of Materials, 2006, 18: 2740.

[43] LERF A, HE H, FORSTER M, et al. Structure of graphite oxide revisited [J]. The Journal of Chemical Physics B, 1998, 102: 4477.

[44] CAI W W. Synthesis and solid-state NMR structural characterization of C-13-labeled graphite oxide[J]. Science, 2008, 321: 1815.

[45] FERRARI A C, ROBERTSON J. Interpretation of Raman spectra of dis-

ordered and amorphous carbon [J]. Physical Review B, 2000, 61: 14095.
[46] KUDIN K N. Raman spectra of graphite oxide and functionalized graphene sheets[J]. Nano Letters,2007,8:36.
[47] ELIAS D C. Control of graphene's properties by reversible hydrogenation: evidence for graphane[J]. Science,2009,323:610.
[48] WEI Z Q, BARLOW D E, SHEEHAN P E. The assembly of single-layer graphene oxide and graphene using molecular templates[J]. Nano Letters,2008,8:3141.
[49] MATTEVI C. Evolution of electrical, chemical, and structural properties of transparent and conducting chemically derived graphene thin films [J]. Advanced Functional Materials,2009,19:2577.
[50] YANG D. Chemical analysis of graphene oxide films after heat and chemical treatments by X-ray photoelectron and micro-Raman spectroscopy[J]. Carbon,2009,47:145.
[51] TUINSTRA F, KOENIG J L. Raman spectrum of graphite[J]. Journal of Chemistry Physics,1970,53:1126.
[52] TUNG V C, ALLEN M J, YANG Y, et al. High-throughput solution processing of large-scale graphene[J]. Nature Nanotechnology,2009,4:25.
[53] GAO W, ALEMANY L B, CI L J, et al. New insights into the structure and reduction of graphite oxide[J]. Nature Chemistry,2009,1:403.
[54] LEE V. Large-area chemically modified graphene films: electrophoretic deposition and characterization by soft X-ray absorption spectroscopy [J]. Chemistry of Materials,2009,21:3905.
[55] WANG S. High mobility, printable, and solution-processed graphene electronics[J]. Nano Letters,2010,10:92.
[56] GÓMEZ-NAVARRO C. Atomic structure of reduced graphene oxide [J]. Nano Letters,2010.
[57] ISHIGAMI M, CHEN J H, CULLEN W G, et al. Atomic structure of graphene on SiO_2[J]. Nano Letters,2007,7:1643.
[58] MKHOYAN K A. Atomic and electronic structure of graphene oxide[J]. Nano Letters,2009,9:1058.
[59] GÓMEZ-NAVARRO C. Atomic structure of reduced graphene oxide [J]. Nano Letters,2010,10:1144.

[60] WILSON N R. Graphene oxide: structural analysis and application as a highly transparent support for electron microscopy[J]. ACS Nano,2009, 3:2547.

[61] MEYER J C. The structure of suspended graphene sheets[J]. Nature, 2007,446:60.

[62] MEYER J C,GIRIT C O,CROMMIE M F,et al. Imaging and dynamics of light atoms and molecules on graphene[J]. Nature,2008,454:319.

[63] GASS M H. Free-standing graphene at atomic resolution[J]. Nature Nanotechnology,2008,3:676.

[64] PACILÉ D. Electronic properties and atomic structure of graphene oxide membranes[J]. Carbon,2011,49:966.

[65] PANDEY D,REIFENBERGER R,PINER R. Scanning probe microscopy study of exfoliated oxidized graphene sheets[J]. Surface Science,2008, 602:1607.

[66] LI J L. Oxygen-driven unzipping of graphitic materials[J]. Physics Review Letters,2006,96:176101.

[67] NOVOSELOV K S. Electric field effect in atomically thin carbon films [J]. Science,2004,306:666.

[68] LEEEDUARDO J H,BALASUBRAMANIAN K,WEITZ R T,et al. Contact and edge effects in graphene devices[J]. Nature Nanotechnology, 2008,3:486.

[69] XIA F. Photocurrent imaging and efficient photon detection in a graphene transistor[J]. Nano Letters,2009,9:1039.

[70] SUNDARAM R S,GOMEZ-NAVARRO C,LEE E J H,et al. Noninvasive metal contacts in chemically derived graphene devices[J]. Applied Physics Letters,2009,95:223507.

[71] GILJE S,HAN S,MINSHENG W,et al. A chemical route to graphene for device applications[J]. Nano Letters,2007,7:3394.

[72] GENGLER R Y N. Large-yield preparation of high-electronic-quality graphene by a Langmuir-Schaefer approach[J]. Small,2010,6:35.

[73] EDA G,MATTEVI C,YAMAGUCHI H,et al. Insulator to semimetal transition in graphene oxide[J]. The Journal of Chemical Physics C, 2009,113:15768.

[74] LUO Z T,LU Y,SOMERS L A,et al. High yield preparation of macro-

scopic graphene oxide membranes[J]. Journal of the American Chemistry Society,2009,131:898.

[75] SQUE S J,JONES R,BRIDDON P R,et al. The transfer doping of graphite and graphene[J]. Physica Status Solidi A,2007,204:3078.

[76] ELLISON M D,GOOD A P,KINNAMAN C S,et al. Interaction of water with single-walled carbon nanotubes: reaction and adsorption[J]. The Journal of Chemical Physics B,2005,109:10640.

[77] SCHEDIN F. Detection of individual gas molecules adsorbed on graphene [J]. Nature Materials,2007,6:652.

[78] MOTT N F,DAVIS A E. Electronic processes in non-crystalline materials [M]. 2nd ed. Oxford:Oxford University Press,1979.

[79] RAVI S, KAISER A B, BUMBY C W, et al. Improved conduction in transparent single walled carbon nanotube networks drop-cast from volatile amine dispersions[J]. Chemical Physics Letters,2010,496:80.

[80] KAISER A B,GÓMEZ-NAVARRO C,SUNDARAM R S,et al. Electrical conduction mechanism in chemically derived graphene monolayers[J]. Nano Letters,2009,9:1787.

[81] EDA G,CHHOWALLA M. Chemically derived graphene oxide: towards largearea thin-film electronics and optoelectronics[J]. Advanced Materials,2010,22:2392.

[82] PEREIRA V M,DOS SANTOS J,CASTRO A H,et al. Modeling disorder in graphene[J]. Physical Review B,2008,77:115109.

[83] BECERRIL H A. Evaluation of solution-processed reduced graphene oxide films as transparent conductors[J]. ACS Nano,2008,2:463.

[84] ANDREA CHIPMAN. A commodity no more[J]. Nature, 2007, 449 (7159):131.

[85] EDA G. Transparent and conducting electrodes for organic electronics from reduced graphene oxide[J]. Applied Physics Letters, 2008, 92: 233305.

[86] WU J. Organic solar cells with solution-processed graphene transparent electrodes[J]. Applied Physics Letters,2008,92:263302.

[87] WU J. Organic light-emitting diodes on solution-processed graphene transparent electrodes[J]. ACS Nano,2009,4:43.

[88] AL-DAHOUDI N,BISHT H,GÖBBERT C,et al. Transparent conduc-

ting, anti-static and anti-static-anti-glare coatings on plastic substrates [J]. Thin Solid Films,2001,392:299.
[89] GORDON R G. Criteria for choosing transparent conductors[J]. MRS Bull,2000,25:6.
[90] NAIR R R. Fine structure constant defines visual transparency of graphene[J]. Science,2008,320:1308.
[91] KASRY A,KURODA M A,MARTYNA G J,et al. Chemical doping of large-area stacked graphene films for use as transparent,conducting electrodes[J]. ACS Nano,2010,4:3839.
[92] KIM K K. Enhancing the conductivity of transparent graphene films via doping[J]. Nanotechnology,2010,21:285205.
[93] WATCHAROTONE S. Graphene-silica composite thin films as transparent conductors[J]. Nano Letters,2007,7:1888.
[94] WANG X,ZHI L J,MULLEN K. Transparent conductive graphene electrodes for dye-sensitized solar cells[J]. Nano Letters,2008,8:323.
[95] LI D,MULLER M B,GILJE S,et al. Processable aqueousdispersions of graphene nanosheets[J]. Nature Nanotechnology,2008,3:101.
[96] SU Q. Composites of graphene with large aromatic molecules[J]. Advanced Materials,2009,21:3191.
[97] GEIM A K. Graphene: status and prospects[J]. Science,2009,324:1530.
[98] WANG X,ZHI L,MULLEN K. Transparent,conductive graphene electrodes for dye-sensitized solar cells[J]. Nano Letters,2008,8:323.
[99] ZHU Y,CAI W,PINER R D,et al. Transparent self-assembled films of reduced graphene oxide platelets[J]. Applied Physics Letters,2009,95:103103.
[100] YANYU L. Transparent,highly conductive graphene electrodes from acetylene-assisted thermolysis of graphite oxide sheets and nanographene molecules[J]. Nanotechnology,2009,20:434007.
[101] TUNG V C. Low-temperature solution processing of graphene-carbon nanotube hybrid materials for high-performance transparent conductors[J]. Nano Letters,2009,9:1949.
[102] KIM Y-K,MIN D-H. Durable large-area thin films of graphene/carbon nanotube double layers as a transparent electrode[J]. Langmuir,2009,

25:11302.

[103] WASSEI J K, KANER R B. Graphene, a promising transparent conductor[J]. Materials Today, 2010, 13:52.

[104] YIN Z. Electrochemical deposition of ZnO nanorods on transparent reduced graphene oxide electrodes for hybrid solar cells[J]. Small, 2010, 6:307.

[105] CZERW R. Substrate-interface interactions between carbon nanotubes and the supporting substrate[J]. Physical Review B, 2002, 66:033408.

[106] PARK Y, CHOONG V, GAO Y, et al. Work function of indium tin oxide transparent conductor measured by photoelectron spectroscopy[J]. Applied Physics Letters, 1996, 68:2699.

[107] KAVAN L, YUM J-H, NAZEERUDDIN M K, et al. Graphene nanoplatelet cathode for Co(Ⅲ)/(Ⅱ) mediated dye-sensitized solar cells [J]. ACS Nano, 2011, 5:9171.

[108] CHOI H, KIM H, HWANG S, et al. Graphene counter electrodes for dye-sensitized solar cells prepared by electrophoretic deposition[J]. Journal of Materials Chemistry, 2011, 21:7548.

[109] YIN Z. Organic photovoltaic devices using highly flexible reduced graphene oxide films as transparent electrodes[J]. ACS Nano, 2010, 4: 5263.

[110] LI S-S, TU K-H, LIN C-C, et al. Solution-processable graphene oxide as an efficient hole transport layer in polymer solar cells[J]. ACS Nano, 2010, 4:3169.

[111] KYMAKIS E, STRATAKIS E, STYLIANAKIS M M, et al. Spin coated graphene films as the transparent electrode in organic photovoltaic devices[J]. Thin Solid Films, 2011, 520:1238.

[112] LIU Z. Organic photovoltaic devices based on a novel acceptor material: graphene[J]. Advanced Materials, 2008, 20:3924.

[113] KYMAKIS E, AMARATUNGA G A J. Single-wall carbon nanotube/conjugated polymer photovoltaic devices[J]. Applied Physics Letters, 2002, 80:112.

[114] TARASCON J M, ARMAND M. Issues and challenges facing rechargeable lithium batteries[J]. Nature, 2001, 414:359.

[115] SINGH V. Graphene based materials: past, present and future[J].

Progress in Materials Science,2011,56:1178.
[116] LAHIRI I. High capacity and excellent stability of lithium ion battery anode using interface-controlled binder-free multiwall carbon nanotubes grown on copper[J]. ACS Nano,2010,4:3440.
[117] GUO Z P,ZHAO Z W,LIU H K,et al. Electrochemical lithiation and de-lithiation of MWNT-Sn/SnNi nanocomposites[J]. Carbon,2005,43:1392.
[118] DE LAS CASAS C,LI W. A review of application of carbon nanotubes for lithium ion battery anode material[J]. Journal of Power Sources, 2012,208:74.
[119] DIKIN D A. Preparation and characterization of graphene oxide paper [J]. Nature,2007,448:457.
[120] ABOUIMRANE A,COMPTON O C,AMINE K,et al. Non-annealed graphene paper as a binder-free anode for lithium-ion batteries[J]. The Journal of Chemical Physics C,2010,114:12800.
[121] YANG S,FENG X,IVANOVICI S,et al. Fabrication of graphene encapsulated oxide nanoparticles: towards high-performance anode materials for lithium storage [J]. Angewandte Chemie International Edition, 2010,49:8408.
[122] WU Z-S. Graphene anchored with Co_3O_4 nanoparticles as anode of lithium ion batteries with enhanced reversible capacity and cyclic performance[J]. ACS Nano,2010,4:3187.
[123] WANG H,et al. Mn_3O_4-graphene hybrid as a high-capacity anode material for lithium ion batteries[J]. Journal of the American Chemistry Society,2010,132:13978.
[124] POIZOT P,LARUELLE S,GRUGEON S,et al. Nano-sized transition-metal oxides as negative-electrode materials for lithium-ion batteries [J]. Nature,2000,407:496.
[125] WANG D. Ternary self-assembly of ordered metal oxide-graphene nanocomposites for electrochemical energy storage[J]. ACS Nano,2010,4:1587.
[126] ZHOU G. Graphene-wrapped Fe_3O_4 anode material with improved reversible capacity and cyclic stability for lithium ion batteries [J]. Chemistry of Materials,2010,22:5306.

[127] WANG D. Self-assembled TiO$_2$-graphene hybrid nanostructures for enhanced Li-ion insertion[J]. ACS Nano,2009,3:907.

[128] LEE J K,SMITH K B,HAYNER C M,et al. Silicon nanoparticles-graphene paper composites for Li ion battery anodes[J]. Chemical Communications,2010,46:2025.

[129] YOO E. Large reversible Li storage of graphene nanosheet families for use in rechargeable lithium ion batteries[J]. Nano Letters,2008,8:2277.

[130] PAN D. Li storage properties of disordered graphene nanosheets[J]. Chemistry of Materials,2009,21:3136.

[131] SUNDARAM R S. Electrochemical modification of graphene[J]. Advanced Materials,2008,20:3050.

[132] JUNG I. Effect of water vapor on electrical properties of individual reduced graphene oxide sheets[J]. The Journal of Chemical Physics C,2008,112:20264.

[133] FOWLER J D. Practical chemical sensors from chemically derived graphene[J]. ACS Nano,2009,3:301.

[134] LU G,OCOLA L E,CHEN J. Gas detection using low-temperature reduced graphene oxide sheets[J]. Applied Physics Letters,2009,94:083111.

[135] SHAFIEI M. Platinum/graphene nanosheet/SiC contacts and their application for hydrogen gas sensing[J]. The Journal of Chemical Physics C,2010,114:13796.

[136] LANGE U,HIRSCH T,MIRSKY V M. Hydrogen sensor based on a graphene-palladium nanocomposite[J]. Electrochimica Acta,2011,56:3707.

[137] MOHANTY N,BERRY V. Graphene-based single-bacterium resolution biodevice and DNA transistor: interfacing graphene derivatives with nanoscale and microscale biocomponents[J]. Nano Letters,2008,8:4469.

[138] MAO S,LU G,YU K,et al. Specific protein detection using thermally reduced graphene oxide sheet decorated with gold nanoparticle-antibody conjugates[J]. Advanced Materials,2010,22:3521.

[139] SHAN C S. Direct electrochemistry of glucose oxidase and biosensing

for glucose based on graphene[J]. Analytical Chemistry,2009,81:2378.

[140] KANG X H. Glucose oxidase-graphene-chitosan modified electrode for direct electrochemistry and glucose sensing[J]. Biosensors and Bioelectronics,2009,25:901.

[141] LIU Y,YU D,ZENG C,et al. Biocompatible graphene oxidebased glucose biosensors[J]. Langmuir,2010,26:6158.

[142] KURKINA T,VLANDAS A,AHMAD A,et al. Label-free detection of few copies of DNA with carbon nanotube impedance biosensors[J]. Angewandte Chemie International Edition,2011,50:3710.

[143] ZHOU M,ZHAI Y M,DONG S J. Electrochemical sensing and biosensing platform based on chemically reduced graphene oxide[J]. Analytical Chemistry,2009,81:5603.

[144] HE Q. Centimeter-long and large-scale micropatterns of reduced graphene oxide films:fabrication and sensing applications[J]. ACS Nano,2010,4:3201.

第4章 电化学剥离制备石墨烯

本章介绍了电化学剥离制备石墨烯,并深入讨论了在特定电解质溶液中采用电化学剥离石墨和(或)电化学还原剥离的氧化石墨大批量制备高品质石墨烯纳米片的方法。探讨了该过程涉及的潜在问题,并对电化学剥离石墨烯的应用进行了研究。

4.1 引　言

石墨烯,由 sp^2 键合的碳原子组成一个原子厚度的层,以二维(2D)蜂窝晶格形式紧密堆叠。石墨烯可以包裹形成 0 维(0D)的富勒烯[1],卷曲成为 1 维(1D)碳纳米管(CNTs)[2],或堆叠成三维(3D)的石墨。由于石墨烯具有一系列优异性能,如高载流子迁移率[3-4]、独特的传输特性[5-6]、高的机械强度[7-8]和高的导热系数[9-10],使其成为 21 世纪最有魅力的材料之一。上述特性使石墨烯适合应用于许多科技领域,包括石墨烯纳米电子[11-13]、聚合物纳米复合材料[7-14]、传感器[15]和能源存储[16-18]。为了满足这些应用,高品质的 2D 石墨烯的制备是首要也是最重要的一步。至今为止,制备石墨烯有如下方法:

(1)石墨的微机械剥离(胶带法),这是最初发现石墨烯的方法[19]。该方法广泛用于实验室内获取石墨烯进行基础科学研究;然而石墨烯的产率低,因此不适于大范围使用。

(2)单晶二氧化硅[20]、钌[21]和铜[22]上的超薄石墨烯的外延生长是制备具有图案化结构石墨烯的有效方法,适合于电子领域应用。

(3)使用催化金属基板化学气相沉积法(CVD)可生长获得大面积石墨烯[23-24]。用这种方法制备的石墨烯可用作透明度高、柔韧性强的导电膜,但是该方法的可行性却受到生产成本的限制。使用催化金属基板还会将缺陷引入制备的石墨烯材料。

(4)石墨或氧化石墨(GO)进行热膨胀会生成石墨烯,但这一过程很少能做到石墨的完整剥离并得到原子尺度的独立石墨烯片层[20,25,26]。

(5)采用挥发性试剂对石墨进行插层和膨胀处理[31-32],再利用超声进行液相分离[27-30]，也可得到石墨烯；然而通过液相剥离或石墨插层得到的石墨烯片层的尺寸通常小于 1 μm，石墨烯产量太低，不适于技术应用。

(6) 化学剥离法，即石墨氧化成为氧化石墨烯，之后进行化学还原或热还原的方法。由于成本低、易制备的特点，该方法已经引起众多关注[33-39]。然而，由于过量氧化及后续还原，使石墨烯片层的蜂窝晶格受到严重破坏。GO 的化学还原过程包含强还原剂（如肼[34,40]和硼氢化钠）的使用[41]，但这些药品的推广及使用受到它们的危险性和毒性限制。其他还原剂（如对苯二酚[36]、亚硫酸氢钠[42]、氢卤酸[43]和菌类[44]）也已经用于 GO 的还原，但它们各自都有不利因素，会影响 GO 向石墨烯的转变。此外，也出现了采用几种环保的还原剂来制备石墨烯，如 L-抗坏血酸[45]和还原性糖[46]。然而 L-抗坏血酸亲水性强、价格昂贵，不适合做还原剂。热还原方法还原 GO 需要很高的温度才能恢复石墨的结构。GO 氧化物并不能通过化学还原剂完全去除，从而会导致电性能的下降，进一步限制了此方法用来制备石墨烯。

(7) 电化学剥离法是制备石墨烯的环保方法。电化学法通过调整外部能量，改变了石墨烯的电子状态。通过改变电压，可以保证产物稳定。电化学剥离的石墨烯的边缘尺寸很大，极大地减少了层间连接点的数量。剥离石墨烯膜具有优良的导电性和场效应迁移率。在透光率低于 80% 时，化学氧化-还原得到的 GO 石墨烯片层的电阻为 1～100 $k\Omega \cdot sq^{-1}$，在透光率为 95% 时为 31～18 $M\Omega \cdot sq^{-1}$。相比之下，在透光率为 96% 时，电化学剥离石墨烯的电阻为 0.015～0.21 $k\Omega \cdot sq^{-1}$。此外，通过电化学剥离得到的石墨烯的缺陷密度比化学氧化-还原制得 GO 石墨烯少很多。因此，电化学剥离法可制得品质卓越、透明性好、导电优良的石墨烯片层。

本章将讨论从石墨/GO 进行电化学剥离制备石墨烯，并对电化学剥离的石墨烯的应用进行深度分析。

4.2　电化学剥离制备石墨烯相关基本概念

在特定电解质溶液中，可以通过石墨电化学剥离和（或）剥离 GO 并电化学还原的方法，大规模制备高品质石墨烯。通常电化学剥离在窄电化学窗口（如水）和宽电化学窗口液体的混合溶剂中进行。电化学剥离制备石墨烯的原理如下所述：在电极上发生水电解并产生羟基和氧自由基；氧自由基会引起石墨阳极发生腐蚀，并导致石墨层边缘打开；具有大电化学窗口的液体插入边层引起电极膨胀；最后，将片层沉淀获得分散于溶液中的石墨烯。

4.2.1 石墨的电化学剥离

通过电化学剥离法制备高品质、大面积石墨烯,并在有效电解质(聚4-苯乙烯磺酸钠溶液)中成功包覆在柔性石墨箔片表面[47]。在电解剥离过程中,箔片电极的附近有黑色粉末出现,表明石墨箔片腐蚀明显。剥离的石墨箔片用去离子水和乙醇清洗,然后在60 ℃真空烘箱中干燥2 h。电解剥离之前,石墨箔片的表面光滑,剥离后表面微结构则变成卷曲状态。石墨箔片的表面粗糙度可以通过原子力显微镜(AFM)观察到(图4.1)。

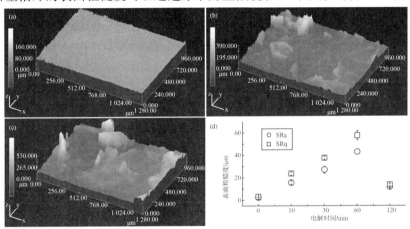

图4.1 电解过程中(a)接收层石墨和剥离石墨在(b)10 min 和(c)1 h 的图。(d)剥离石墨层的表面粗糙度与电解时间的函数关系。其中,SRa 表示平均值;SRq 表示均方根值[47]

石墨箔片表面粗糙度很小,而经过电解剥离后,表面变得非常粗糙。剥离石墨越粗糙,表面积越大。1 h 的剥离可形成最大的表面积,之后表面积急剧下降。很明显剥离箔片阴极电流比阳极电流大。锂与石墨烯纳米层里残余含氧基团相互作用以及固态电解质界面的形成,可能引起初始电流损耗[48-49]。更重要的是,每个循环中剥离石墨箔层的电流密度与接收层的相比急剧增加,由此证明剥离石墨层的锂电活性较高。实际上作为接收的石墨层的层状结构阻碍了电解质和离子的迁移,限制了典型石墨片层作为电池电极的有效性。由于剥离石墨与液态电解质接触面积很大,所以电极具有更高的存储容量。

Wang 等报道了采用电解剥离从石墨上制备石墨烯,并很容易扩大规模化生产[50]。在典型合成中,两个石墨棒放置于充满电解质的电解池中,电极电位恒定在5 V。制备过程如图4.2所示。

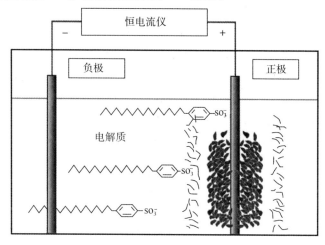

图4.2 电解剥离法制备石墨烯示意图[50]

电解 20 min 后,正极逐渐产生黑色产物。剥离 4 h,将产物从电解池取走,低速离心以去除大的团聚体。底层弃掉,收集上层分散物。这种石墨烯-聚苯乙烯磺酸(PSS)悬浮液非常稳定,即使过了 6 个月也没有沉淀出现。最后分散体用去离子水和乙醇清洗,然后在 80 ℃的真空干燥箱干燥得到石墨烯粉末。用该方法可以获得 15% 的产率。石墨烯在高能量电子束下是透明的,看上去像起皱的丝绸面料。出现的褶皱和卷曲是石墨烯层固有的特点[51]。

由于石墨烯表面和 PSS 芳香环之间的边-面相互作用,使 PSS 吸附到石墨烯表面。PSS 是一种吸水性材料,在石墨的电解剥离中起重要作用。一旦溶解在水里,PSS 就会电离成钠离子和聚苯乙烯磺酸盐阴离子。在电解过程中,聚苯乙烯磺酸盐阴离子在电场力作用下转移到正的石墨电极上,与石墨相互作用,引起石墨棒的电解剥离。

石墨柱通过氧化-还原循环,可电化学剥离产生自立的少层石墨烯(FFG)[52]。FFGs 通过以下步骤制备:石墨柱电极(GCE)置于 0.1 mol/L 的 Na_2SO_4 水溶液里,在 -1.0 ~ +3.0 V、扫描速率为 500 mV·s^{-1} 状态下循环。电解质溶液的颜色从无色到黄色,最后随循环数增加而变成深棕色。深棕色溶液在 4 000 r/min 转速下离心 30 min,经超纯水洗涤沉淀,在氮气下干燥得到石墨烯。上层清液最后经过渗析除掉多余的 Na_2SO_4。得到

FFG 的宽度和长度分别是 500 nm 和 800 nm。FFG/GCE 电极的电荷转移电阻比纯 GCE 电极的低,表明 FFG/GCE 的电化学活性较高,从而使 FFG/GCE 电极可应用于生物传感。烟酰胺腺嘌呤二核苷酸(NADH)阳极电位为 0.45 V,在特定 pH 范围内,具有比 GCE 高的电流信号,这与碳纳米管改性电极[53]或石墨烯改性电极[54]相似。FFG 的边-面类位点,有利于电子转移到生物分子[55]。

Wei 等研究了在含有室温亲水离子液体(RTIL)、1-丁基-3-甲基咪唑四氟硼酸盐和 BMIM-BF$_4$ 的水溶液中,用电化学剥离石墨制备石墨烯[56]。实验结束时,石墨被剥离到海绵状电极上,溶液的颜色由透明变成深棕色。得到的石墨烯片分离并分散在水中。从石墨烯溶液中可以看到紫外诱导发光,证明 RTIL 通过化学键连接到石墨烯上,254 nm 的紫外光激发的蓝色荧光可能来自 RTIL 基团。去离子水对减小溶液黏度起了重要作用,从而促使 RTIL 离子扩散进入石墨。将锂盐 LiClO$_4$ 加入到[BMIM][BF$_4$]和去离子水的混合溶液中,溶液的离子电导率增加会进一步促进电化学剥离。

用磷酸三甲酯(TMP)基的电解液进行电化学剥离可制备亚微米厚度的石墨片层[57]。为了实现石墨电化学剥离,电池恒流放电至 0 V,随后充电到 2.5 V。在放电过程中,溶剂 Li^+-$(TMP)_n$ 的插入和还原分解引起石墨颗粒的剥离。从热动力学观点看,低温下(0 ℃)溶剂向石墨层的转移和插层还是有问题的;而温度升高,还原分解程度又会加剧。

有报道称,石墨的阳极和阴极在高氯酸水溶液中电化学插入,之后通过微波辐射膨胀来制备高品质少层石墨烯[58]。预制备过程包括石墨电极的电化学预膨胀,之后微波辐射辅助热处理。第一步,将工作电极进行双电位阶跃实验(double potential step experiments)。首先采用 0 V 初始电位保持 10 s;然后采用扰动电位并保持一段时间 $t(0 \sim 1\ 200\ s)$,最后设置到初始电位保持 10 s。施加双电位阶跃后,电极开始发生剥离,并在电解池底部出现固体。第二步,电化学制备的预膨胀电极被立即放入微波炉中,在氮气中,800 W 功率下处理 5~10 s。最后,剥离的石墨分散到 N-甲基-1-吡咯烷酮(NMP)中超声处理 5 min,2 000 r/min 下离心 30 min 并去除离心残余得到高品质的石墨烯。反应过程中生成的氢分子插入石墨层中,并引起工作电极膨胀。实际上,氢分子会加剧碳/电解质溶液界面的机械应力,这样有助于克服层间的范德瓦耳斯力。石墨烯的质量好坏可以通过 D 带和 G 带的强度比来评定。D 带的低强度和 D′带的消失表明高品质石墨烯的生成。

Su 等提出一个简单快捷的电化学方法剥离石墨得到薄的石墨烯层,主要是 AB 堆叠的具有大横向尺寸($1\sim40~\mu m$)的双层石墨烯[59]。电化学剥离过程用硫酸作为电解质。图 4.3 是实验装置,高度定向热解石墨(HOPG)作为电极和电化学剥离石墨烯的来源。选择铂线作为接地电极。对石墨实施低偏置电压+1 V,时间为 5 min;再增加到+10 V,时间为 1 min。低偏置有助于加湿试样,并可能引起 SO_4^{2-} 离子进入石墨的晶界[60]。高偏置之前,石墨是整片,一旦加+10 V 的高偏置电压,石墨就会快速解离成小片,并分散到溶液表面。用过滤法收集剥离的石墨烯,在二甲基甲酰胺(DMF)中再分散。用硫酸溶液进行石墨烯的电化学剥离非常有效,该过程可能在几分钟内完成,但是硫酸本身会引起石墨的氧化,因此会生产有高度缺陷的薄层。为了解决这一问题,可向硫酸溶液中加入 KOH 以降低电解质溶液的酸度,得到的石墨烯的产率是 5% ~ 8%。由自组装石墨烯制成的薄膜具有优良的电导率和透明性,这些特性使其在柔性电子领域非常有用。

石墨负电极进行电化学充电,在高电流密度下,以锂盐和聚碳酸酯(PC)作为电解质,可被有效地剥离成少层石墨烯。分散的石墨烯可以以墨水状涂刷于纸张表面,制成高度贴合的导电涂层[61]。采用高电位(-15 V)激发 Li/PC 共同插入石墨。高压和高电流下的电化学充电过程中,有机溶液(PC)在阴极分解,产生丙烯气体。该过程也会引起石墨电极的膨胀。膨胀石墨在溶解于 PC 和 DMF 的高浓度 LiCl 中进行超声处理。超声空化作用产生热冲击从而引起石墨烯片层的剥离和切割[62]。用酸和水冲洗去除插入的 Li^+/PC。电化学充电方法可以与微波辐射法相结合,实现从石墨到分散石墨烯片的大量(克数量级)制备。

Lu 等提出了用离子液体辅助电化学剥离方法从石墨电极上产生荧光碳纳米带、纳米颗粒和石墨烯的便捷途径[63]。由于离子液体具有独特性能,如可忽略的蒸汽压、热稳定性、宽泛的电化学电位窗口、低黏度、突出的离子电导率以及可循环性,因此离子液体(ILs)可作为"绿色"溶剂替代传统溶剂。此外,ILs 的高介电常数使它可以作为堆叠相互作用的屏障,有助于纳米材料有效分散。在确定剥离的纳米颗粒的分布形状和尺寸时,水起到了很重要的作用。在电化学实验中,高压会引起水解离产生羟基和氧自由基,它们会攻击阳极,导致电化学反应过程中阳极腐蚀。为了克服这一缺点,当电化学剥离石墨时,采用 IL(1-丁基-3-甲基咪唑四氟硼酸盐,[BMIM][BF_4])与水以不同比例混合。将水加入到 ILs 中,会扰乱 ILs 内部结构并形成新的氢键网络,导致内聚能降低和黏度下降。用高含水量的

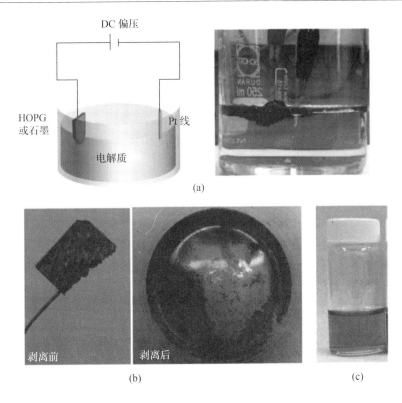

图4.3 (a)石墨电化学剥离的示意图和照片。(b)电化学剥离前后的石墨片。(c)DMF溶液中分散的石墨烯片[59]

ILs作为电解质,可制得水溶性氧化碳纳米材料。而通过使用浓ILs(含水量低)作为电解质,可以获得IL功能化的碳纳米材料。电化学剥离过程包括3个阶段(图4.4)。

阶段Ⅰ:在发生显著剥离之前有一个诱导期。电解质溶液的颜色变化从无色到黄色,然后是深棕色。

阶段Ⅱ:观察到石墨电极的膨胀。

阶段Ⅲ:膨胀的薄片从电极上剥离下来,在电解质溶液中产生黑色胶体。

剥离机理和相关反应步骤概括如下:

(Ⅰ)水发生阳极氧化产生羟基和氧自由基。石墨的羟基化或氧化通常发生在边缘,从而引起电极上荧光碳纳米晶的溶解。

$$H_2O \xrightarrow{-e} OH^{\bullet} + H^+ \xrightarrow{-e} O^{\bullet} + H^+$$

（Ⅱ）氧化反应为阴离子 BF_4^- 的插入提供了通道，导致石墨阳极的去极化和膨胀[64-65]。

$$C_x + BF_4^- \rightarrow BF_4C_x + e^- \xrightarrow{H_2O} C_xOH + HBF_4$$

（Ⅲ）膨胀石墨烯片的氧化剥离产生石墨烯纳米带和一些膨胀片层沉淀为石墨烯片。

这种方法为大量生产用于生物标记和成像领域的荧光生物相容性碳纳米材料提供了新的可能性。

图4.4 高定向热解石墨（HOPG）电极在60%水/[BMI_m][BF_4]的电解质中剥离过程随时间变化。阶段Ⅰ、Ⅱ和Ⅲ分别对应图中（b）、（c）和（d）。（f）剧烈膨胀的 HOPG[63]

Liu 等在类似实验中，提出了一步 IL 辅助直接从石墨电化学制备石墨烯的方法[66]。1-丁基-3-甲基咪唑四氟硼酸盐、[C_8MIM]$^+$[PF_6]$^-$ 和水（质量比为1∶1）的混合物用作电解质溶液。两个石墨棒之间施加 15 V 的电位。室温下腐蚀 6 h 后，在反应器底部得到了黑色石墨烯沉淀。

4.2.2 剥离氧化石墨的电化学还原

有报道称,阴极电位存在时,通过电化学方法还原剥离氧化石墨前驱体,可实现高品质石墨烯纳米层的大量制备[67]。在酸性介质中,通过改进的 Humers 方法采用石墨片制备剥离氧化石墨[68]。与 GO 表面连接的氧化基团增加了电容量,因此提高了 GO 在水性介质中的分散度。剥离 GO 的电化学还原反应,在采用磁力搅拌不同阴极电位下的石墨工作电极进行。GO 改性的玻碳电极(GCE)在-1.2 V 具有大的阴极电流峰,说明 GO 表面的含氧基团发生了还原反应。在第二个周期,阴极电流峰的强度在负电位快速降低,经过几次电位扫描后阴极电流峰消失。这表明在负电位下采用电化学还原的方法,可以快速不可逆地还原 GO 和剥离 GO 的表面氧化基团。由于还原速度非常快,因此电化学还原的石墨烯会有很多缺陷,但如果电化学还原在高温下进行或对产物进行退火,这些缺陷就会消除。

在乙酰胺-尿素-硝酸铵三元共熔体里,剥离 GO 的电化学还原会产生少层石墨烯薄膜[69]。对作为原材料的硫酸氢盐插层的天然石墨进行热膨胀,运用预膨胀石墨采用传统改进的 Hummers 方法制备剥离 GO(EGO)。由于带有含氧基团,EGO 水性分散态带有大量负电荷。自组装胱胺单层的胺基的酸度系数 pKa=7.6,所制备的 EGO 的 pH≈2.5。因此,胱胺单层的胺基团以质子化形式存在。胱胺单层的带正电荷胺基团通过静电吸附于金表面带有负电荷的 EGO 纳米层,获得有序的 EGO 层。随后,EGO 片层上未补偿的负电荷会彼此排斥。

石墨烯改性的电极(ERG)通过在磷酸盐缓冲溶液 PBS 中的 GCE 表面进行 GO 电化学还原制得。典型实验中,取一定量的 GO 水性溶液滴到 GCE 表面,之后真空条件下干燥电极得到改性 GO 电极。电极被浸入 PBS 中,在电位为 0~-1.5 V 下进行五次循环电位扫描。第一次循环,出现-1.2 V 的宽阴极电流峰,它与氧化石墨烯中的 C—O 键还原有关。四个循环后,-1.2 V 处阴极峰完全消失,这表明 GO 的电化学还原完成,形成石墨烯。制备的电极对一氧化氮(NO)响应快速,即使六周后,电极的响应信号仍保持不变,证实了电极对 NO 具有高敏感性和稳定性。

Shao 等报道了通过 GO 的电化学还原反应制备 ERG[70]。电位为 -1.0~1.0 V,采用 Hg/Hg_2SO_4 和铂参比电极的标准三电极体系,以 0.1 $mol·L^{-1}$ 硫酸钠溶液作为电解质,通过扩展伏安法进行 GO 的电化学还原反应。阴极峰在施加电位时先出现增长,然后随着电位循环的进行而消失,表明 GO 的还原性。该还原过程容易被原位监测和控制。与碳纳米管和化学还原石墨烯相比,ERG 表现出高电化学电容和循环持久性。

4.3 石墨烯和石墨烯基材料的应用

石墨烯作为轰动 21 世纪的材料,具有广泛的应用领域:

(1)石墨烯研究最广泛的应用之一是场效应晶体管(FET)。二维石墨烯表现出金属性,并具有强的双极性场效应。

(2)石墨烯相关材料可用作记忆器件或存储器件已被广泛认可。在弹道电流下,同时存在自旋极化和自旋非极化态,在二进制记忆装置中通过施加偏置,二者可以进行转换。记忆装置具有高通断比和低转换阈值电压,而这些都与石墨烯的薄膜电阻有关。

(3)由于石墨烯具有光学透明性,仅能吸收 2.3% 的可见光,并且是一种高导电材料,所以在柔性电子领域可以完全替代氧化铟锡(ITO)。ITO 薄膜的脆性、高成本和原材料不足均限制了它在韧性设备中的应用。

(4)由于石墨烯具有良好的电子迁移率,因此它也是一种新的电子受体材料。

(5)石墨烯的可调带宽和高吸光率,使其作为高效捕光材料具有很大优势。

(6)石墨烯基的化学和生物传感器具有高度敏感性、低噪声、制作简单和生物相容性等特点。石墨烯气体传感器也具有良好的发展前景。石墨烯基生物传感器利用石墨烯和细菌或细胞之间相互作用,探测生物分子(如 DNA、蛋白质和生物信号),这方面已处于领先地位。

(7)石墨烯基材料比表面积大、导电性好、表面可功能化。这些对于电化学应用都是非常有利的。由于石墨烯独特的 2D 结构和边缘缺陷的存在,使石墨烯在电化学传感器应用方面可完美替代其他碳材料,如 CNTs 和石墨。

(8)石墨烯和石墨烯基导电聚合物纳米复合材料可用于获得具有高比电容和优良循环稳定性的高能量存储器件。

(9)制备具有可调电子特性的图案化石墨烯是当前研究的热点。

更特别的是,电化学剥离的石墨烯电极对一氧化氮响应迅速。如此制备的电极具有更高的稳定性,对于 NO 的氧化具有更强的催化活性。因此电化学剥离石墨烯为制造电化学传感平台提供了一个新的视角。电化学剥离技术具有成本优势并有潜力制备出具有极高电导率的石墨烯。还原程度、剥离程度和石墨烯尺寸均可以通过改变电化学参数控制。剥离溶剂中引入锂盐会获得比容量相对纯石墨电极提高4倍的电极。因此电化学剥离石墨烯可以在原电池电极区进行剥离。此外,电化学剥离石墨烯的高电导率和巨大活性区域使其可应用于制造锂离子电池和电子导体。石墨烯片的透明性使其在透明和半透明储能器件方面具有极大的发展潜力。

4.4 结　　论

作为构建各维数碳材料的基本单元,石墨烯具有优良的电学、力学、光学和热学性能以及独特的化学结构。为满足薄膜、复合材料和器件等领域对石墨烯的需求,迄今为止,已有几种制备石墨烯的方法,如微机械剥离(胶带法)、不同基板的外延生长法、CVD法、液相剥离、化学氧化还原法和电化学剥离法。本书重点关注了采用电化学剥离石墨和GO制备石墨烯的研究,这一过程的研究对于石墨烯大规模的制备非常有用。未来的研究仍需对现有制造工艺进行改进以满足新应用领域对石墨烯的要求和需求。

4.5　参考文献

[1] KROTO H W,HEATH J R,OBRIEN S,et al. C_{60}: buckminsterfullerene [J]. Nature,1985,318:162.

[2] IIJIMA S. Helical microtubules of graphitic carbon[J]. Nature,1991, 354:56.

[3] ZOMER P J,DASH S P,TOMBROS N,et al. A transfer technique for high mobility graphene devices on commercially available hexagonal boron nitride[J]. Applied Physics Letters,2011,99:232104.

[4] NOVOSELOV K S,GEIM A K,MOROZOV S V,et al. Two-dimensional gas of massless Dirac fermions in graphene[J]. Nature,2005,438:197.

[5] ZHANG Y B,TAN Y W,STORMER H L,et al. Experimental observation of the quantum Hall effect and Berry's phase in graphene[J]. Nature, 2005,438:201.

[6] HEERSCHE H B,JARILLO-HERRERO P,OOSTINGA J B,et al. Bipolar supercurrent in graphene[J]. Nature,2007,446:56.

[7] STANKOVICH S,DIKIN D A,DOMMETT G H B,et al. Graphene-based

composite materials[J]. Nature,2006,442:282.
[8] LEE C,WEI X,KYSAR J W,et al. Measurement of the elastic properties and intrinsic strength of monolayer graphene[J]. Science,2008,321: 385.
[9] YU A P,RAMESH P,ITKIS M E,et al. Graphite nanoplatelet let-epoxy composite thermal interface materials[J]. The Journal of Physical Chemistry C,2007,111:7565.
[10] BALANDIN A A,GHOSH S,BAO W Z,et al. Superior thermal conductivity of single-layer graphene[J]. Nano Letters,2008,8:902.
[11] LI X L,WANG X R,ZHANG L,et al. Chemically derived ultrasmooth graphene nanoribbon semiconductors[J]. Science,2008,319:1229.
[12] AVOURIS P,CHEN Z,PEREBEINOS V. Carbon-based electronics[J]. Nature Nanotechnology,2007,2:605.
[13] SON Y W,COHEN M L,LOUIE S G. Half-metallic graphene nanoribbons[J]. Nature 2006,444:347.
[14] WATCHAROTONE S,DIKIN D A,STANKOVICH S,et al. Graphene-silica composite thin films as transparent conductors[J]. Nano Letters, 2007,7:1888.
[15] SCHEDIN F,GEIM A K,MOROZOV S V,et al. Detection of individual gas molecules adsorbed on graphene[J]. Nature Materials,2007,6:652.
[16] YOO E J,KIM J,HOSONO E,et al. Large reversible Li storage of graphene nanosheet families for use in rechargeable lithium ion batteries [J]. Nano Letters,2008,8:2277.
[17] STROLLER M D,PARK S,ZHU Y,et al. Graphene-based ultracapacitors[J]. Nano Letters,2008,8:3498.
[18] BONG S,KIM Y R,KIM I,et al. Graphene supported electrocatalysts for methanol oxidation[J]. Electro chemical Communications, 2010, 12: 129.
[19] NOVOSELOV K S,GEIM A K,MOROZOV S V,et al. Electric field in atomically thin carbon films[J]. Science,2004,306:666.
[20] BERGER C,SONG Z M,LI X B,et al. Electronic confinement and coherence in patterned epitaxial graphene[J]. Science,2006,312:1191.
[21] SUTTER P W,FLEGE J,SUTTER E A. Epitaxial graphene on ruthenium [J]. Nature Materials,2008,7:406.

[22] GAO L,GUEST J R,GUISINGE N P. Epitaxial graphene on Cu(111)[J]. Nano Letters,2010,10:3512.

[23] LI X S,CAI W W,AN J H,et al. Large-area synthesis of high quality and uniform graphene films on copper foils[J]. Science,2009,324:1312.

[24] KIM K S,ZHAO Y,JIANG H,et al. Large-scale pattern growth of graphene films for stretchable transparent electrodes[J]. Nature,2009,457:706.

[25] SCHNIEPP H C,LI J L,MCALLISTER M J,et al. Functionalized single graphene sheets derived from splitting graphite oxide[J]. The Journal of Chemical Physics B, 2006,110:8535.

[26] MCALLISTER M J,LI J L,ADAMSON D H,et al. Single sheet functionalized graphene by oxidation and thermal expansion of graphite[J]. Chemistry of Materials,2007,19:4396.

[27] HERNANDEZ Y,NICOLOSI V,LOTYA M,et al. High-yield production of graphene by liquid-phase exfoliation of graphite[J]. Nanotechnology,2008,3:563.

[28] BISWAS S,DRZAL L T. A novel approach to create a highly ordered monolayer film of graphene nanosheets at the liquid-liquid interface[J]. Nano Letters,2009,9:167.

[29] GU W T,ZHANG W,LI X M,et al. Graphene sheets from worm-like exfoliated graphite[J]. Journal of Materials Chemistry,2009,19:3367.

[30] BOSE S,KUILA T,MISHRA A K,et al. Preparation of non-covalently functionalized graphene using 9-anthracene carboxylic acid[J]. Nanotechnology,2011,22:405603.

[31] LI X,ZHANG G,BAI X,et al. Highly conducting graphene sheets and Langmuir-Blodgett films[J]. Nanotechnology,2008,3:538.

[32] VALLES C,DRUMMOND C,SAADAOUI H,et al. Solutions of negatively charged graphene sheets and ribbons[J]. Journal of the American Chemistry Society,2008,130:15802.

[33] STANKOVICH S,PINER R D,CHEN X Q,et al. Stable aqueous dispersions of graphitic nanoplatelets via the reduction of exfoliated graphite oxide in the presence of poly(sodium 4-styrenesulfonate)[J]. Journal of Materials Chemistry,2006,16:155.

[34] STANKOVICH S, DIKIN D A, PINER R D, et al. Synthesis of graphene-based nanosheets via chemical reduction of exfoliated graphite oxide[J]. Carbon, 2007, 45: 1558.

[35] LI D, MULLER M B, GILJE S, et al. Processable aqueous dispersions of graphene nanosheets[J]. Nanotechnology, 2008, 3: 101.

[36] WANG G X, YANG J, PARK J, et al. Facile synthesis and characterization of graphene nanosheets[J]. The Journal of Chemical Physics C, 2008, 112: 8192.

[37] PARK S, RUOFF R S. Nature chemistryical methods for the production of graphenes[J]. Nanotechnology, 2009, 4: 217.

[38] PARK S, AN J, JUNG I, et al. Colloidal suspensions of highly reduced graphene oxide in a wide variety of organic solvents[J]. Nano Letters, 2009, 9: 1593.

[39] GOMEZ-NAVARRO C, WEITZ R T, BITTNER A M, et al. Electronic transport properties of individual chemically reduced graphene oxide sheets[J]. Nano Letters, 2009, 9: 2206.

[40] TUNG V C, ALLEN M J, YANG Y, et al. High-throughput solution processing of large-scale graphene[J]. Nature Nanotechnology, 2009, 4: 25.

[41] SHIN H J, KIM K K, BENAYAD A, et al. Efficient reduction of graphite oxide by sodium borohydride and its effect on electrical conductance[J]. Advanced Functional Materials, 2009, 19: 1987.

[42] ZHOU T, CHEN F, LIU K, et al. A simple and efficient method to prepare graphene by reduction of graphite oxide with sodium hydrosulfite [J]. Nanotechnology, 2011, 22: 045704.

[43] PEI S, ZHAO J, DU J, et al. Direct reduction of graphene oxide films into highly conductive and flexible graphene films by hydrohalic acids[J]. Carbon, 2010, 48: 4466.

[44] SALAS E C, SUN Z, LÜTTGE A, et al. Reduction of graphene oxide via bacterial respiration[J]. ACS Nano, 2010, 4: 4852.

[45] ZHANG J, YANG H, SHEN G, et al. Reduction of graphene oxide via L-ascorbic acid[J]. Chemical Communications, 2010, 46: 1112.

[46] ZHU C, GUO S, FANG Y, et al. Reducing sugar: new functional molecules for the green synthesis of graphene nanosheets[J]. ACS Nano, 2010, 4: 2429.

[47] LEE S-H,SEO S-D,JIN Y-H,et al. A graphite foil electrode covered with electrochemically exfoliated graphene nanosheets[J]. Electrochemical Communications,2010,12:1419.

[48] LIAN P,ZHU X,LIANG S,et al. Large reversible capacity of high quality graphene sheets as an anode material for lithium-ion batteries[J]. Electrochimica Acta,2010,55:3909.

[49] WANG C,LI D,TOO C O,et al. Electrochemical properties of graphene paper electrodes used in lithium batteries[J]. Chemistry of Materials, 2009,21:2604.

[50] WANG G,WANG B,PARK J,et al. Highly efficient and large scale synthesis of graphene by electrolytic exfoliation[J]. Carbon, 2009, 47: 3242.

[51] MEYER J C,GEIM A K,KATSNELSON I,et al. The structure of suspended graphene sheets[J]. Nature,2007,446:60.

[52] QI B,HE L,BO X,et al. Electrochemical preparation of free-standing few-layer graphene through oxidation-reduction cycling[J]. Chemical Engineering Journal,2011,171:340.

[53] AGÜÍ L,PEÑA-FARFAL C,YÁNEZ-SEDEÑO P,et al. Poly-(3-methylthiophene)/carbon nanotubes hybrid composite-modified electrodes[J]. Electrochimica Acta,2007,52:7946.

[54] SHAN C S,YANG H F,SONG J F,et al. Direct electrochemistry of glucose oxidase and biosensing for glucose based on graphene[J]. Analytical Chemistry,2009,81:2378.

[55] ZHOU M,ZHAI Y M,DONG S J. Electrochemical sensing and biosensing platform based on chemically reduced graphene oxide[J]. Analytical Chemistry,2009,81:5603.

[56] WEI D,GRANDE L,CHUNDI V,et al. Graphene from electrochemical exfoliation and its direct applications in enhanced energy storage devices [J]. Chemical Communications,48,9:1239.

[57] XIANG H F,SHI J Y,FENG X Y,et al. Graphitic platelets prepared by electrochemical exfoliation of graphite and their application for Li energy storage[J]. Electrochimica Acta,2011,56:5322.

[58] MORALES G M,SCHIFANI P,ELLIS G,et al. High-quality few layer graphene produced by electrochemical intercalation and microwave-assis-

ted expansion of graphite[J]. Carbon,2011,49:2809.
[59] SU C-Y,LU A-Y,XU Y,et al. High-quality thin graphene films from fast electrochemical exfoliation[J]. ACS Nano,2011,5:2332.
[60] KANG F,LENG Y,ZHANG T Y,In fluencies of H_2O_2 on synthesis of H_2SO_4-GICs[J]. The Journal of Chemical Physics Solids, 1996, 57: 889.
[61] WANG J,MANGA K,BAO Q,et al. High-yield synthesis of few-layer graphene flakes through electrochemical expansion of graphite in propylene carbonate electrolyte[J]. Journal of the American Chemistry Society,2011,133:8888.
[62] SUSLICK K S,PRICE G. Application of ultrasound to materials chemistry[J]. Annual Review of Materials Science,1999,29:295.
[63] LU J,YANG J. X,WANG J,et al. One-pot synthesis of fluorescent carbon nanoribbons,nanoparticles,and graphene by the exfoliation of graphite in ionic liquids[J]. ACS Nano,2009,3:2367.
[64] SEELA J A,DAHN J R. Electrochemical intercalation of PF_6 into graphite[J]. Journal of the Electrochemical Society,2000,147:892.
[65] KATINAONKUL W,LERNER M M. Graphical abstract: chem asian J 3/2007[J]. Fluorine Chemistry,2007,128:332.
[66] LIU N,LUO F,WU H,et al. One-step ionic-liquid-assisted electrochemical synthesis of ionic-liquid-functionalized graphene sheets directly from graphite[J]. Advanced Functional Materials,2008,18:1518.
[67] GUO H L,WANG X-F,QIAN Q-Y,et al. A green approach to the synthesis of graphene nanosheets[J]. ACS Nano,2009,3:2653.
[68] HUMMERS W S,OFFEMAN R E. Preparation of graphitic oxide[J]. Journal of the American Chemistry Society,1958,80:1339.
[69] DILIMON V S,SAMPATH S. Electrochemical preparation of few layer-graphene nanosheets via reduction of oriented exfoliated graphene oxide thin films in acetamide-urea-ammonium nitrate melt under ambient conditions[J]. Thin Solid Films,2011,519:2323.
[70] SHAO Y,WANG J,ENGELHARD M. Facile and controllable electrochemical reduction of graphene oxide and its applications[J]. Journal of Materials Chemistry,2010,20:743.

第 2 部分 石墨烯的表征

第 5 章 透射电子显微镜与石墨烯

本章综述了石墨烯表征过程中所涉及的透射电子显微镜(TEM)基本知识。讨论了如何采用 TEM 表征石墨烯结构和石墨烯缺陷,并进行了实例分析。对高分辨透射电镜和电子衍射技术进行了详细介绍。

5.1 引 言

透射电子显微镜(TEM)以及扫描透射电子显微镜(STEM)是在原子尺度上研究材料最常用的方法之一。现阶段 TEM 设备已经实现了与成像技术、衬度机理以及分析工具等手段的结合并应用[1,2]。因此,通过简单的介绍或一本书的方式无法囊括与 TEM 相关的知识。下面将介绍材料研究中最常用的三种手段:高分辨明场成像技术、微区电子衍射和扫描透射电子显微镜。

图 5.1 为三种操作模式下电子射线示意图。为了获得 TEM 图像,如图 5.1(a)所示,样品的某一区域被照亮,然后电子束通过光学元器件在显示屏、CCD 相机或胶片上形成放大的图像。为了使图像清晰,成像系统需要轻微的离焦以提高衬度。因此,在探头上获得了稍高于或稍低于样品位置处的电子波强度的清晰图像。对于电子衍射分析,目镜聚焦在物镜的后聚焦面上。这时所成的像为样品后某一距离电子束强度的图像,这一距离为相机常数(一般为 0.3~1.5 m)。可在样品上方通过选区光阑选择成像区域(nanobeam electron diffraction,纳米束电子衍射),或者在样品下方中间镜位置通过选区光阑选择成像区域(selected area electron diffraction,选区电子衍射)。对于 STEM,在偏转线圈的作用下高精度探针扫过样品产生散射电子,该散射电子被多种探头收集[3]。在一定程度上,STEM 中电子光路设置和电子与样品的相互作用与高分辨透射电镜(HRTEM)(图

5.8(a))相似,所不同的在于光源和探头的选择。

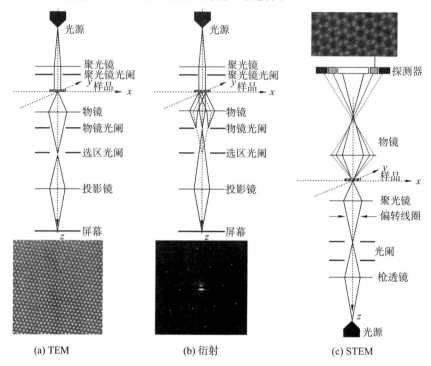

图 5.1　TEM、电子衍射和 STEM 中电子光路示意图以及所获得的单层石墨烯形貌 (a)采用 FEI Titan 高分辨透射电镜,经像差校正(image-side aberration corrector)后单层石墨烯形貌。(b)采用 Zeiss EM912 获得的石墨烯电子衍射图。(c)采用 Nion Ultra STEM100 中角环形暗场探测器获得的形貌[5,27,73]

对于 TEM 分析,首要的问题就是针对不同目的而正确制备所需样品。由透射电子显微镜名称即可看出,TEM 需要电子穿透样品到达探头。对于薄样品和原子序数小的样品,例如石墨烯或者碳纳米管,任何支撑膜都会在所成图像中形成比材料本身强的衬度。因此,样品必须具有自立性(free-standing)。实现石墨烯片层的良好铺展和自立是一项具有挑战性的工作。第一种解决该问题的方法如下,采用电子束印刷技术在石墨烯片层上制备金属栅格,然后将支撑衬底去除[4-5]。该方法不仅用于电镜分析,还被发展为制备达到 25 μm 的自立性单晶膜的方法[6-7]。之后人们又发明了一种将石墨烯片层机械剥离下来并置于 TEM 栅格上的方法[8]。但随着化学气相沉积法(CVD)在金属衬底上制备大面积石墨烯方法的推广[9-11],提出了将 CVD 法制备石墨烯转移到标准 TEM 栅格上的方法[12],并成为制备自立性薄膜样品最有效的方法。

从理论方面理解石墨烯、碳纳米管及相关材料的 TEM 形貌和电子衍射相对而言比较简单。该类材料在沿电子束方向仅有一个或几个碳原子层厚度,电子束与样品间的相互作用可以近似看成样品静电势对电子波的轻微干扰。在电子能量大于等于 60 kV 的情况下(在 20 kV 即使石墨烯也是很强的散射物质[13]),该类材料非常适合应用弱相位物体近似(a weak phase object approximation)进行讨论。如果进一步近似在埃瓦尔德球上所观察区域为平面(散射角很小),可以将样品后方电子波的相移表示为

$$T(x,y) = \exp[-\mathrm{i}\frac{eK}{2E}\int V(x,y,z)\mathrm{d}z]$$

其中,$V(x,y,z)$ 为样品的电势;e 为电荷数;E 为电量;K 为波矢,$T(x,y)$ 为与原电子波相乘(一般为平面)后的透射方程。可以看出,经上述近似后透射波是投影电子电势的函数。这种单片近似于单原子层是合理的(或可作为少层样品的近似),但是对于稍厚的样品需采用多层计算。例如双层样品采用单层模型所引起的误差已有研究者进行了描述[14]。

样品电势 $V(x,y,z)$ 通常看成由元素周期表中单个原子电势的叠加。单个原子电势的求和可表示为

$$V(\boldsymbol{r}) = \sum_n V_n(\boldsymbol{r}-\boldsymbol{R}_n)$$

其中,R_n 为原子 n 所处位置;V_n 为原子 n 的原子电势。上述近似忽略了由于固体材料中电子结构的变化(在实验研究二维材料、六方氮化硼、N 掺杂石墨烯的过程中,该模型的误差可以得到验证[15])。对于单一元素组成的样品,上述方程可以简化为

$$V(\boldsymbol{r}) = V_0(\boldsymbol{r}) \times \sum_n \partial(\boldsymbol{r}-\boldsymbol{R}_n)$$

其中,$V_0(\boldsymbol{r})$ 为碳原子电势;求和为原子位置 R_n 的求和。

5.2 石墨烯结构基础

现在考虑二维材料在三维倒易空间的表示。所有原子在一个平面上(即 $z=0$),而且所有原子在二维空间内规则排布,原子电势 $V(r)$ 的傅里叶变换可以表示为

$$\tilde{V}(\boldsymbol{q}) = f(\boldsymbol{q})\sum_n F_n \partial(q_x - Q_{n,x})\partial(q_y - Q_{n,y})$$

其中,$f(q)$ 为原子散射因子(原子电势的傅里叶变换);$(Q_{n,x}, Q_{n,y})$ 为二维

倒易晶格上倒易点;F_n 为结构因子。在 3D 倒易空间里,这就出现了沿 q_z 方向无限延伸的非零强度带(在晶体学上,零阶劳厄区是无限大的)。倒易空间的三维形貌如图 5.2(a)所示,平面处 $q_z=0$。图 5.2(b)给出了真实形貌内原始晶格间距。上述数值可以作为 TEM 和 STEM 研究石墨烯过程中的参考标尺。需要注意的是,石墨烯中最近邻原子面(0.142 nm)和六边形中心距离(0.246 nm)不产生布拉格衍射。

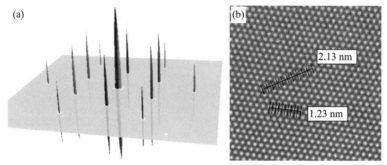

图 5.2　(a)石墨烯倒易空间示意图,非零强度由垂直于平面的连续棒表示。(b)[10-10]和[11-20]布拉格衍射所对应的晶格间距,图中显示晶格间距为 0.213 nm 和 0.123 nm 的 10 倍

图 5.3 为电子衍射图中计算强度的定量分解。在 $q_z=0$ 处倒易空间的结构因子如图 5.3(a)所示。图 5.3 为假设原子为理想点状条件下相对强

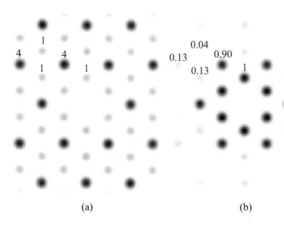

图 5.3　(a)石墨烯电子衍射图的强度,仅考虑结构因子。
　　　　(b)根据 Doyle 和 Turner 独立原子散射因素计算所得的强度

度。由于原子电势是有限的,减少了高阶反射的可能。图5.3(b)为基于Doyle和Turner独立原子散射因素(the independent-atom(IAM) scattering factors)计算获得的相对强度[16]。该结果可以看成是结构因子和原子散射因子共同作用,且[11-20]/[10-10]相对比值近似等于1。由IAM计算所得数值0.9与实验数值(1.0±0.05)接近;若采用密度泛函理论(DFT)计算石墨烯晶体电势则完全吻合[15]。

5.3 石墨烯的电子衍射分析

电子衍射分析是验证单层石墨烯的一种有效可靠的方法。图5.4给出了单层石墨烯和少层石墨烯的电子衍射图。图5.4(a)为单层石墨烯的电子衍射图,衍射相对强度与计算数值(图5.3(b))吻合得非常好。图5.4(b)为AB叠层结构的双层石墨烯,一般而言,[11-20]强度高于[10-10]强度;而对于少层石墨烯样品(ABA或ABC叠层结构),[11-20]强度高于[10-10]强度。AB叠层结构的石墨烯样品常见于机械剥离石墨,该石墨烯保留了石墨原材料的结构。CVD法制备的少层石墨烯往往为无序堆叠,表现为所得石墨烯层层间以完全随机的角度堆叠。因此,该类样品的电子衍射图出现多层现象,如图5.4(c)所示[17]。

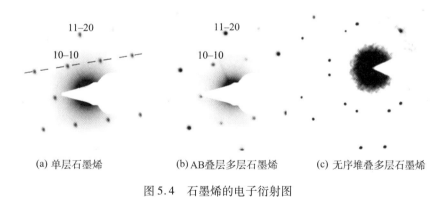

(a) 单层石墨烯　　(b) AB叠层多层石墨烯　　(c) 无序堆叠多层石墨烯

图5.4　石墨烯的电子衍射图

电子衍射图还可以提供缺陷密度的信息。样品原子晶格与理想石墨烯晶格间的偏差,可以用Debye-Waller因子描述,该不一致性导致高阶反射强度的降低。图5.5所示对比了高结晶度石墨烯(机械剥离制备)和带有缺陷石墨烯(单层氧化石墨烯,GO)[18]。图中给出了前者和后者的反射强度(图5.4(a)中虚线所示)。由于氧化石墨烯中具有较高的缺陷密度,导致其[11-20]/[10-10]强度之比仅为 c.0.5。这可以由静态 Debye-Waller 因子进行描述,缺陷引起的晶格与理想晶格间的偏差导致高阶峰强度的下降。

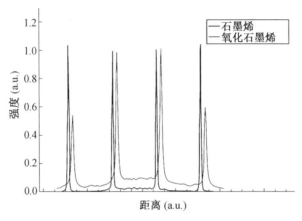

图 5.5 石墨烯和氧化石墨烯的电子衍射强度对比([10-10]峰进行归一化处理)

即使样品表现出图 5.4(a)所示的电子衍射图,也不能完全说明样品为单层石墨烯。理论上而言,AA 叠层的多层石墨烯也会产生上述衍射图(关于 AA 叠层石墨烯的报道很少)。此外,高缺陷密度的少层石墨烯也容易被误判为单层石墨烯,这是由于缺陷引起[11-20]强度下降到与[10-10]相似。若需要利用电子衍射图确定样品为单层石墨烯,须将样品在不同角度获得一系列衍射图[4-5]。这就需要在一个较大的倒易空间内进行。单层石墨烯样品所产生的电子衍射图的强度受角度变化影响很小。然而,石墨烯片层的起伏将导致衍射斑点的宽化,如图 5.6 所示。

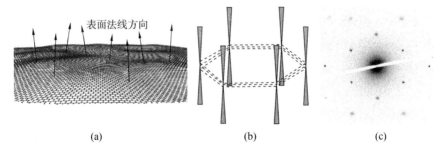

图 5.6 (a)悬浮石墨烯(suspended graphene sheets)的粗糙度。(b)石墨烯表面法线的变化导致倒易空间内非零强度为锥体而非带状。(c)倾斜样品(与水平轴线成 15°角)的电子衍射图表现为宽化的斑点[5]

5.4 石墨烯及其缺陷的像差校正 TEM 和 STEM 分析

接下来介绍如何利用更直接的图像(如真实空间的图像)来研究石墨烯及其缺陷。需要强调的是,TEM 探头上所获得图像并不一定是原子空间位置的图像或其投影像。总体而言,TEM 图像的获得可分为两步:样品对电子波的影响,以及电子光学系统对穿过样品电子波的影响。对于 STEM 而言,过程恰恰与上述相反,但获得结果相同的明场 STEM 形貌。

对于只有单原子厚度且原子序数小的石墨烯而言,样品对电子波的影响相对比较简单。电子光学系统的影响可以认为是穿过样品的电子波受(复杂)傅里叶滤波器的影响。按照 Buseck 等人的方法[2],该过程可以表示为

$$\varphi(\rho, z_{Det}) = \varphi(\rho, 0)\exp[\mathrm{i}(\pi\lambda z\rho^2 + \frac{1}{2}\pi\lambda^3\rho^4 C_s)] \cdot$$
$$[E_t(\rho)E_s(\rho)E_i(\rho)A(\rho)]^{\frac{1}{2}}$$

其中,$\varphi(\rho, z_{Det})$ 为电子波在探头处的波函数;$\varphi(\rho, 0)$ 为穿过样品后电子波的波函数;λ 为电子波长;z 和 C_s 为离焦和球面像差系数;E_t、E_s 和 E_i 分别为时间、空间和仪器阻尼系数;A 为物镜光阑相关系数。将上述方程进行变换,可获得以 ρ 为变量的表达式:

$$Q(\rho) = \sin(\pi\lambda z\rho^2 + \frac{1}{2}\pi\lambda^3\rho^4 C_s)E_t(\rho)E_s(\rho)E_i(\rho)A(\rho)$$

其中,$Q(\rho)$ 是对比传递函数(contrast transfer function,CTF),它受离焦和球面像差系数 z 和 C_s 以及其他像差系数的影响,其他像差系数在此不展开讨论。这些参数均可在像差校正透射电镜中进行调整。由于这些参数随时间发生漂移,所以必须保证图像是在设定的参数下获得的。在最佳条件下,设定 Scherzer 离焦等于 Lichte's 离焦[19],像差因素调节可以使 CTF 函数从零至显微镜极限(microscope's information limit),而 CTF 未与横坐标相交(图 5.7)。上述操作条件下所获得图像不存在离域的信息,获得 HR-TEM 图像能直接反映样品原子结构。

电子光学系统像差校正[20-21]使 TEM 晶格分辨率接近或低于 sp^2 杂化碳材料的阈值[22-24]。此外,碳材料研究方面的需求也在一定程度上促进了透射电子显微镜在低电压领域的研究进展[25-27]。像差校正后 TEM 图像中避免了信息的误差,从而为缺陷结构[23-24]、位错[24,28]、掺杂[15]以及非晶态[28-29]的研究提供了直接证据。像差校正 TEM 和 STEM 可提供其他研究

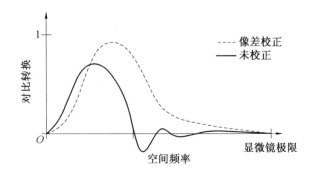

图 5.7 TEM 中像差校正和未校正所对应的对比传递函数(CTF)

手段无法提供的样品信息。例如,在上述的非六元环结构中,键结构与六元环晶格非常接近,仅利用分光镜的方法无法确认。还有一些碳同素异形体的弯曲利用扫描探针类设备也难以表征。

图 5.8 给出了一个 HRTEM 图像的例子,该图像是由 FEI Titan 在 80 kV 下获得的像差校正图像。该图像中包含了双层石墨烯、单层石墨烯、孔洞以及污染物。样品的层数在图 5.8(a)中用 0、1、2 进行了标记。单层区域内,与较暗区域形成衬度对比的六边形结构为原子的排列。在双层区域内,第一层和第二层可以被区分开(AB 叠层,白色和黑色)。该分辨率下(c. 0.2 nm),双层区域的图像在两个碳原子重叠的区域形成黑点。因此,双层区域显现出与单层区域完全相反的衬度。图像中的孔洞为层数的确定提供了进一步的证据。这些由光束刻蚀形成的孔洞,从真空到厚层的过渡过程呈现为台阶(如图从孔洞到第一层,从第一层到第二层)。石墨烯边缘处的高原子序数污染物如箭头所示。此外,石墨烯样品上无定型污染物如图 5.8(a)中虚线圆内所示。

在样品观测过程中,距离目标区域较近目标的污染物可以被用来验证和调整成像条件,以获得高质量形貌。图 5.8(b)~5.8(d)给出了污染物区域的傅里叶变换。该过程的目的是获得单通道带,以获得最佳信息,如图 5.8(b)所示。经过调节仪器参数达到该条件后,会获得图 5.8(b)所示的一个没有黑圈和黑线的圆形区域,此时可获得最佳信息。很小的失谐都会导致图 5.8(c)和图 5.8(d)所示的不对称形状区域或者区域内出现黑色圈,出现这些情况导致所获得图像存在或多或少的问题。需要指出的是,并非所有的失谐都可以通过上述过程发现,尤其是存在彗形像差(B2)或者三级象散(A2)的情况下。

接下来讨论在不同分辨率下石墨烯片层上的拓扑缺陷。例如多个碳

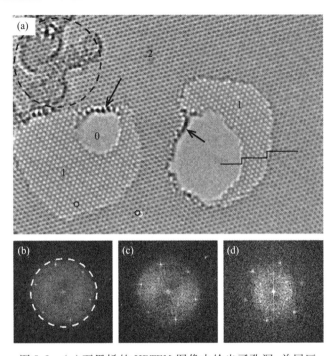

图 5.8 (a) 石墨烯的 HRTEM 图像中给出了孔洞、单层区域和双层区域。(b)~(d) 为不同成像条件下的傅里叶变换;其中(b)为最佳的成像条件,可以获得石墨烯晶格分辨率的形貌。该傅里叶变换是由图(a)中虚线区域内的无定形区域获得的。(c)和(d)所示傅里叶变换存在非圆像差时的状态

原子形成的环状局部拓扑结构,该结构是描述碳纳米材料结构的重要手段。纯拓扑缺陷被定义为在不增加和减少原子的条件下与六角晶格间的偏差,Stone-Wales(SW)缺陷就是最典型的例子[30]。空位和多重空位往往以非六元环的结构存在[29,31],五元环-七元环不匹配可形成位错中心[28,32-33]。此外,石墨烯内晶界是由五元环和六元环晶格不匹配所形成的[34-35]。由上述内容可以看出,非六元环结构的研究是理解 sp^2 杂化碳材料的关键。

图 5.9 给出了石墨烯和五元-七元结构(SW 缺陷)在不同分辨率下的形貌。在分辨率为 c.0.3 nm 条件下(常规透射电子显微镜,80~100 kV),在石墨烯原始晶格背景上可以观察到衬度很小的缺陷,但是无法准确描述。在约 0.2 nm 分辨率下(石墨烯[10-10]衍射晶格分辨率为

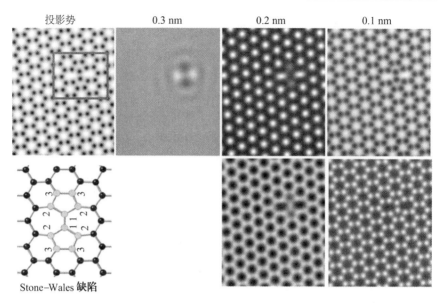

图 5.9 石墨烯片层上 Stone-Wales 缺陷不同分辨率的形貌。左上图像为投影势，其中黑色衬度表明具有较高的投影势。左下图像为 Stone-Wales 缺陷模型[72]。其他图像为分辨率是 0.3 nm、0.2 nm 和 0.1 nm 时的投影势，上排图像暗色为原子，下排图像亮色为原子

0.213 nm)，石墨烯的六角中心可以清晰地看到，根据成像条件不同表现为亮点或暗点。此时，五元环和七元环可以根据中心亮度弱和强进行区分，也可通过组成环状的原子个数进行区分。在约 0.1 nm 的分辨率下（石墨烯[20-20]衍射晶格分辨率为 0.108 nm)，所有原子可以准确地被区分开，因此可以区分组成环状结构的原子个数。原则上而言，点分辨率为 0.2 nm 时可以通过对比碳原子环中心的亮度，简便地表征出不同的拓扑结构。

碳纳米管的拓扑缺陷图像是由 Suenaga 首次报道[22]的，而石墨烯膜的拓扑缺陷图像由 Meyer 首次报道[23]。图 5.10 汇总了基本的缺陷类型，其中包括五元环、七元环或八元环。如图 5.10 所示，环的形状可以根据以下方式区分：五个原子组成的五元环中心亮度很弱或无亮度，而六元环中心亮度则较高，随着原子个数增加到七元环或八元环，环内亮度进一步增强（需要注意的是，图 5.12(c) 和图 5.14 中衬度相反）。

第 5 章　透射电子显微镜与石墨烯

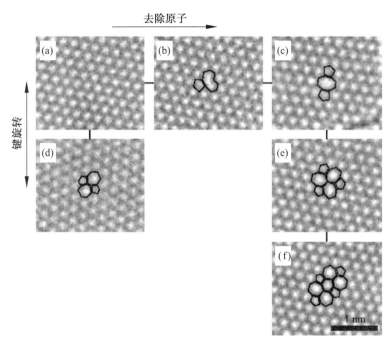

图 5.10　石墨烯中缺少 0、1、2 个原子时的基本缺陷形貌,并采用了非六元环进行表示[29]

5.5　石墨烯电子显微分析的启示

石墨烯的电子显微分析是 sp^2 杂化类碳纳米材料电子显微分析的一类。碳纳米管[36-38]、富勒烯[39]、石墨烯[40] 以及石墨均为 sp^2 杂化类碳材料,所不同的是它们的结构不同。碳纳米管和富勒烯可以看成由石墨烯卷曲而成的,而石墨可以看成是由石墨烯叠层而成的[41]。将结构进一步演化,尤其是非六元环的出现,可得到活性炭[42]、五肽(pentaheptide)[43]、二维非晶碳[29]以及其他[47]。当考虑到多层结构时,除了石墨和多层碳纳米管,还可形成洋葱状[44-45]、锥状[46]和卷轴状碳材料。在所有这些碳材料中,其碳原子均与三个相邻原子键合,这三个原子共平面或近似平面;所不同的是,它们的局部拓扑结构(环内原子个数)和总体拓扑结构不同,例如石墨烯为平坦的,碳管是弯曲的。此外,由实验手段所制备的平坦的单层石墨烯[40](且石墨烯片层具有自立性,即无须衬底的支撑[4,8,12,34])为我们研究 sp^2 杂化类碳材料提供了便利。如单层石墨烯的投影图像清晰地给出

101

了所有单原子的位置(而不是显示为一列原子),而且可以研究 sp^2 键合形式的碳原子。因此,石墨烯薄膜结构及其缺陷研究对于碳纳米材料这类具有极大科学和应用前途的新材料而言,具有重要意义。此外,显微结构研究实现了洞察 sp^2 杂化形式结合的碳原子。

前面论述的倒易空间方法(衍射方法)是研究晶体结构最准确的方法,也可描述晶体结构与标准晶格的偏差。显微图像的优势在于揭示了原子的非规则结构,如点缺陷、晶界、官能团、边缘等。最后一部分综述了电子显微图像研究中典型的例子,当然无法实现该领域研究中出现的所有结果和现象。首先要介绍与理想六角晶格相比所有由于偏差形成的缺陷。正如前面指出的,研究非六元环对于碳纳米材料而言非常重要,这是由于非六元环结构涉及拓扑缺陷、重构空位、晶界、边缘和其他很多相似问题。上述结构的识别前面已经进行描述,这些结构主要出现在石墨烯的早期电子显微镜研究过程,该类结构认为是由于辐射引起的缺陷[23-24]。图 5.10[29] 给出了重构空位的例子。

非六元环结构的发现实例来源于还原的氧化石墨烯(RGO)。虽然众多研究者对 RGO 进行了大量光谱研究,但是在电子显微研究手段之前,研究者并未描述 RGO 内部由于氧化-还原过程所形成的大量拓扑缺陷。正如 Gómez-Navarro 等报道[28],结晶良好的 RGO 片层上包含集群分布的尺寸几纳米的拓扑缺陷(图 5.11)。这些拓扑缺陷群主要是由碳五元环、七元环和旋转的六元环组成的纳米尺度区域,而这些拓扑缺陷群被尺寸为几十纳米的六角晶格隔离。然而,所有碳原子均与面内三个相邻原子以 sp^2 方式键合。机械剥离制备石墨烯不具备该类缺陷,因此该特征成为鉴定氧化-还原过程的有效手段。总体而言,氧化-还原过程如下:在氧化过程中,强氧化的非晶团簇形成,并与石墨烯完整晶格脱离[49]。还原过程中,这些团簇被还原成 sp^2 杂化的网状结构(由之前的光谱结果证实[50])。然而电子显微研究发现,上述网状结构并未与 sp^2 杂化的完整六角晶格结合,而是形成随机的、准非晶的单层 RGO 薄膜。

晶界被认为是由非六元环形成的位错结构[51-52],但是该位错结构无确切周期性。CVD 方法实现了在金属表面制备大面积石墨烯薄膜,该石墨烯薄膜内片层与片层间存在由于取向不同形成的晶界[9,10,17]。采用 CVD 方法可以在铜衬底上制备多晶的单层石墨烯薄膜[11]。该类单层石墨烯内晶界是由 Huang 等[34] 和 Kim 等[35] 首次报道的。图 5.12 给出了晶界的 STEM 环形暗场像(ADF-STEM),样品为铜衬底制备的多晶石墨烯并转移

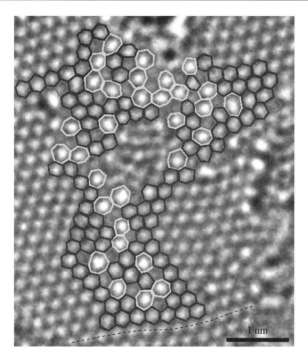

图 5.11 还原的氧化石墨烯(RGO)的结构缺陷[28]。图中给出了五元环、六元环、七元环或更高阶碳原子环。灰色虚线表示晶格扭曲变形至 TEM 用样品栅格之上[34]。图 5.12(c)和图 5.12(d)可以清晰地看出晶界的结构:碳五元环和七元环构成了晶界,环与环之间采用一定角度进行键合实现 sp^2 杂化。不同于简单的直线型晶界,该晶界并非是一条直线[53]。为了更清晰地研究 CVD 法制备石墨烯的晶界结构,需要结合高分辨和暗场像,前者可描述纳米尺度原子结构,后者可清晰描述晶体在微米尺度的取向。

二维材料的边缘是一维原子链形成的。石墨烯的边缘可以由高分辨电子显微镜观察到。Girit 等给出了第一张像差校正的高分辨形貌[54],以及石墨烯边缘的形成和动态重排过程视频。Chuvilin 等进一步研究了石墨烯边缘[55],并得到了边缘内某区域的动态重构过程(图 5.13)。可以发现一个有趣的现象是,在 5 nm 长的边缘内,某区域内存在从"纯"锯齿状边缘(即六元环形成的锯齿状边缘)到五元环-七元环重构锯齿状边缘的转换。

埃(1 Å(埃)= 10^{-10} m)(亚埃)大小的 STEM 探针不仅提供了获得高分辨图像的方法,同时为单原子尺度或单原子体积(single-atomic-column)尺

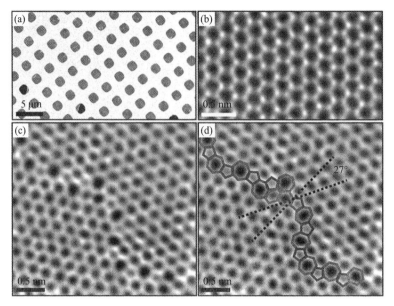

图 5.12　石墨烯及石墨烯晶界的环形暗场 STEM 形貌[36]。(a) CVD 法制备石墨烯膜的低倍形貌。(b) 高倍下石墨烯的晶格形貌。(c) 典型的石墨烯晶界形貌。(d) 五元环、七元环以及几个六元环所形成的线形结构

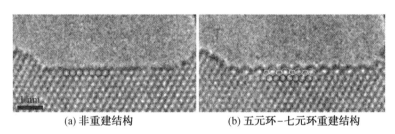

(a) 非重建结构　　　　　　　(b) 五元环-七元环重建结构

图 5.13　石墨烯锯齿状边缘由非重建结构到五元环-七元环重建结构之间的转换[55]

度光谱研究提供了可行性。图 5.14 给出了石墨烯边缘的高分辨形貌和图中所标记位置的电子能量损失谱[56]。值得注意的是，该结果可以给出边缘碳原子和内部碳原子的光谱。

严格控制样品的几何形状、样品低电压稳定性以及样品表面无非晶层，是确保获得高信噪比图像的前提条件。上述条件所获得高质量图像为我们采用 HRTEM 图像研究电子成键（如氮掺杂的石墨烯和六方氮化硼）提供了直接的实验基础[15]。众所周知，从基础化学而言，当孤立的原子通过化学键结合成为化合物后，其电子分布情况将发生变化。这种电荷密度

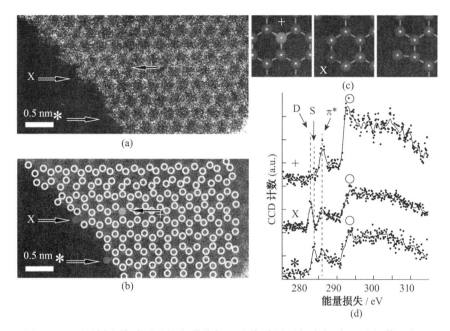

图 5.14 石墨烯边缘碳原子的光谱分析。边缘碳原子与内部碳原子光谱具有显著的不同[56]

的变化以及该变化所引起的电位和电子散射因子的变化,与其原数值相比很小[57]。然而,孤立的原子散射电位差是可以被检测到的[15]。根据独立原子模型和 DFT 计算,图 5.15 给出了石墨烯内由于氮原子点缺陷所形成的电荷密度变化(图 5.15(a))和电位变化(图 5.15(b))的投影。值得注意的是,可以看出氮原子与其最近邻碳原子存在键合作用。换句话说,与石墨烯内所有与碳原子相连的碳原子相比,邻近氮原子的碳原子具有不同的电子散射因子,而该差异足以采用 HRTEM 检测到。若采用 DFT 模型来模拟计算上述过程[15],其结果与实验数据非常吻合。因此,通过对比模拟和实验数据,可以分析单原子尺度上电荷的重新分布。

同时,该分析方法为我们深入了解材料提供了可能。首先,通过该方法可以直观地观察到石墨烯内单个氮原子的掺杂,即使二者的原子序数相差仅为 1。其次,HRTEM 的数据与 DFT 计算的对比明确了氮原子对与其相近碳原子电子结构的改变。有关氮掺杂石墨烯的相似结果也被之后的扫描隧道显微镜(STM)研究结果证实[58]。

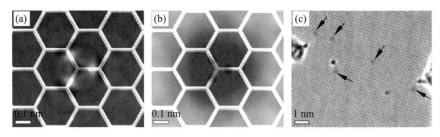

图 5.15 （a）独立原子模型（IAM）和 DFT 计算所表现出的电荷密度不同。与独立原子模型（IAM）不同，DFT 模型中白色表示低电荷密度,黑色表示高电荷密度。(b) DFT 和 IAM 投影电位的不同。由于键合原子间投影电位的增加,颜色由白色转变为黑色。由此获得石墨烯周期性晶格结构,如图(a)和(b)中白线所示。(c)单层石墨烯膜中氮掺杂原子的形貌[15]

5.6 结 论

电子显微镜是一种非常适合研究碳纳米材料的高度有效方法,特别是当电子束能量低于 80 kV 或者低于引起样品损害的能量[59]。现阶段像差校正的电子显微镜可以在相对较低的能量下获得晶格尺度甚至原子尺度分辨率的图像,可以显示出原子、点缺陷或晶界。需要指出的是,虽然石墨烯晶格对于低于 60~80 kV 的电子束相对而言比较稳定,然而其内部包含的缺陷、石墨烯边沿、吸附物、官能团在原子分辨率图像所需的束流下变换非常快。随着研究的发展,若能采用更低电子束能量获得上述材料信息,是非常有意义的。

本章综述了采用电子显微方法研究石墨烯所涉及的基础知识和应用实例。为了知识的完整性,在这里简要介绍其他几种与石墨烯相关的电子显微研究。(1)作为由原子序数低的元素组成的最薄的结晶材料,石墨烯具有良好的导电性,因此其可以作为其他材料电子显微分析的支撑材料[4,6,60-69]。(2)石墨烯是电子显微设备发展研究中理想的"测试标准样",这归因于它规整的结构(特别是厚度)、无吸附物和对低压电子束的高稳定性[13,14,26,27]。(3)电子显微技术本身即可用于石墨烯结构的改性,且该过程可在极高分辨率下提供直接的反馈[8,70,71],也就为实现新的应用提供了可能。总体而言,电子显微技术的发展和石墨烯相关物理研究的发展为相关研究提供了广阔的发展和探索空间,包括制备技术发展、缺陷和掺杂控制、晶界和表面等。

5.7 参考文献

[1] SPENCE J C H. High-resolution electron microscopy[M]. New York: Oxford University Press, 2009.

[2] BUSECK P, COWLEY J M, EYRING L. High-resolution transmission electron microscopy and associated techniques[M]. New York: Oxford University Press, 1992.

[3] PENNYCOOK S J. Scanning transmission electron microscopy: imaging and analysis[M]. Germany: Springer, 2011.

[4] MEYER J C, GEIM A K, KATSNELSON M I, et al. The structure of suspended graphene sheets[J]. Nature, 2007, 446(7131): 60.

[5] MEYER J C, GEIM A K, KATSNELSON M I, et al. On the roughness of single-and bi-layer graphene membranes[J]. Solid State Communications, 2007, 143: 101.

[6] BOOTH T J, BLAKE P, NAIR R R, et al. Macroscopic graphene membranes and their extraordinary stiffness[J]. Nano Letters, 2008, 8(8): 2442.

[7] NAIR R R, BLAKE P, GRIGORENKO A N, et al. Fine structure constant defines visual transparency of grapheme[J]. Science, 2008, 320(5881): 1308.

[8] MEYER J C, GIRIT C O, CROMMIE M F, et al. Hydrocarbon lithography on graphene membranes[J]. Applied Physics Letters, 2008, 92: 123110.

[9] YU Q, LIAN J, SIRIPONGLERT S, et al. Graphene segregated on Ni surfaces and transferred to insulators[J]. Applied Physics Letters, 2008, 93(11): 113103.

[10] REINA A, JIA X, HO J, et al. Large area, few-layer graphene films on arbitrary substrates by chemical vapor deposition[J]. Nano Letters, 2009, 9(1): 30.

[11] LI X, CAI W, AN J, et al. Large-area synthesis of high-quality and uniform graphene films on copper foils[J]. Science, 2009, 324(5932): 1312.

[12] LIN Y-C, JIN C, LEE J-C, et al. Clean transfer of graphene for isolation and suspension[J]. ACS Nano, 2011, 5(3): 2362.

[13] LEE Z, MEYER J C, ROSE H, et al. Optimum HRTEM image contrast at 20 kV and 80 kV-exemplified by graphene[J]. Ultramicroscopy, 2011, 112(1):39.

[14] JINSCHEK J R, YUCELEN E, CALDERON H A, et al. Quantitative atomic 3D imaging of single/double sheet graphene structure[J]. Carbon, 2011, 49(2):556.

[15] MEYER J C, KURASCH S, PARK H J, et al. Experimental analysis of charge redistribution due to chemical bonding by high-resolution transmission electron microscopy[J]. Nature Materials, 2011, 10(3):209.

[16] DOYLE P A, TURNER P S. Relativistic Hartree-Fock X-ray and electron scattering factors[J]. Acta Crystallographica Section A, 1968, 24(3):390.

[17] PARK H J, MEYER J, ROTH S, et al. Growth and properties of fewlayer graphene prepared by chemical vapor deposition[J]. Carbon, 2010, 48(4):108.

[18] PACILÉ D, MEYER J, RODRIGUEZ A F, et al. Electronic properties and atomic structure of graphene oxide membranes[J]. Carbon, 2011, 49:966.

[19] LENTZEN M. Progress in aberration-corrected high-resolution transmission electron microscopy using hardware aberration correction[J]. Microscopy and Microanalysis, 2006, 12(3):191.

[20] HAIDER M, UHLEMANN S, SCHWAN E, et al. Electron microscopy image enhanced[J]. Nature, 1998, 392(6678):768.

[21] BATSON P E, DELLBY N, KRIVANEK O L. Sub-ångstrom resolution using aberration corrected electron optics[J]. Nature, 2002, 418(6898):617.

[22] SUENAGA K, WAKABAYASHI H, KOSHINO M, et al. Imaging active topological defects in carbon nanotubes[J]. Nature Nanotechnology, 2007, 2(6):358.

[23] MEYER J C, KISIELOWSKI C, ERNI R, et al. Direct imaging of lattice atoms and topological defects in graphene membranes[J]. Nano Letters, 2008, 8(11):3582.

[24] GASS M H, BANGERT U, BLELOCH A L, et al. Free-standing graphene at atomic resolution[J]. Nature Nanotechnology, 2008, 3(11):676.

[25] SAWADA H, SASAKI T, HOSOKAWA F, et al. Higher-order aberration corrector for an image-forming system in a transmission electron microscope[J]. Ultramicroscopy,2010,110(8):958.

[26] KAISER U, BISKUPEK J, MEYER J C, et al. Transmission electron microscopy at 20 kV for imaging and spectroscopy[J]. Ultramicroscopy, 2011,111(8):1239.

[27] KRIVANEK O L, DELLBY N, MURFITT M F, et al. Gentle STEM: ADF imaging and EELS at low primary energies[J]. Ultramicroscopy,2010, 110(8):935.

[28] GóMEZ-NAVARRO C, MEYER J C, SUNDARAM R S, et al. Atomic structure of reduced graphene oxide[J]. Nano Letters,2010,10(4): 1144.

[29] KOTAKOSKI J, KRASHENINNIKOV A, KAISER U, et al. From point defects in graphene to two-dimensional amorphous carbon[J]. Physical Review Letters,2011,106(10):105505.

[30] STONE A, WALES D. Theoretical studies of icosahedral C60 and some related species[J]. Chemical Physics Letters,1986,128(5):501.

[31] LEE G-D, WANG C, YOON E, et al. Diffusion, coalescence, and reconstruction of vacancy defects in graphene layers[J]. Physical Review Letters,2005,95(20).

[32] HASHIMOTO A, SUENAGA K, GLOTER A, et al. Direct evidence for atomic defects in graphene layers[J]. Nature,2004,430:17.

[33] JEONG B, IHM J, LEE G-D. Stability of dislocation defect with two pentagon-heptagon pairs in graphene[J]. Physical Review B,2008,78(16):1.

[34] HUANG P Y, RUIZ-VARGAS C S, VAN DER ZANDE A M, et al. Grains and grain boundaries in single-layer graphene atomic patchwork quilts[J]. Nature,2011,469(7330):389.

[35] KIM K, LEE Z, REGAN W, et al. Grain boundary mapping in polycrystalline graphene[J]. ACS Nano,2011,5(3):2142.

[36] IIJIMA S. Helical microtubules of graphitic carbon[J]. Nature,1991, 354(6348):56.

[37] IIJIMA S, ICHIHASHI T. Single-shell carbon nanotubes of 1 nm diameter[J]. Nature,1993,363(6430):603.

[38] BETHUNE D S, KLANG C H, de VRIES M S, et al. Cobalt-catalysed growth of carbon nanotubes with single-atomiclayer walls[J]. Nature, 1993,363(6430):605.

[39] KROTO H W, HEATH J R, O'BRIEN S C, et al. C60: buckminsterfullerene[J]. Nature,318(6042):162.

[40] NOVOSELOV K S, GEIM A K, MOROZOV S V, et al. Electric field effect in atomically thin carbon films[J]. Science,2004,306(5696): 666.

[41] GEIM A K, NOVOSELOV K S. The rise of graphene[J]. Nature Materials,2007,6(3):183.

[42] HARRIS P J F, LIU Z, SUENAGA K. Imaging the atomic structure of activated carbon [J]. Journal of Physics: Condensed Matter, 2008, 20 (36):362201.

[43] CRESPI V, BENEDICT L, COHEN M, et al. Prediction of a pure-carbon planar covalent metal[J]. Physical Review B, Condensed Matter,19996, 53(20):R13303.

[44] UGARTE D. Curling and closure of graphitic networks under electron-beam irradiation[J]. Nature,359(6397):707.

[45] SAITO Y, YOSHIKAWA T, INAGAKI M, et al. Growth and structure of graphitic tubules and polyhedral particles in arc-discharge[J]. Chemical Physics Letters,1993,204(3-4):277.

[46] GE M, SATTLER K. Observation of fullerene cones[J]. Chemical Physics Letters,1994,220(3-5):192.

[47] VICULIS L M, MACK J J, KANER R B. A chemical route to carbon nanoscrolls[J]. Science,2003,299(5611):1361.

[48] PARK S, RUOFF R S. Chemical methods for the production of graphenes [J]. Nature Nanotechnology,2009,4(4):217.

[49] LERF A, HE H, FORSTER M, et al. Structure of graphite oxide revisited [J]. Journal of Physical Chemistry B,1998,102(23):4477.

[50] CAI W, PINER R D, STADERMANN F J, et al. Synthesis and solid-state NMR structural characterization of ^{13}C-labeled graphite oxide[J]. Science,2008,321(5897):1815.

[51] YAZYEV O V, LOUIE S G. Electronic transport in polycrystalline graphene[J]. Nature Materials,2010,9(10):806.

[52] READ W, SHOCKLEY W. Dislocation models of crystal grain boundaries [J]. Physical Review, 1950, 78(3):275.

[53] KOTAKOSKI J, MEYER J. Mechanical properties of polycrystalline graphene based on a realistic atomistic model [J]. Physical Review B, 2012, 85(19):195447.

[54] GIRIT C O, MEYER J C, ERNI R, et al. Graphene at the edge: stability and dynamics [J]. Science, 2009, 323(5922):1705.

[55] CHUVILIN A, MEYER J C, ALGARA-SILLER G, et al. From graphene constrictions to single carbon chains [J]. New Journal of Physics, 2009, 11(8):083019.

[56] SUENAGA K, KOSHINO M. Atom-by-atom spectroscopy at graphene edge [J]. Nature, 2010, 468(7327):1088.

[57] DENG B, MARKS L D. Theoretical structure factors for selected oxides and their effects in high-resolution electron-microscope (HREM) images [J]. Acta Crystallographica Section A: Foundations of Crystallography, 2006, 62, 208.

[58] ZHAO L, HE R, RIM K T T, et al. Visualizing individual nitrogen dopants in monolayer graphene [J]. Science, 2011, 333(6045):999.

[59] MEYER J, EDER F, KURASCH S, et al. Accurate measurement of electron beam induced displacement cross sections for singlelayer graphene [J]. Physical Review Letters, 2012, 108(19):1.

[60] IIJIMA S. Observation of single and clusters of atoms in bright field electron microscopy [J]. Optik, 1977, 48(2):193.

[61] MEYER J C, GIRIT C O, CROMMIE M F, et al. Imaging and dynamics of light atoms and molecules on graphene [J]. Nature, 2008, 454(7202):319.

[62] WILSON N R, PANDEY P A, BEANLAND R, et al. Graphene oxide: structural analysis and application as a highly transparent support for electron microscopy [J]. ACS Nano, 2009, 3(9):2547.

[63] PANTELIC R S, MEYER J C, KAISER U, et al. Graphene oxide: a substrate for optimizing preparations of frozen-hydrated samples [J]. Journal of Structural Biology, 2010, 170(1):152.

[64] LEE Z, JEON K-J, DATO A, et al. Direct imaging of soft-hard interfaces enabled by graphene [J]. Nano Letters, 2009, 9(9):3365.

[65] MCBRIDE J R, LUPINI A R, SCHREUDER M A, et al. Few-layer graphene as a support film for transmission electron microscopy imaging of nanoparticles[J]. ACS Applied Materials & Interfaces, 2009, 1(12): 2886.

[66] PANTELIC R S, SUK J W, MAGNUSON C W, et al. Graphene: substrate preparation and introduction[J]. Journal of Structural Biology, 2011, 174(1): 234.

[67] NAIR R R, BLAKE P, BLAKE J R, et al. Graphene as a transparent conductive support for studying biological molecules by transmission electron microscopy[J]. Applied Physics Letters, 2010, 97(15): 153102.

[68] WESTENFELDER B, MEYER J C, BISKUPEK J, et al. Transformations of carbon adsorbates on graphene substrates under extreme heat[J]. Nano Letters, 2011, 11, 5123.

[69] SCHÄFFEL F, WILSON M, WARNER J H. Motion of light adatoms and molecules on the surface of few-layer graphene[J]. ACS Nano, 5(12): 9428.

[70] FISCHBEIN M D, DRNDIC M. Electron beam nanosculpting of suspended graphene sheets[J]. Applied Physics Letters, 2008, 93(11): 113107.

[71] SONG B, SCHNEIDER G F, XU Q, et al. Atomic-scale electron-beam sculpting of near-defect-free graphene nanostructures[J]. Nano Letters, 2011, 11(6): 2247.

第6章 石墨烯的扫描隧道显微镜分析

本章综述了采用扫描隧道显微镜和光谱学研究石墨烯在不同衬底上的生长或沉积。讨论了衬底对石墨烯形貌和电子结构的影响。描述了如何采用实验手段表征石墨烯网格内缺陷,介绍了如何制备石墨烯纳米带以及其边缘电子结构的研究。

6.1 引　　言

石墨烯最早是1947年 P. R. Wallace 提出的理论假设,用以计算石墨的能带结构[1]。近70年来,实验过程中石墨烯被认为是碳的单原子层,常见于超高真空条件下制备过渡金属单晶膜的过程[2]。标准的金属膜制备工艺包括高温退火过程,在该过程中金属内部掺杂的碳原子常扩散至表面。之前已经存在在多种衬底上采用化学气相沉积的方法(CVD)有目的地制备"单原子层石墨"的报道[3],然而实验手段限制了该薄膜的表征。1982年,Gerd Binnig 和 Heinrich Rohrer 发明扫描隧道显微镜(STM)[4],实现了原子尺度表征不同衬底上碳单原子层,如 Ru(0001) 和 Ru(11-20)[5]、Ni(111)[6]、Pt(111)[7-9]。Geim 和 Novoselov 等于2004年采用机械剥离法从石墨剥离获得石墨烯并转移至 SiO_2 衬底之上,而后引发了科学界的关注[10]。在此之后,研究者采用了诸多方法实现石墨烯制备,如机械剥离石墨、SiC 衬底外延生长、金属衬底上 CVD 法。本章主要介绍采用扫描隧道显微镜 STM 和扫描隧道谱 STS 表征不同衬底上沉积或生长的石墨烯片或单层石墨烯,并研究不同衬底对石墨烯性能的影响。

6.2 不同惰性衬底上沉积石墨烯片的形貌、完整性和电子结构

衬底与所沉积的石墨烯片层间相互作用比较弱,因此其电子结构与衬底剥离后石墨烯或独立石墨烯类似。

6.2.1 石墨衬底上的石墨烯

将高定向热解石墨(HOPG)经空气气氛裂解,在其表面会自发形成一些石墨烯片层,这些石墨烯与石墨衬底不存在电荷耦合相互作用,因此其电子结构与独立石墨烯非常接近[11]。该片层的 STM 图像(图 6.1)显现出了由原子层形成的阶梯结构。原子尺度分辨率的 STM 形貌中存在两种情况:一种为由于 Bernal 堆叠结构石墨晶格断裂所形成的三角结构,另一种是自立性石墨烯表现出的蜂窝状结构(图 6.1 中 A 区域)。层与层间距离也进一步验证了上述结果:具有蜂窝状结构区域的层间距大于石墨层间距平均值(0.34 nm),而具有三角形结构区域层间距与 Bernal 堆叠结构层间距接近。

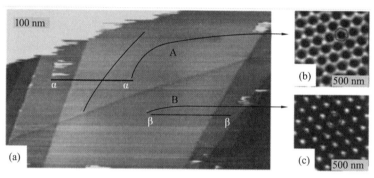

图 6.1 石墨表面石墨烯层的形貌

(a)低倍率下形貌。底部存在两个缺陷:最上方两层下存在一个倾斜的长脊,第一层下存在一条细微的垂直线。该长脊上方具有蜂窝状结构(b)的 A 区域,下方具有三角形结构(c)的 B 区域。图中箭头所指为原子分布图所研究的位置。(b)和(c)所示原子分布图给出了 A 区域的蜂窝状结构(六边形的六个顶点原子均可见),以及 B 区域的三角形结构(仅三个顶点原子可见,并对应一个晶格)[11]。样品测试条件为(a)偏置电压为 300 mV、隧穿电流为 9 pA,(b)偏置电压为 200 mV、隧穿电流为 22 pA,(c)偏置电压为 300 mV、隧穿电流为 55 pA

STS 最简单的表达方式为隧道电导与偏置电压的函数,与局部态密度(LDOS)大致成比例。图 6.2(a)给出了低温零磁场下该类石墨烯片层的 STS 数据。可以看出,石墨烯的 DOS 在费米能级附近(± 150 mV)呈 V 形,在狄拉克点接近零。狄拉克点位于费米能级以上 16 meV 附近,对应于 10^{10} cm^{-2} 能级的轻空位掺杂。作为对比,石墨区域的 STS 曲线如图 6.2(b)

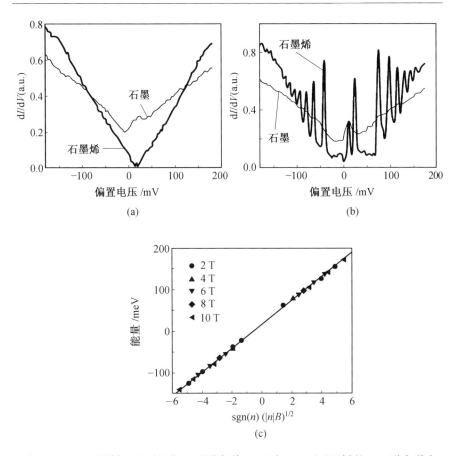

图 6.2 (a) 石墨烯以及石墨的 V 形隧穿谱。(b) 在 4 T 下石墨烯的 STS 隧穿谱表现出明显的朗道能级结构,而石墨不存在该现象。(c) 朗道能级峰位置与 $(|n|B)^{1/2}$ 作图,与狄拉克费米子的预期值非常吻合[12]

所示,表现为偏置电压为 0 附近存在一定的电导,反映出第一层碳原子层与石墨衬底间的耦合作用[11-12]。

在垂直层的方向施加一个磁场,可获得石墨烯层的非耦合程度。在磁场的作用下,石墨烯的 DOS 出现一系列具有不寻常能量的尖锐朗道能级,该能量由如下公式表示:

$$E_n = \mathrm{sgn}(n)\sqrt{2e\hbar v_F^2 |n| B} \quad n = \cdots, -2, -1, 0, 1, 2, \cdots$$

该能量由与磁场无关的零能量态以及磁场和能级因子平方根项组成: $n=0$ 能级的出现是狄拉克费米子存在手征性的结果,其不存在于其他已知的二维电子体系。该能级具有与其他能级相同的简并,但该能级为唯一与

施加磁场无关的能级。在 0 ~ 10 T 磁场下,如果朗道能级的能量与 $(|n|B)^{1/2}$ 作图,其数据变为一条直线,如图 6.2(c)所示,揭示出单个朗道能级序列具有狄拉克费米子的特征[11]。由直线的斜率可知费米速率为 $0.79 \times 10^6 \text{ m} \cdot \text{s}^{-1}$。所表现出的费米速率降低 20% 归因于电子-声子相互作用[11]。根据密度泛函理论(DFT)计算,电子-声子耦合导致零磁场下 dI/dV 曲线在光子能量($\hbar\omega_{ph}$)出现额外的扭结(kink)。数据在 ±150 meV 处的确出现了扭结的特征。

6.2.2 SiO$_2$ 衬底上的石墨烯

2004 年,Geim 和 Novoselov 首先在具有一层无定型 SiO$_2$ 层的 Si 晶片上测量了石墨烯的电子特性。在 STM 过程中使用该类衬底的目的在于 SiO$_2$ 低的电导率。但是,另一方面,硅基片有利于实现石墨烯的掺杂。

图 6.3(a)给出了 SiO$_2$ 层上石墨烯的 STM 形貌,从图中可以看出存在横向尺寸在几个纳米和垂直尺寸在 0.15 nm 左右的随机波纹,这是由衬底上 SiO$_2$ 层的粗糙度引起的[13]。由于杂质引起的电荷不均匀性也可清晰地看到。尽管如此,在表面波纹上仍可以清晰地看出石墨烯的蜂窝状结构(图 6.3(b))。STM/STS 结果显示出,SiO$_2$ 衬底上的石墨烯片层在零栅压时费米能级出现一个间隙[13]。如图 6.3(c)所示,曲线在 130 mV 出现以费米能级为中心的间隙,这由于弹性隧道到狄拉克锥之间线状态密度(可在石墨衬底上石墨烯检测到)显著不同。隧道电导在 138 mV 处出现一个额外的最低点,导致谱线对于费米能级不对称。该局部最低点被认为是由于衬底电子掺杂石墨烯引起的狄拉克点偏离费米能级。将低偏置处谱线放大,如图 6.3(c)所示,可以看出隧道电导在间隙处并未线性趋于零。间隙的位置与样品的具体位置无对应关系,虽然 V_D 的位置与探针的位置有关,而探针位置反映出掺杂的空间变化。

当改变栅极电压 V_g 时,石墨烯的狄拉克点 E_D 相对于 E_F 出现偏移,反映出二维电荷载流子密度 $n = \alpha V_g$,这里 $\alpha = 7.1 \times 10^{10} \text{ cm}^{-2} \cdot \text{V}^{-1}$。在石墨烯的相同位置上,在 $V_g = -60 \sim 60$ V 范围内,采用 STM 获得的 dI/dV 曲线如图 6.4(a)所示。中心间隙的宽度和能量位置不受栅极电压影响。但是,最小电导 V_D 随栅极电压发生显著变化,甚至表现出正负值的变化。该间隙特征未发现具有温度依赖性[13]。

该间隙的存在以及其与载流子密度无关,可能是由于声子参与下较高能量电子的非弹性隧穿,抑制了 E_F 附近的弹性隧穿。位于倒易空间内

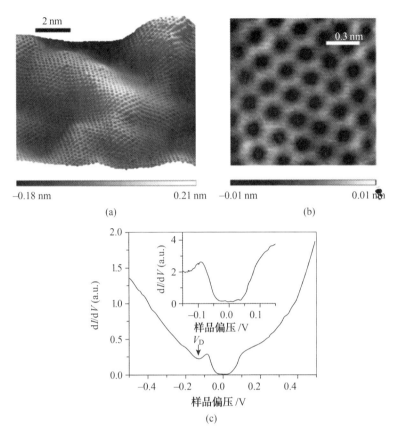

图 6.3 （a）SiO_2 衬底上石墨烯片的恒电流 STM 形貌（1 V,50 pA）。（b）恒电流 STM 所获得的石墨烯蜂窝晶格形貌（0.15 V,40 pA）。（c）零栅压下石墨烯的 dI/dV 曲线,阻抗为 5 GΩ（0.5 V,100 pA）。在阻抗为 1~100 GΩ,间隙宽度和相邻电导最小位置 V_D 对 STM 针尖高度不敏感。插图中给出了 dI/dV 谱图中央位置处的特征[13]

K/K' 点附近的 67 meV 出石墨烯平面的声子,被认为是该间隙存在的证据[13]。能量低于该隧道阈值的电子将在 E_F 点（K 点附近）弹性地进入石墨烯并表现出较低的迁移率,这是由于电子波矢较大地降低了隧穿概率。一旦偏置电压达到声子的能量,K 点的隧穿由于形成新的非弹性通道而增强。在这个新机制中,一个电子首先隧穿进入倒易空间内 Γ 点附近的石墨烯 σ^* 带,形成一个虚拟的过渡状态,然后在 K' 点释放出一个出平面声子进入 π 带中的 K 点。采用 DFT 理论计算结果也支持该解释[14]。由于 K/K'

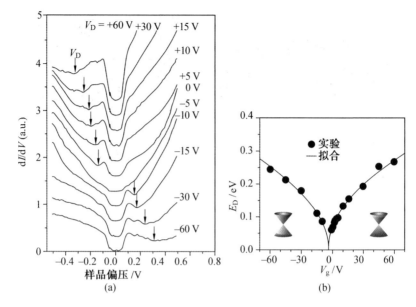

图 6.4 (a) 相同阻抗 5 GΩ(0.5 V,100 pA)条件下,石墨烯上固定位置处 dI/dV 曲线随不同栅极电压的变化。图中曲线进行了一定的垂直位移以便对比观察。箭头给出了相邻电导最小值的位置 V_D。(b) 狄拉克点的能量 E_D 与栅极电压的关系(E_D 由(a)图中电导最小值 V_D 获得,$E_D = e|V_D| - \hbar\omega_0$)。拟合曲线表明,$E_D$ 与栅极电压的平方根有关。当栅极电压一定时,图中给出了线形的石墨烯能带,其中占有态为深灰色[13]

处的出平面声子,导致非弹性隧道电流增加,从而石墨烯的 σ 和 π 电子能带出现明显的混合。

石墨烯的电子性质取决于载流子与多种因素(等离子体、声子)的相互作用。这些相互作用使石墨烯的能带结构归一化并产生新的非弹性通道,而这些相互作用主要受石墨烯衬底表面状态的影响。由 SiO_2 衬底上背栅单层石墨烯的扫描隧道谱还可以看出,谱线受多体作用(many-body excitations)影响。根据石墨烯载流子密度的不同可以区分等离子体或声子所导致的不同变化特征。一种特征为能量随栅极电压发生偏移,且对于给定的掺杂该现象只发生在费米能级的一侧,另一种特征为与栅极(和载流子密度)无关。第一性原理计算结果表明:第一种特征是由于准粒子自能量的多体化(many-body renormalization of the quasiparticle self-energy),而第二种特征是由于声子的非弹性隧穿。

6.3 SiC 和金属衬底上外延生长石墨烯的形貌、完整性及电子结构

6.3.1 SiC 衬底上石墨烯外延生长

采用 SiC 表明升华 Si 制备石墨层在 1975 年被 Van Bommel 等人发现[15]。碳化硅存在多种晶型,例如,4H 和 6H 属于六方结构,具有能带为 3 eV 的半导体性质。这种晶体具有两种不同的终端面,即 C 终端面或 Si 终端面。SiC 衬底上生长的石墨烯的物理和电子性质很大程度上受终端面的影响。在 C 终端面上,石墨烯以多层结构存在,层与层间具有无序的取向角度,因此不存在电子耦合[16]。另一方面,在 Si 终端面的石墨化过程中,碳化硅表面出现复杂的重构,而重构后碳化硅表面在石墨烯层连续生长过程中保持不变。该重构层的结构以及其对石墨烯生长和性能的影响现在还不够清楚。

(1)SiC 的 C 终端面($000\bar{1}$)。在碳化硅 C 终端面上,多层石墨烯可以与原有表面重建层共存,而不需要界面层存在。石墨烯层间无规律的层叠已经被衍射技术和 STM 手段证实,该结构被称为"moiré"超结构[17]。这类样品形貌内包含周期性为 2.5~3.8 nm 的 moiré 超晶格,超晶格中波纹具有埃米数量级的间距。

当 SiC($000\bar{1}$)面上石墨烯层厚至少为 3 层时,由 STM 图像可以获得层排列和局部应变的信息[18]。图 6.5 给出了 STM 获得的两个干涉 moiré 结构的例子。从图 6.5(a)和(b)可以看出,moiré 结构在不同的尺度出现(较大者为 26.5 nm,较小周期为 3 nm),该尺度仍大于原子晶格间距的数量级。由于出现了不止一个的 moiré 现象,可以断定至少存在三个原子层。利用快速傅里叶变换,图 6.5(c)给出了两组倒易空间格点。通过两组倒易空间格点的相对位置,可以获得形成特定大小的 moiré 结构所需的旋转角度。

这种无序的旋转导致层与层间的耦合作用减弱,导致该种结构电学性质与单层石墨烯相似。将层与层间旋转使其不存在贝纳尔堆叠结构,即使出现新的 moiré 超结构,保证了隔离的石墨烯层具有线性的能带。一项采用 STS 研究 SiC 衬底上六层石墨烯的研究发现,最上层石墨烯具有单层石墨烯的特性(图 6.6)[19]。数据显示最上层石墨烯的载流子迁移率与单层

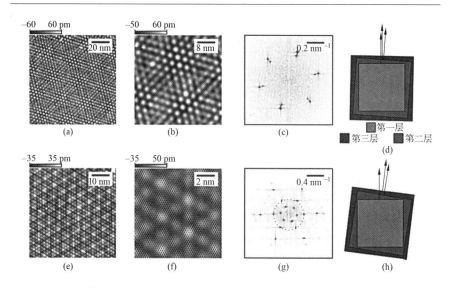

图6.5 SiC外延生长多层石墨烯采用STM获得的两个moiré结构（样品偏压 $V_s=0.5$ V，隧道电流 $I=100$ pA）。(a)两个大小相近的moiré结构共同组成一个较大的超晶格。第一和第二层、第二和第三层具有相似的旋转角度，但旋转方向相反，如图(d)所示。(b)为图(a)的放大形貌。(c)为图(a)的傅里叶变换。(d)为图(a)和(b)中moiré结构形成的示意图。(e)STM所获得moiré超晶格。(f)为相同区域的高分辨形貌。(g)为图(e)的傅里叶变换，图中组成圆环的斑点为两个moiré结构（三层石墨烯）相互作用的结果。(h)为图(e)中moiré结构形成示意图[18]

石墨烯相近，10^6 m·s^{-1}。研究磁场对朗道能级的影响表明，朗道能级的能级位置与磁场有关系。进一步研究发现，在高磁场下朗道能级出现四重分裂。第一次分裂是valley简并打破（valley degeneracy）的结果，第二次为自旋简并（spin degeneracy）打破引起的，能量较小。

(2) SiC的Si终端面(0001)。对于SiC(0001)表面，控制石墨化过程可以在连续的石墨烯层与SiC(0001)晶体间形成具有 $R30(6R3\times6R3)$ 结构的缓冲层，该缓冲层为非金属的富碳层。该缓冲层的结构非常复杂，且该缓冲层与其上层石墨烯、下层SiC间相互作用研究非常重要，这将有利于理解所制备石墨烯的性质[20]。

图6.7给出了6R3缓冲层的复杂结构，其中包括6×6、5×5和 $R3\times R3$ 的近似周期性重建。对比STM图像的傅里叶变换和 $6R3\times6R3$ 的低能电子衍射图（low energy electron diffraction, LEED），后者可以通过表面扫描与重建共同作用进行很好的解释[21-23]。

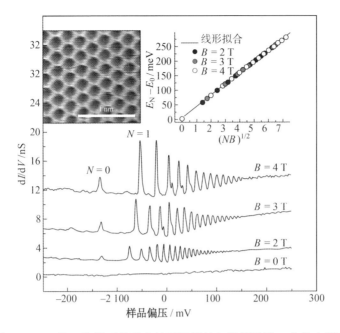

图6.6 SiC 的 C 终端面外延生长石墨烯的扫描隧道谱。曲线表明不同样品偏压表现出不同的 dI/dV 值。参数:设定电流为 400 pA,样品偏压为−300 mV,调制电压为 1 mV。左侧插图为石墨烯蜂窝状结构的 STM 形貌。参数:设定电流为 100 pA,样品偏压,250 mV,T = 35 mK。右侧插图为朗道能级峰位置与 NB 平方根的关系[19]

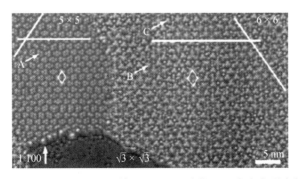

图6.7 SiC(0001)面经 1 250 ℃处理 1.5 min 后的 STM 占有电子态像,针尖电压为2.5 V,恒定电流为 60 pA。图中给出了 6×6 以及 5×5 近似周期性重构,对称方向如图中线所示[23]

如图6.8所示,给出了SiC(0001)衬底上单层石墨烯的STM形貌,图像可以看成是由SiC的界面特征与石墨烯晶格叠加的结果。单层石墨烯层的隧穿透明性为我们研究石墨烯层下面缓冲层的结构特征提供可能,如图6.8所示。低偏置STM图像揭示了类石墨的结构[24]。石墨烯层显示出2 nm的周期性。在更高分辨率下,结构的非规整性以凸起或点缺陷的形式存在。

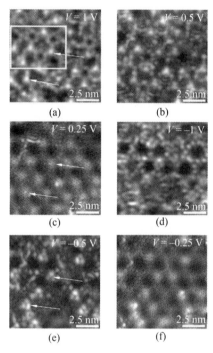

图6.8　偏压对石墨烯/SiC(0001)形貌的影响,在高偏压时可获得SiC界面的结构,低偏压时获得石墨烯薄膜的形貌。隧道电流固定为100 pA,图(a)～(f)中给出了偏压的数值。同一位置处采用不同偏压所表现出的不同特征如箭头所指区域:(a)四聚体(tetramers),(c)石墨烯的6×6结构,(e)三聚体(trimers)[28]

石墨烯/SiC体系为电子掺杂体系,其E_D值为E_F以下200～400 meV。关于SiC(0001)表面外延生长的单层和双层石墨烯的电子结构研究仍然是热点,且已发表的研究结果充满了争议。图6.9所示的SiC(0001)衬底上单层和双层石墨烯表现出100 meV的间隙特性[25]。在间隙以上,空间分辨光谱(spatially resolved spectroscopy)显示出明显的空间不均匀性。该间隙特性可能存在以下三种解释:石墨烯下方SiC层的电子态;STM探针电场引起的充电/带弯曲;与表面激发的非弹性耦合。如前面所述,该课题组之后报道了在 SiO_2 衬底上剥离石墨烯具有类似的间隙结构,归因于增强的非弹性(声子)通道[13]。

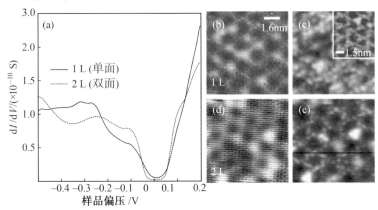

图 6.9　(a)单层石墨烯区域(实线)和双层石墨烯区域(虚线)的平均空间 dI/dV 曲线。低偏压时单层石墨烯区域(b)和双层石墨烯区域(d)的形貌(-0.05 V,0.025 nA)。高偏压时相同位置处单层石墨烯(c)和双层石墨烯(e),(c)中插图为单层石墨烯区域更高偏压时形貌(-1.0 V,0.003 nA)[25]

Vitali 及其合作者 2008 年报道了 SiC 衬底剥离石墨烯在狄拉克能量具有空间调制(与厚度有关)的间隙。对于单层石墨烯,该间隙的宽度可以通过缓冲层进行调制[26]。作者认为,这种可调制性并不是由于缓冲层导致的对称性破坏,而是由于 6R3 超结构在界面所形成的结构周期性变化。相反,文章认为双层石墨烯的间隙宽度与空间无关。另外,Rutter 等人报道结果表明单层石墨烯具有该间隙,而双层石墨烯不存在该间隙[27]。

Lauffer 等人报道了多层石墨烯的厚度与隧道谱的相关性,同时根据 STS 结果发现了双层石墨烯电荷分布的不均匀性[27]。所有的 dI/dV 曲线在零偏压处出现最小值,但是 Brar 等人发现了与之完全相反的结果[25]。除了零偏压处最低值,2 ML、3 ML 和 4 ML 的 dI/dV 曲线局部最低值在负偏压处随着厚度的增加逐渐向零偏压偏移。Lauffer 等人认为该第二最低点给出了狄拉克点的位置,而厚度引起的偏移是由于电荷转移至不同层引起的电荷减少[27]。

6.3.2　金属表面石墨烯

金属表面石墨烯的发现已有 40 多年,之前它被称为"单层石墨"或

"碳单原子层"[29]。这些碳层来源于高温下材料退火所引起的材料内碳原子的表面偏析[30]。现阶段金属表面外延生长的石墨烯一般采用分解含碳氢分子(常采用乙烯)[31]获得。在这种情况下,分子前驱体常温下首先吸附于金属表面,然后经高温退火分解前驱体,或直接将前驱体分解在热的金属表面[5]。在以上过程中,石墨烯的生长过程自发结束,且第二层或多层很难生长。当考虑到热金属表面催化分解含碳分子的作用时,上述过程很容易理解。

(1)石墨烯的生长、结构及其与衬底相互作用。高度完美外延生长的单层大尺寸石墨烯是该材料在诸多领域应用的前提条件。现阶段已经实现在金属表面制备该类石墨烯。在超高真空下,通过软化学气相沉积方法,可以在众多单晶金属上外延生长单层石墨烯,如 Ir(111)[32-35]、Ru(0001)[31,36-41]、多晶 Ru[42]、Pt(111)[43-44]、Ni(111)[45-46]、Cu(111)[47]、Rh(111)[48]、Re(0001)[49]或 Co(0001)[50-51]。通过表面状态的精确控制,超高真空的制备条件,以及扫描隧道显微镜所提供的原子尺度分辨率,使该体系达到了原子尺度。

C 1s 的 X 射线光电子能谱表明金属衬底不同,导致单层石墨烯与衬底键合作用不同,同时其电子结构也不同[52]。对于强相互作用的金属,如 Ru(0001),C 1s 的 XPS 图谱表现出两个峰,表明存在两种结合态的碳原子。因此,石墨烯的电子结构由于金属与石墨烯相互作用发生显著变化[31,37,53,54]。此外,对于弱相互作用金属衬底,如 Ir(111),C 1s 图谱仅表现出一个峰,表明所有碳原子其化学结合状态是一致的。这种石墨烯的电子结构仅仅表现出与自立性石墨烯轻微的改变[52]。

石墨烯晶格参数(0.246 nm)与金属衬底晶格参数的不同(Ni(111)0.249 nm、Rh(111)0.269 nm、Ru(0001)0.271 nm、Ir(111)0.272 nm 和 Pt(111)0.277 nm),导致出现了一系列与石墨烯结构有关的周期性 moiré 结构。Ni(111)和 Co(0001)例外,这是由于它们与石墨烯的晶格失配足够小。

在强相互作用金属衬底上,如 Ru(0001)衬底,外延生长的石墨烯表现为尺寸为几个微米大小的均匀单层石墨烯,可以再现表面阶梯、位错及其他结构缺陷[54,55]。图 6.10(a)给出了 Ru(0001)表面单层石墨烯的 STM 形貌,可以发现,石墨烯表现出间距为 3 nm 的三角形周期性点阵,且点阵间存在约为 0.1 nm 的纵向波纹。这些由 moiré 结构出现的三角形点阵是由于外延生长的石墨烯与钌衬底晶格常数不同形成的。假设未发生晶格变形的石墨烯覆盖于 Ru(0001)表面,那么石墨烯晶格将与钌晶格出现不匹配现象。根据晶格常数大小,碳 11 的蜂窝(2.707 nm)可以与 Ru—Ru 的 10 晶格(2.706 nm)相匹配,表现出一个可以被忽略的 0.05% 的压缩应变。需要指出的是,(12×12) 超结构,如 12(C—C)(2.953 nm) 和 11(Ru—Ru)(2.976 nm),仅为 0.78% 的拉伸应变。在此之前,Grant 和 Haas 等报道了 Ru 晶体表面 C 偏析形成的(9×9) 超结构[56],Goodman 及其合作者报道了热解甲烷并加热至 1 300 K 获得的(11×11) 超结构。Marchini 等 2007 年通过 STM 和 LEED 分析结果,提出了 3 nm 周期性的(12×12) 超结构[36]。Martoccia 等基于 X 射线衍射数据,提出了一个(25×25) 的更大超级结构,认为其是由于 Ru(0001) 表面石墨烯引起的 0.02 nm 的体变形和 0.01 nm 的横向弛豫[57]。

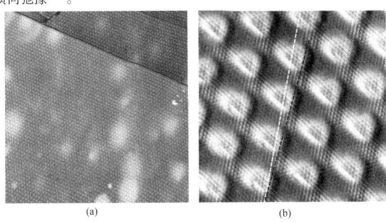

图 6.10 (a) Ru(0001)表面单层石墨烯的 STM 形貌(200 nm×200 nm)。对不同高度石墨烯层的颜色进行了调整以便于观察。白色圆点对应于 moiré 超晶格的最大值。(b) 无缺陷区域的原子尺度分辨率的 STM 形貌(13 nm×13 nm)(参数设定 V_s = 1 mV 和 I_t = 1 nA)。图中白色虚线为石墨烯内 C 原子的[1120]方向,黑色虚线为 moiré 超晶格的方向[54]

图 6.10(b) 给出了外延生长石墨烯的高分辨 STM 拓扑结构,图中可以看出原子分布与 moiré 超结构的叠加。不同于石墨的 STM 形貌仅表现出两个碳原子中的一个,石墨烯的蜂窝状结构被清晰地表达出来,C—C 间距为 0.14 nm。图中白色虚线为高度对称的碳原子晶格[1120]晶向,而黑色虚线表示六角 moiré 结构的晶向。显然在没有缺陷的区域二者是无法对齐的。二者间的角度为 $\phi_{gr,moire}$ = 4.5°±0.5°。这种碳原子晶格与 moiré 结构间的偏差通过 LEED 图谱也可以被发现[39,58]。由于 moiré 结构的高精度,

我们可以通过 moiré 超结构的研究获得 C 和 Ru 间的晶格不匹配程度，$\phi_{gr,Ru}=0.5°\pm0.05°$。这个小旋转角度可以很好地解释，XRD 所获得钌表面的(25×25)周期性结构和 STM 图像所获得 moiré 结构间的矛盾。当考虑到 Ru(0001)原子层，石墨烯与 Ru(0001)晶格(无变形)[1010]晶向出现 0.5°的旋转，表现出(24×24)的周期性。这个大的周期性的存在减少了石墨烯生长层与 Ru(0001)间的应力，也可能减少了石墨烯层的褶皱。

在弱相互作用衬底上石墨烯样品的 STM 研究中也发现了此类 moiré 结构，如 Pt(111)和 Ir(111)，然而在此类样品中由于相互作用较弱，导致形成几种不同周期的超结构。对于 Pt(111)存在周期为 2.2 nm 的无旋转相，旋转角度为 1.5°的周期为 2 nm 的相及旋转角度为 90°的周期为 0.5 nm 的相[7]。最近，出现了六种周期性 0.5~2.1 nm 的原子分辨率的 moiré 超结构[44]。后来，一个($R3\times R3$)的 30°超结构被发现[43]。

在 Ir(111)上，石墨烯形成一个周期为 9.32 Ir 晶格常数和 0.02 nm 垂直波纹的 moiré 超结构[32,34]。当采用高温分解乙烯制备薄膜时，可获得上述高度有序的相[33]。图 6.11 中 STM 图像显示出了高度有序的六角 moiré 超结构。石墨烯表现为地毯似的覆盖于多个衬底的台阶之上。作为对比，在低温时制备石墨烯在 Ir(111)上表现为旋转取向的多相区域。图 6.11(d)标记出了六角 moiré 图形的旋转角度。周边区域间的角度差异很明显(这是由于 moiré 现象的放大效应)，但是实际上衬底的高度对称方向和石墨烯晶格原子间角度偏差较小。

(2)金属表面石墨烯的电子结构。由于石墨烯与金属衬底间晶格常数的差别，外延生长的石墨烯与 Ru(0001)衬底间相互作用表现出空间周期性变化，从而使石墨烯可分为纳米大小的 H 区域("高"区域)和 L 区域("低"区域)，进而具有不同的电子结构。图 6.12 显示了 LDOS 在费米能级以上和以下的空间分布。实验所获得的图像在图中上排显示。图中明亮区域 LDOS 较大，图(e)为 $dI/dV-V$ 的隧道谱，该隧道谱与 LDOS 大致呈比例[31]。不同空间区域表现出不同的扫描隧道谱：波纹层内"高"区域的占据态 LDOS 较大，而波纹层"低"区域的空态 LDOS 较大。上述不同即使在 300 K 的条件下也会存在。此外，在波纹区域内费米能级两侧还存间隔为 0.3 eV 的能量相同的位置。这种不均匀性可以通过一个简单的紧束缚模型进行理解，该模型是由结构波纹周期性势场和石墨烯与 Ru 衬底周期性相互作用共同组成的[31,54]。该模型与实验结果相吻合，模型计算可得占据态 LDOS 在超晶格的"高"区域较大，这里的势场最小，而空态 LDOS 在石墨烯"低"区域较大。这些结果都表明，在费米能级附近，这种空间周

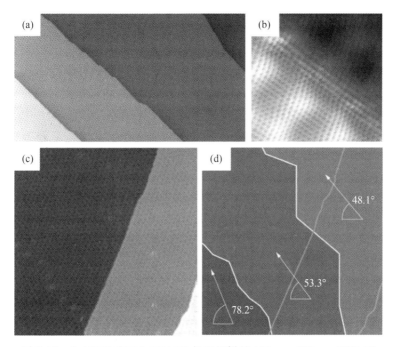

图6.11 (a)1320 K下Ir(111)生长石墨烯的125 nm×250 nm STM形貌(0.10 V, 30 nA),石墨烯覆盖了多个Ir台阶。(b)台阶处石墨烯原子的分布(5 nm×5 nm, 0.04 V, 30 nA)。(c) 1 120 K下Ir(111)生长石墨烯的108 nm×108 nm的STM形貌(-0.05 V, 30 nA),存在两个台阶以及三个不同取向的moiré区域。(d) 三个不同取向的moiré区域和取向关系[33]

期性和化学相互作用所调制出的电子结构形成了一系列电子口袋(electron pockets)阵列[31,37]。

石墨烯和Ru(0001)间moiré结构和空间调制的化学相互作用的存在还对表面静电电位发生影响。该表面静电电位变化可以通过STM的共振场发射方法表征[59]。图6.13(a)给出了Ru(0001)表面外延生长石墨烯的形貌。局部的电子结构往往通过强度和隧道结电压的关系确定,而这里可以通过隧穿距离Z与电压V给出相同的结果。图6.13(b)内插图给出了Ru(0001)区域的dZ/dV图。从图中可以看出存在三个峰,第一个FER峰的能量略高于Ru(0001)的功函数(5.4 eV)[60]。图6.13(b)给出了石墨烯moiré结构的"高"和"低"区域的dZ/dV曲线,分别由实线和虚线表示。moiré结构"高"区域(即石墨烯与金属衬底间距离较远)的测试曲线,其第一个FER峰出现在4.4 eV附近,与功函数(4.5 eV)接近[60]。

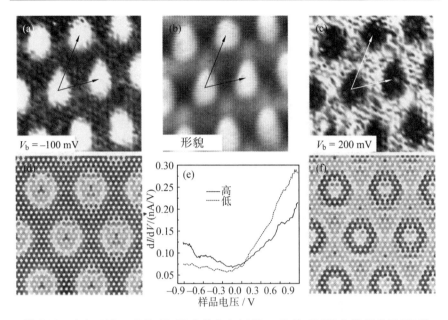

图6.12 (a)-100 meV 的 dI/dV 曲线和(c)200 meV 的 dI/dV 曲线反映了 LDOS 在费米能级以下和以上的空间分布。(b) 所对应的 STM 形貌(topography)。(d)和(f)按照石墨烯(11×11)周期性波纹所计算的 LDOS 空间分布,其中浅灰色表示高 LDOS,深灰色表示低 LDOS[31],(e)300K 下 Ru(0001)表面单层石墨烯波纹高低不同所对应的 dI/dV 曲线,其中黑色实线为石墨烯的高处,虚线为石墨烯的低处。

moiré 结构"低"区域的测试曲线(即石墨烯与钌结合力较强,间距小)表现出了意想不到的特点。首先,在费米能级之上出现了 3 eV 的新峰,该峰在"高"区域不会出现。其次,第一个 FER 峰出现在 4.8 eV,高于 H 区域("高"区域)的数值。L 区域和 H 区域的所有其他高阶的 FERs 均位于不同的能量,表明其局部功函数不同[41],与其他体系的理论和实验相符。L 区域的第一 FER 峰位置与 H 区域的位置存在 0.4 eV 的偏移,偏移方向与局部功函数方向相反。第一 FER 峰的这种相反方向的偏移不受实验过程中探针尖锐程度、样品温度(4.6~300 K)或隧道电流的影响,虽然每个实例中能量偏移与探针和样品间的电场有关,而电场与探针尖锐度和隧道电流有关。

总之,该体系中发现了三个独特的现象:①石墨烯的第一态是分裂的,在 H 区域出现能量低的分裂,L 区域出现能量高的分裂;②不同于其他态,分裂后态未显示出 0.25 eV 的间隔,这是由于 L 区域和 H 区域的局部态密

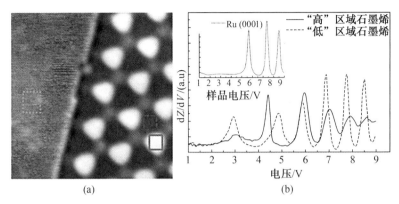

图6.13 （a）4.6K下Ru(0001)衬底上石墨烯岛边缘的STM形貌（15 nm×15 nm，$V_s = -0.5$ V）。左边为未被石墨烯覆盖的Ru，右边为石墨烯岛的三角形moiré结构。（b）石墨烯岛的moiré超晶格上所获得的dZ/dV曲线。实线为图（a）中高处，虚线为图（a）moiré区域的低处。（b）中插图为Ru(0001)表面的dZ/dV曲线[61]

度变化；③在L区域3 eV出现新的态，归因于石墨烯超晶格的拓扑结构。理论计算结果认为以上现象是准二维带的分裂和空间限制的结果，这是由于石墨层与金属表面相互结合强度的调制引起的[54,61]。石墨烯/Rh(111)具有相似的实验结果[48]。

对于Ir(111)表面石墨烯，其moiré单胞内能量不存在变化，结果表现为单套峰。这种moiré单胞内的均匀性是由于石墨烯和Ir(111)间弱相互作用[62]。由于表面电位的不同，石墨烯的第一层和第二层的FER表现出轻微的能量位置不同。

对于金属衬底与石墨烯间存在强相互作用的体系，由于几何学和电子结构的混合作用，导致STM所获得moiré超结构表现得相当复杂。Ru(0001)上生长石墨烯就是一个典型的例子。moiré结构的图像状态取决于探针和样品间的偏置电压，例如对于电压高于2.6 V，甚至出现和图6.14所示相反衬度的结果。这种moiré结构与样品偏压的关系可以通过DFT计算进行解释[63]。计算过程采用了一个可以容纳moiré结构特征的足够大单胞。图6.15显示了不同偏置电压下计算结果与实验结果的吻合。从图中可以清晰地看出，当考虑范德瓦耳斯力时计算结果与实验结果很好地吻合，表明表面存在较强的电子效应。对于Ni(111)上生长石墨烯，采用插入Au[46]、Ag[64]或Cu[65]的方法可以减弱石墨烯层与衬底的强

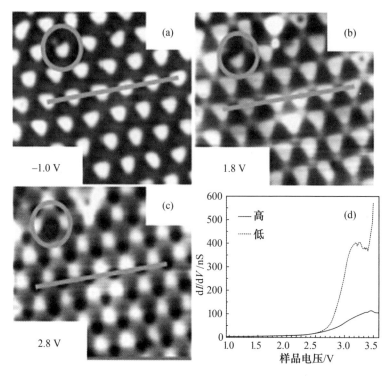

图 6.14 石墨烯/Ru(0001)的 STM 形貌及衬度变化。(a)~(c)三个不同样品偏压下的 STM 形貌。(d)高于费米能级时图(a)中 moiré 结构高处(实线)及低处(虚线)的局部隧道谱[54]

烈相互作用。

不同衬底上石墨烯薄膜的曲率是一个重要的参数。最近的理论研究认为,应变可以产生伪磁场,进而用于调节石墨烯的电子态[66]。这种效应是石墨烯所特有的,这是由于石墨烯无质量的类似于费米能级的狄拉克能带结构和晶格完美性。利用石墨烯与金属衬底间热膨胀系数的不同引入应变是一种有效的方法。采用有机分子为原料,在高温下金属衬底退火制备石墨烯薄膜,当样品冷却时,由于金属衬底和石墨烯的热膨胀系数不同,表现出不同的收缩。由此产生的应变导致石墨烯层中存在纳米气泡,例如,Ir(111)[67]和 Pt(111)[66]上就出现了纳米气泡。研究者采用 STM/STS 研究了 Pt(111)上石墨烯纳米气泡。研究结果认为是由于应变引起的大于 300 T 的伪磁场所形成的朗道能级[66]。

第6章 石墨烯的扫描隧道显微镜分析

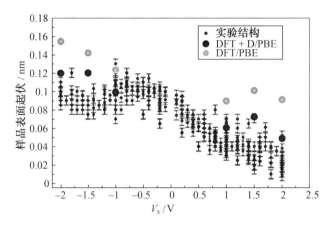

图6.15 石墨烯/Ru(0001)样品表面起伏随偏置电压的变化。黑色小圆点为不同实验条件下获得的数据(针尖、隧道电流、样品、温度)。未采用范德瓦耳斯修正的理论结果由浅灰色点表示,采用范德瓦耳斯修正的理论结果采用黑色大圆点表示。对STM图像进行理论计算时采用了$1.69×10^{-1}$ nm^{-3}的电子密度[63]

6.4 点缺陷的扫描隧道显微镜(STM)和扫描隧道谱(STS)分析

了解石墨烯点缺陷的传输性质对于石墨烯的应用和基础研究十分重要。石墨烯蜂窝状晶格的完整性是决定石墨烯独特电子结构的重要因素。石墨烯的晶格对称性产生低能电子结构,具有线性能量–动量色散的特征。晶格的缺陷(点缺陷)打破了晶格的对称性,从而为研究石墨烯的量子特性提供可能。

掺杂是石墨烯内引入点缺陷的一种方法。这种有效的改变电子性质的方法对于石墨烯二维材料具有不同的结果。2011年,Zhao等研究了石墨烯内掺杂N原子附近的结构,测量了纳米尺度态密度和载流子浓度[68]。该N掺杂的石墨烯样品采用化学气相沉积法在铜箔衬底上生长。图6.16(a)给出了STM的形貌,可以看出在石墨烯晶格上存在一些亮点。这些亮点是由于石墨烯层上N原子掺杂引起的。这个结论可通过X射线和拉曼光谱进行支持。图6.16(b)给出了某个N掺杂位置附近的dI/dV曲线。其最突出的特点为,在费米能级以下零偏压附近和300 mV附近存在两个凹点。这些特征在之前单层石墨烯中已经被观察到[13]。在−300 meV

131

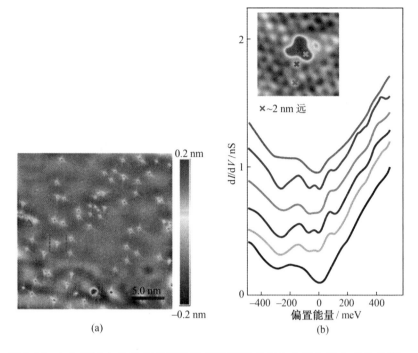

图6.16 (a) 铜箔上 N 掺杂石墨烯的 STM 图像表现出大量与掺杂有关的信息($V_{bias}=0.8$ V, $I=0.8$ nA)。(b) N 原子(最低处曲线)及其附近的 dI/dV 曲线,对曲线进行了一定的垂直位移。最高处曲线为距离杂原子2 nm 位置处的 dI/dV 曲线($V_{bias}=0.8$ V, $I=1.0$ nA)[68]

处特征与电子掺杂石墨烯的狄拉克点有关。从狄拉克点的能量位置,可以计算出载流子的密度,结合 N 掺杂浓度数据,可以得出石墨烯内每个 N 原子提供 0.42 个载流子。N 掺杂位置附近的 dI/dV 图表明 N 掺杂所引起的电子扰动仅在原子附近存在。研究者对 SiC(0001) 生长石墨烯的点缺陷也进行了研究。采用 STS 方法研究准粒子干涉图[28]发现,LDS 显示出两个不同的调制波长,作者认为其归因于谷内散射和谷间散射。

6.5 石墨烯纳米带的 STM/STS 研究

石墨烯纳米带(graphene nanoribbons, GNR)有望获得一些独特的电子性质,使其在纳米电子器件领域存在潜在的应用。尤其是,石墨烯纳米带能带间隙与纳米带宽度成反比[69],并具有不寻常磁结构的一维边缘态[70-72]。然而,石墨烯纳米带的研究受到其制备的局限。

石墨烯纳米带的制备已出现了一些方法。例如采用碳纳米管切开的方法,然而制备宽度为 10 nm 边缘整齐的石墨烯纳米带仍然是个挑战。最近,Cai 等[73]在 2010 年报道了一种自下而上的制备石墨烯纳米带的方法,可以原子尺度制备各种拓扑结构和宽度的石墨烯纳米带。该方法包括表面辅助聚合分子前驱体获得线性聚合物。接下来是退火使之脱氢环化反应生成石墨烯纳米带。石墨烯纳米带的拓扑结构、宽度、边缘结构与有机前驱体的关系如图 6.17 所示。图 6.17(a)给出了采用 10,10'-二溴-9,9'-二蒽为前驱体制备的石墨烯纳米带的 STM 形貌。石墨烯纳米带的分子模型在图的右下方,通过 DFT 计算所得的 STM 模拟纳米带见图中灰色部分。图 6.17(b)给出了采用 6,11-二溴-1,2,3,4-四苯基三亚苯前驱体单体制备石墨烯纳米带的高分辨率 STM 图。在这种情况下,所制备的波浪形石墨烯纳米带具有类似扶手椅的边缘。插图给出了分子模型与 DFT 计算所得 STM 模拟形貌对比。

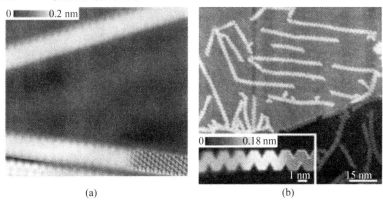

图 6.17　(a) Au(111)上石墨烯纳米带的高分辨 STM 形貌,并包含了纳米带的原子模型($T=5$ K, $U=-0.1$ V, $I=0.2$ nA)。右下角为基于 DFT 模拟计算所得的 STM 形貌。(b) Au(111)表面波浪形石墨烯纳米带的总体形貌($T=35$ K, $U=-2$ V, $I=0.02$ nA)。插图为高分辨率的 STM 形貌($T=77$ K, $U=-2$ V, $I=0.5$ nA)以及基于 DFT 模拟计算所获得的 STM 形貌(灰色)[73]

Tao 等[74]2011 年报道了采用石墨烯纳米管切开的方法制备石墨烯纳米带的边缘电子结构。STS 测试揭示了石墨烯纳米带一维边缘态的存在,其能级位置和宽度取决于边缘态能隙。GNR 边缘能带由于宽度引起分裂,可能是理论预测所描述的其磁结构的证据,但是至今为止未发现直接的证据。

边缘结构被认为可以改变石墨烯纳米带和量子点的电子性质。据报道,横向尺寸在 7~8 nm 的石墨烯量子点具有金属性质,石墨烯纳米带也具有这一特点。石墨烯纳米带边缘为锯齿型,与扶手椅型相比,具有较小的间隙[75]。值得指出的是,石墨烯的结构不规则,这就导致很难确切地说明电子结构受石墨烯量子体大小和结构的影响。

另一重要方面是石墨烯边缘的电子散射,而该现象被认为通过量子干涉对石墨烯量子体内电子传输有重要贡献。研究 6H-SiC(0001) 衬底上生长的单层石墨烯的不同边缘结构表明,沿 C—C 键的电子态密度导致了沿石墨烯碳键网络的量子干涉,而其形状仅受到边缘结构的影响,不受电子能量影响[76]。

6.6 结　论

本章总结了利用 STM 和 STS 研究不同衬底上生长或分解制备石墨烯的一些结果。STM 为我们提供了研究石墨烯结构完整性和石墨烯电子结构以及不同衬底对石墨烯电子性能影响的直接手段。由于 STM 的空间原子分辨能力,使其成为研究石墨烯晶格上单原子缺陷电子性质的独特手段。对石墨烯边缘及石墨烯纳米带的电子结构也进行了论述。现阶段石墨烯的研究正处于快速发展阶段中,在本章撰写过程中出现了一些重要的新发现,本章中所描述的某些未确定问题也可能出现定论,此外,也会出现一些新的研究领域。

6.7 参考文献

[1] WALLACE P R. The band theory of graphite[J]. Physical Review,1947, 71:622.

[2] OSHIMA C,NAGASHIMA A. Ultra-thin epitaxial films of graphite and hexagonal boron nitride on solid surfaces[J]. Journal of Physics:Condensed Matter,1997,9:1.

[3] HU Z-P,OGLETREE D F,VAN HOVE M A,et al. LEED theory for incommensurate overlayer:application to graphite on Pt(111)[J]. Surface Science,1987,180:433.

[4] BINNIG G,ROHRER H,GERBER C,et al. Surface studies by scanning

tunneling microscopy[J]. Physical Review Letters,1982,49:57.
[5] WU M-C,XU Q,GOODMAN D W. Investigation of graphitic overlayers formed from methane decomposition on Ru(0001) and catalysts with scanning tunneling microscopy and high resolution electron energy loss spectroscopy[J]. Journal of Physics Chemistry,1994,98:5104.
[6] KLINK C,STENSGAARD I,BESENBACHER F,et al. An STM study of carbon-induced structures on Ni(111): evidence for a carbidic-phase clock reconstruction[J]. Surface Science,1995,342:250.
[7] LAND T A,MICHELY T,BEHM R J,et al. STM investigation of single layer graphite structures produced on Pt(111) by hydrocarbon decomposition[J]. Surface Science,1992,264:261.
[8] SASAKI M,YAMADA Y,OGIWARA Y,et al. Moiré contrast in the local tunneling barrier height images of monlayer graphite on Pt(111)[J]. Physical Review B,2000,61:15653.
[9] UETA H,SAIDA M,NAKAI C,et al. Highly oriented monolayer graphite formation on Pt(111) by a supersonic methane beam[J]. Surface Science,2004,560:183.
[10] NOVOSELOV K S,GEIM A K,MOROZOV S V,et al. Electric field effect in atomically thin carbon films[J]. Science,2004,306:666.
[11] LI G,LUICAN A,ANDREI E Y. Scanning tunneling spectroscopy of graphene on graphite[J]. Physical Review Letters,2009,102:176804-1.
[12] LUICAN A,LI G,ANDREI E Y. Scanning tunneling microscopy and spectroscopy of graphene layers on graphite[J]. Solid State Communications,2009,149:1151.
[13] ZHANG Y,BRAR V W,WANG F,et al. Giant phonon-induced conductance in scanning tunneling spectroscopy of gate-tunable graphene[J]. Nature Physics,2008,4:627.
[14] WEHLING T O,GRIGORENKO I,LICHTENSTEIN A I,et al. Phonon-mediated tunneling into graphene[J]. Physical Review Letters,2008, 101:216803-1.
[15] VAN BOMMEL A J,CROMBEEN J E,VAN TOOREN A. LEED and auger electron observations of the SiC(0001) surface[J]. Surface Science,1975,48:463.
[16] MILLER D L,KUBISTA K D,RUTTER G M,et al. Observing quantiza-

tion of zero mass carriers in graphene[J]. Science,2009,324:924.
[17] VARCHON F, MALLET P, MAGAUD L, et al. Rotational disorder in few-layer graphene films on $6H-SiC(000\bar{1})$: a scanning tunneling microscopy study[J]. Physical Review B,2008,77:165415-1.
[18] MILLER D L, KUBISTA K D, RUTTER G M, et al. Structural analysis of multilayer graphene via atomic moiré interferometry[J]. Physical Review B,2010,81:125427-1.
[19] SONG Y J, OTTE A F, KUK Y, et al. High-resolution tunnelling spectroscopy of a graphene quartet[J]. Nature,2010,467:185.
[20] MALLET P, VARCHON F, NAUD C, et al. Electron states of mono-and bilayer graphene on SiC probed by scanning tunneling microscopy[J]. Physical Review B,2007,76:041403-1.
[21] RUTTER G M, CRAIN J N, GUISINGER N P, et al. Scattering and interference in epitaxial graphene[J]. Science,2007,317:219.
[22] RIEDL C, STARKE U, BERNHARDT J, et al. Structural properties of the graphene-SiC(0001) interface as a key for the preparation of homogeneous large-terrace graphene surfaces[J]. Physical Review B,2007,76:245406-1.
[23] MÄTENSSON P, OWMAN F, JOHANSSON L I. Morphology, atomic and electronic structure of 6H-SiC(0001) surfaces[J]. Physica Status Solidi B,1997,202:501.
[24] HIEBEL F, MALLET P, MAGAUD L, et al. Atomic and electronic structure of monolayer graphene on $6H-SiC(000\bar{1})(3\times3)$: a scanning tunneling microscopy study[J]. Physical Review B,2009,80:235429-1.
[25] BRAR V W, ZHANG Y, YAYON Y, et al. Scanning tunneling spectroscopy of inhomogeneous electronic structure in monolayer and bilayer graphene on SiC[J]. Applied Physics Letters,2007,91:122102-1.
[26] VITALI L, RIEDL C, OHMANN R B, et al. Spatial modulation of the Dirac gap in epitaxial graphene[J]. Surface Science,2008,602:L127.
[27] LAUFFER P, EMTSEV K V, GRAUPNER R, et al. Atomic and electronic structure of few-layer graphene on SiC(0001) studied with scanning tunneling microscopy and spectroscopy[J]. Physical Review B,2008,77:155426-1.

[28] RUTTER G M, GUISINGER N P, CRAIN J N, et al. Imaging the interface of epitaxial graphene with silicon carbide via scanning tunneling microscopy[J]. Physical Review B,2007,76:235416-1.

[29] MAY J W. Platinum surface LEED rings[J]. Surface Science,1969,17:267.

[30] GRANT J. Auger electron spectroscopy studies of carbon overlayers on metal surfaces[J]. Surface Science,1971,24:332.

[31] VáZQUEZ DE PARGA A L, CALLEJA F, BORCA B, et al. Periodically rippled graphene:growth and spatially resolved electronic structure[J]. Physical Review Letters,2008,100:056807-1.

[32] N'DIAYE A T, BLEIKAMP S, FEIBELMAN P J, et al. Two-dimensional Ir cluster lattice on a graphene moiré on Ir(111)[J]. Physical Review Letters,2006,97:215501-1.

[33] CORAUX J, N'DIAYE A T, BUSSE C, et al. Structural coherency of graphene on Ir(111)[J]. Nano Letters,2008,8:565.

[34] N'DIAYE A T, CORAUX J, PLASA T N, et al. Structure of epitaxial graphene on Ir(111)[J]. New Journal of Physics,2008,10:043033-1.

[35] LOGINOVA E, NIE S, THÜRMER K, et al. Defects of graphene on Ir (111): rotational domains and ridges[J]. Physical Review B,2009,80:085430-1.

[36] MARCHINI S, GÜNTHER S, WINTTERLIN J. Scanning tunneling microscopy of graphene on Ru(0001)[J]. Physical Review B,2007,76:075429-1.

[37] VÁZQUEZ DE PARGA A L, CALLEJA F, BORCA B, et al. Vázquez de Parga reply[J]. Physical Review Letters,2008,101:99704-1.

[38] PAN Y, ZHANG H, SHI D, et al. Highly ordered, millimeter-scale, continuous, single-crystalline graphene monolayer formed on Ru(0001)[J]. Advanced Materials,2009,21:2777.

[39] SUTTER P W, FLEGE J-I, SUTTER E A. Epitaxial graphene on Ru [J]. Nature Materials,2008,7:406.

[40] ZHANG H, FU Q, DALI Y C, et al. Growth mechanism of graphene on Ru(0001) and O_2 adsorption on the graphene/Ru(0001) surface[J]. Journal of Physical Chemistry C,2009,113:8296.

[41] BRUGGER T, GÜNTHER S, WANG B, et al. Comparison of electronic

structure and template function of single-layer graphene and a hexagonal boron nitride nanomesh on Ru(0001)[J]. Physical Review B,2009, 79:045407-1.

[42] SUTTER E A, ALBRECHT P M, SUTTER P W. Graphene growth on polycrystalline Ru thin films[J]. Applied Physics Letters,2009,95: 133109-1.

[43] OTERO G, GONZáLEZ C, PINARDE A L, et al. Ordered vacancy network induced by the growth of epitaxial graphene on Pt(111)[J]. Physical Review Letters,2010,105:216102-1.

[44] GAO M, PAN Y, HUANG L, et al. Epitaxial growth and structural property of graphene on Pt(111)[J]. Applied Physics Letters,2011,98: 033101-1-3.

[45] DEDKOV Y S, FONIN M, LAUBSCHAT C. A possible source of spin-polarized electrons:the inert graphene/Ni(111) system[J]. Applied Physics Letters,2008,92:052506-1.

[46] VARYKHALOV A, SÁNCHEZ-BARRIGA J, SHIKIN A M, et al. Electronic and magnetic properties of quasifreestanding graphene on Ni[J]. Physical Review Letters,2008,101:157601-1.

[47] GAO L, GUEST J R, GUISINGER N P. Epitaxial graphene on Cu(111) [J]. Nano Letters,2010,10:3512.

[48] WANG B, CAFFIO M, BROMLEY C, et al. Coupling epitaxy, chemical bonding, and work function at the local scale in transition metalsupported graphene[J]. ACS Nano,2010,4:5773.

[49] MINIUSSI E, POZZO M, BARALDI A, et al. Thermal stability of corrugated epitaxial graphene grown on Re(0001)[J]. Physical Review Letters,2011,106:216101-1.

[50] EOM D, PREZZI D, RIM K T, et al. Structure and electronic properties of graphene nanoislands on Co(0001)[J]. Nano Letters,2009,9:2844.

[51] VARYKHALOV A, RADER O. Graphene growth on Co(0001) films and islands and its precise magnetization dependence[J]. Physical Review B,2009,80:035437-1.

[52] PREOBRAJENSKI A B, NG M L, VINOGRADIV A S, et al. Controlling graphene corrugation on lattice-mismatched substrates[J]. Physical Review B,2008,78:073401-1.

[53] BORCA B, BARJA S, GARNICA M, et al. Periodically modulated geometric and electronic structure of graphene on Ru(0001)[J]. Semiconductor Science and Technology, 2010, 25:034001-1.

[54] BORCA B, BARJA S, GARNICA M, et al. Electronic and geometric corrugation of periodically rippled, self-nanostructured graphene epitaxially grown on Ru(0001)[J]. New Journal of Physics, 2010, 12:093018-1.

[55] LOGINOVA E, BARTELT N C, FEIBELMAN P J, et al. Factors influencing graphene growth on metal surfaces[J]. New Journal of Physics, 2009, 11:063046-1.

[56] GRANT J T, HAAS T W. A study of Ru(0001) and Rh(111) surfaces using LEED and Auger electron spectroscopy[J]. Surface Science, 1984, 21:76.

[57] MARTOCCIA D, WILLMOTT P R, BRUGGER T B, et al. Graphene on Ru(0001): a 25×25 supercell[J]. Physical Review Letters, 2008, 101: 126102-1.

[58] LOGINOVA E, BARTELT N C, FEIBELMAN P J, et al. Factors influencing graphene growth on metal surfaces[J]. New Journal of Physics, 2009, 11:063046-1.

[59] BINNIG G, FRANK K H, FUCHS H, et al. Tunneling spectroscopy and inverse photoemission image and field states[J]. Physical Review Letters, 1985, 55:991.

[60] HIMPSEL F J, CHRISTMANN K, HEIMANN P, et al. Adsorbate band dispersions for C on Ru(0001)[J]. Surface Science, 1982, 115:L159.

[61] BORCA B, BARJA S, GARNICA M, et al. Potential energy landscape for hot electrons in periodically nanostructured graphene[J]. Physical Review Letters, 2010, 105:036804-1.

[62] BOSE S, SILKIN V M, OHMANN R, et al. Image potential states as a quantum probe of graphene interfaces[J]. New Journal of Physics, 2010, 12:023028-1.

[63] STRADI D, BARJA S, DÍAZ C, et al. Role of dispersion forces in the structure of graphene monolayers on Ru surfaces[J]. Physical Review Letters, 2011, 106:186102-1.

[64] FARíAS D, SHIKIN A M, RIEDER K-H, et al. Synthesis of a weakly bonded graphite monolayer on Ni(111) by intercalation of silver[J].

Journal of Physics: Condensed Matter,1999,11:8453.

[65] DEDKOV Y S, SHIKIN A M, ADAMCHUK V K, et al. Intercalation of copper underneath a monolayer of graphite on Ni(111)[J]. Physical Review B,2001,64:035405-1.

[66] LEVY N, BURKE S A, MEAKER K L, et al. Strain-induced pseudomagnetic fields greater than 300 teslas in graphene nanobubbles[J]. Science,2011,329:544.

[67] N'DIAYE A T, VAN GASTEL R, MARTíNEZ-GALERA A J, et al. In situ observation of stress relaxation in epitaxial graphene[J]. New Journal of Physics,2009,11:113056-1.

[68] ZHAO L, HE R, RIM K T, et al. Visualizing individual nitrogen dopants in monolayer graphene[J]. Science,2011,333:999.

[69] SON Y-W, COHEN M L, LOUIE S G. Energy gaps in graphene nanoribbons[J]. Physical Review Letters, 2006,97:216803-1-4.

[70] NAKADA K, FUJITA M, DRESSELHAUS G, et al. Edge state in graphene ribbons: nanometer size effect and edge shape dependence[J]. Physical Review B,1996,54:17954.

[71] AKHMEROV A R, BEENAKKER C W J. Boundary conditions for Dirac fermions on a terminated honeycomb lattice[J]. Physical Review B, 2008,77:085423-1.

[72] SON Y-W, COHEN M L, LOUIE S G. Half-metallic graphene nanoribbons[J]. Nature,2006,444:347.

[73] CAI J, RUFFI EUX P, JAAFAR R, et al. Atomically precise bottom-up fabrication of graphene nanoribbons[J]. Nature,2010,466:470.

[74] TAO C, JIAO L, YAZYEV O V, et al. Spatially resolving spin-split edge states of chiral graphene nanoribbons[J]. Nature Physics,2011,7:616.

[75] RITTER K A, LYDING J W. The influence of edge structure on the electronic properties of graphene quantum dots and nanoribbons[J]. Nature Materials,2009,8:235.

[76] YANG H, MAYNE A J, BOUCHERIT M, et al. Quantum interference channeling at graphene edges[J]. Nano Letters,2010,10:943.

第7章 石墨烯的拉曼光谱

本章综述了拉曼散射的原理和石墨烯中声子和电子的性质。重点讨论单层和双层石墨烯的拉曼光谱,以及影响拉曼光谱的关键因素。对不同级的拉曼模式进行了讨论。

7.1 引　　言

石墨烯是众多碳材料家族中的一员,还包括 sp^3 杂化的金刚石、sp^2 杂化的石墨、富勒烯、碳纳米管和各种共轭体系。由于其优越的性能,一直受到了广大科研工作者的注意[1]。

在过去的几十年里,拉曼光谱(Raman spectroscopy)被认为是研究碳材料的理想手段。它能同时提供材料的结构振动和电子性质的信息。此外,拉曼光谱是一种快速、可靠的无损检测方法,所提供信息精度高、分辨率好。

石墨烯本身已受到很多综述文章的关注[2-7],且已出现关于石墨烯拉曼光谱的书籍[8]。Ferrari 和 Basko 等详细列举了迄今为止的相关文献[7]。

本章,我们将主要关注石墨烯中光散射的物理背景。接下来是关于拉曼散射原理的一个简要概述,然后是石墨烯的振动和电子性质分析。最后一部分为石墨烯拉曼光谱及其关键影响因素,例如掺杂、应变、石墨烯层数及取向。重点关注单层和双层石墨烯,因为这二者是被广泛关注的体系。

7.2 拉曼散射原理

光散射具有如下过程,当能量为 $\hbar\omega_i$ 的入射光作用到样品后发生吸收,然后发射出能量为 $\hbar\omega_s$ 的光。当 $\omega_i \neq \omega_s$ 时为非弹性散射。当 $\omega_i > \omega_s$ 时,光能转移到样品;而当 $\omega_i < \omega_s$ 时,光从样品获得能量。经过散射后,样品的最终状态与初始状态不同。散射光相对于 ω_i 的频率位移即为光散射研究所需要测量的数值。非弹性光散射光谱中峰的位置对应样品的不同激发态。

描述光散射的主要参数为微分截面 $d\sigma/d(\Omega d\omega_s)$。微分截面被定义为入射光能量受入射角度 d$\Omega$ 及 ω_s 和 (ω_s+dω_s) 间的频率间隔的影响程度[9-10]：

$$\frac{d\sigma}{d\Omega d\omega_s} = \frac{V^2 \omega_s^2}{8\pi^3 c^4 n_i} \cdot \frac{1}{\tau} \tag{7.1}$$

其中，V 为引起散射的样品体积；c 为光速；n_i 为入射光子的数量；τ 为跃迁速率。它的倒数 $1/\tau$ 为单位时间内系统内初始和最终状态间的跃迁概率。散射微分截面的尺寸为面积除以频率。

值得一提的是，在实际拉曼实验中，散射截面和数量间的关系受多种因素的影响，如散射光检测所用角度（solid-angle）、光谱仪的光学特性和检测器的灵敏度。虽然在较宽的频率范围内很难实现所有条件的定量，但已经实现针对几类材料绝对散射面的测量[11]。人们习惯于将拉曼信号采用相对单位，这是由于其与散射截面具有较为复杂的关系。

由公式(7.1)可以看出，散射微分截面积的定量需要准确计算跃迁速率。为此，可以采用 7.2.1 小节所讨论的微扰理论（perturbation theory）。

7.2.1 弛豫时间

我们考虑 (H_0+W) 的哈密顿量物理系统，其中，H_0 为非扰动部分，而 W 为哈密顿量的微扰部分。如果 $|i\rangle$ 和 $|f\rangle$ 为系统的初始和最终态（也是 H_0 的本征函数），则 W 所引起的由 $|i\rangle$ 到 $|f\rangle$ 转变的概率可以如下表示：

$$P_{fi} = \frac{1}{\tau} = \frac{2\pi}{\hbar} |\langle f|T|i\rangle|^2 \delta(E_f - E_i) \tag{7.2}$$

公式(7.2)中的能量守恒和跃迁矩阵元 $\langle f|T|i\rangle$ 可以扩展为扰动序列：

$$\langle f|T|i\rangle = \langle f|W|i\rangle + \sum_{f'} \frac{\langle f|W|f'\rangle \langle f'|W|i\rangle}{E_i - E_{f'} + i\eta} +$$
$$\sum_{f',f''} \frac{\langle f|W|f''\rangle \langle f''|W|f'\rangle \langle f'|W|i\rangle}{(E_i - E_{f'} + i\eta)(E_i - E_{f''} + i\eta)} + \cdots \tag{7.3}$$

其中，函数 f' 和 f'' 为 H_0 本征函数的中间态，$E_{f'}$ 和 $E_{f''}$ 为相对应的能量本征值。式中 $i\eta$ 因子的引入是为了序列的收敛性。如果 W 足够小，那么公式(7.3)中仅需要考虑其中几项即可描述该转变概率。保留公式(7.3)的第一项，即可得到著名的费米黄金法则（Fermi's golden rule）。

如果体系内存在多种不同类型的相互作用，W 可以为多种扰动的总和。这是拉曼散射中的普遍状况。例如，样品中电子和声子不仅与入射光存在相互作用，彼此间也存在相互作用。显然，即使对于公式(7.3)较低阶数的方程，处理这些微扰都十分困难。因此，针对特定体系选择具有代表性的相互作用是合理的和必要的。

接下来,考虑哈密顿量的微扰部分 W,该部分可以看成电子和声子相互作用及电子和光之间相互作用。电子和声子耦合可以通过下式表示[10]:

$$H_{eR} = -\frac{e}{m}\bm{A}\bm{p} + \frac{e^2}{m}\bm{A}^2 \qquad (7.4)$$

按照第二量子语言(the second-quantisation language),矢势 $\bm{A}$ 为光子的产生和湮灭算符,其或者产生或者湮灭一个光子。矢量 $\bm{p}$ 为电子动量算符,其产生电子态之间的跃迁。

拉曼散射是双光子过程,一个光子被湮灭,另一个光子产生。因此,只有 $\bm{A}^2$ 项对公式(7.3)中第一阶矩阵元产生非零贡献。掺杂的半导体中存在这种散射过程,电子的集体激发(等离子体)能够散射光[12]。对于二阶项,$\bm{A}\bm{p}$ 往往起主要作用。此时,光散射伴随着电子跃迁。首先,一个光子被吸收,电子能态由基态进入一个高能态。然后,释放出一个能量为 $\omega_s <\omega_i$ 的光子(对应谱线中的 Stokes 部分),电子由高能态进入不同于基态的低能态。散射过程因此导致样品电子系统处于激发的状态[10]。

声子包含在公式(7.3)的第三项。这三个矩阵元素中,两个必须包含电子-辐射($\bm{A}\bm{p}$)相互作用:一个为入射光子,另一个为散射光子。一个包含电子-声子相互作用的矩阵 $H_{\text{e-ph}}$ 被引入公式(7.3)的分子。通过入射光子发射(Stokes 过程)或吸收(anti-Stokes 过程)声子,电子被激发进入激发态,然后该电子跃迁回基态释放出一个光子。对于 Stokes 过程,散射光子的能量低于入射光子的能量,能量以声子的形式带走。对于声子波矢 $\bm{q}$,动量守恒要求 $\bm{q} = \bm{k}_i - \bm{k}_f$,表示散射过程光动量变化。

根据量子力学,对于微扰理论中给定的阶需考虑所有可能的时间序列。因此,公式(7.3)的最后一项仅为六个不同三阶项中的一项。对于非共振拉曼散射,所有项对跃迁概率均有影响。对于共振散射,其中某一项将占主导。除了矩阵元素,其他区分共振散射和非共振散射的因素在公式(7.3)的分母处。这些因素中的某些可能较小,但是会引起整项较大。当入射光子能量等于电子激发能量时,或声子能量等于两电子能态差时,将发生上述情况。可以证明,在三阶微扰理论中最重要的项源于如下过程:电子激发→声子发射→光子发射。

$$\frac{1}{\tau} \approx \sum_{f',f''} \frac{\langle f | H_{eR} | f'' \rangle \langle f'' | H_{e\text{-ph}} | f' \rangle \langle f' | H_{eR} | i \rangle}{[\hbar\omega_i - (\varepsilon_f - \varepsilon_i) + i\eta][\hbar\omega_i - \hbar\omega_{ph} - (\varepsilon_{f'} - \varepsilon_i) + i\eta]}$$

$$(7.5)$$

其中，$\hbar\omega_{ph}$ 和 ε 为声子和电子能量。当入射光子与电子激发共振（入射共振，incoming resonance）或散射光子与电子激发共振（$\hbar\omega_s = \hbar\omega_i - \hbar\omega_{ph}$，发射共振，outgoing resonance）时，公式（7.5）中分母实部为零。这种散射称为共振拉曼散射。

可见光在固体中的拉曼散射光子的波矢量与第一布里渊区的大小相比较小。因此，根据拉曼选择定则（selection rules），光子动量 $q \approx 0$，而且只有第一布里渊区（1st BZ）中心的声子发生如公式（7.5）所示的第一阶拉曼过程。此外，为了研究散射光的光谱，涉及 $H_{e\text{-}ph}$ 的矩阵元必须是有限的。基于系统对称性的选择定则可以告诉我们哪个声子参与了拉曼散射过程[9]。

公式（7.5）中的一阶（声子）拉曼项无法影响拉曼散射。对于碳类材料，例如石墨烯，需要利用高阶项来研究拉曼光谱中一些声子线的起源。在二阶项中，如公式（7.6）中存在一个额外的矩阵元和分母中额外的同类项。

$$\frac{1}{\tau} \approx \sum_{f',f'',f'''} \frac{\langle f|H_{eR}|f'''\rangle\langle f'''|H_X|f''\rangle\langle f''|H_{e\text{-}ph}|f'\rangle\langle f'|H_{eR}|i\rangle}{[\hbar\omega_i - (\varepsilon_{f'} - \varepsilon_i) + i\eta][\hbar\omega_i - \hbar\omega_{ph} - (\varepsilon_{f''} - \varepsilon_i) + i\eta]} \cdot \frac{1}{[\hbar\omega_i - \hbar\omega_{ph} - \hbar\omega_X - (\varepsilon_{f'''} - \varepsilon_i) + i\eta]} \quad (7.6)$$

如果这个额外的矩阵元 $\langle f'''|H_X|f''\rangle$ 涉及电子-声子相互作用，两个声子处于系统的最终状态（所说的无缺陷双声子过程，two-phonon defect-free process）。双声子动量的选择规则表示为 $q_1 + q_2 = 0$。此外，这个额外的矩阵元也可能是一个缺陷的电子散射，而不存在声子的参与（即所说的单声子缺陷过程，one-phonon defect-assisted process）[7]。不管哪种情况，第一布里渊区内声子均能在拉曼光谱上反映出来。

显然，声子和电子是拉曼散射中的必然组成。对于石墨烯，具体的细节将在后面进行详细讨论。

7.3 石墨烯内的声子

石墨烯的原胞里包含两个原子 A 和 B。两个原子可以构造出总共六个不同的声子色散分支。第一布里渊区中心三种能量趋于零的模式为声学声子分支，其余三个具有有限能量的为光学声子分支。对于所有的光学和声学分支，在声子传播的波矢 q 的方向存在两个横向声子和一个纵向声子。在蜂窝晶格的二维平面内，传播方向往往与最近的 C—C 方向近似平行,如在 A 和 B 原子之间[2]。对于纵向声子,原子位移与声子传播平行,而与横向声子垂直。这两个横向声子的原子位移也是互相垂直的。对于二维的石墨烯晶格而言,这就意味着一纵一横声子在平面内振动,其他横向声子在垂直于石墨烯层的方向振动(面外)。两个面内纵向分支表示为 LA(纵向声学支)和 LO(纵向光学支)。四个横向声子表示为面内横向声学支(iTA)、面内横向光学支(iTO)、面外横向声学支(oTA 或 ZA)和面外横向光学支(oTO 或 ZO)。这些声子在布里渊区 Γ-K 和 Γ-M 内保持高度对称性。计算所得声子色散关系如图 7.1 所示[2]。

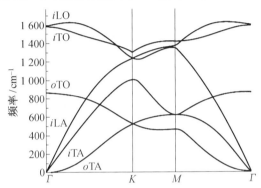

图 7.1　石墨烯计算所得的声子色散关系:LO、iTO、oTO、LA、iTA 和 oTA[2]

声子可以根据石墨烯原胞所属的不可约点群进行分类。单层石墨烯(MLG)的对称性按照 $P6/mmm$ 空间群,与 D_{6h} 点群同构,该点群对应倒易空间内布里渊区中心(Γ 点)波矢。单层石墨烯的六个分支在 Γ 点按照 D_{6h} 点群的 E_{2g}、B_{2g}、E_{1u} 和 A_{2u} 模式[2-3]。其中,两种模式 E_{2g} 和 E_{1u} 为双重简并的,而其他两种为非简并。E_{2g} 和 B_{2g} 为光学模式,E_{1u} 和 A_{2u} 为声学模式。只有 E_{2g} 具有拉曼活性,即石墨烯拉曼光谱中所说的 G 带。E_{2g} 存在双重简并的原因是 iTO 和 LO 在 Γ 点相遇。该简并在布里渊区外消失,见图 7.1 中 Γ 点外 Γ-K 方向的 iTO 和 iLO 分支。

此外,K 点附近的声子对于石墨烯的拉曼光谱具有显著的贡献。在 K 点附近有两个光学支声子,一个为 iTO 分支,另一个为 iLO 和 iLA 的合并分支。前者属于 D_{3h} 点群内有拉曼活性的 A_1。对于完全对称的 A_1 模式,六元环的六个原子在径向方向振动。所对应的拉曼谱线成为 D 带。

对于双层石墨烯,两个六边形网络堆垛在彼此之上。对于 AB 堆垛,上层中的碳原子 B 正好位于下层六元环中心的位置。晶胞由四个原子组成。每层内两个原子是不等价的,因此其模式为单层石墨烯模式的两倍。

在这12个声子模式中,相对于两层内碳原子的振动,六个为对称排列(symmetric combination),六个为反对称排列(antisymmetric combination)。对称和反对称模式表示邻近面碳原子的面内和面外振动。换句话说,单层石墨烯内的每个不可约表示(irreducible representations)均会在双层石墨烯内产生两个不可约表示,分别为对称和反对称。在Γ点,它们将以$2E_{2g} + 2B_{2g} + 2E_{1u} + 2A_{2u}$的方式区分为不同的$D_{3d}$点群不可约表示。双层石墨烯内两个$E_{2g}$模式是唯一的具有拉曼活性的一阶模式。其中一个为$E_{2g}$对称振动,另一个为不对称$E_{1u}$振动。

7.4 石墨烯的电子结构

石墨烯的第一布里渊区六边形,中心存在具有高度对称的Γ点,角落存在两个不对称的K和K'。每个碳原子在面内形成三个键,垂直于平面存在一个轨道。垂直轨道中的电子形成费米能级附近的能带,称为π和π^*。在单层石墨烯的价带结构中,π对应价带,π^*对应导带。

按照紧束缚模型,石墨烯的能带色散关系为线形的,$E^{\pm} = \pm \hbar v_F |\boldsymbol{k}|$,其中(+)和(-)为导带和价带,$v_F \approx 10^6$ m/s 为费米速度。波矢$|\boldsymbol{k}|$的大小是在K点处测量获得。能带在第一布里渊区的角落相遇,所谓的狄拉克点。在零度和非掺杂的情况下,价带为满,而导带为空。费米能级在狄拉克点。能带的能量以参数γ_0(约3 eV)为基本尺度,γ_0即相邻碳原子的π轨道矩阵元。这就意味着,考虑到价带和导带间的线性色散,可以采用可见光范围的激光来进行拉曼分析。

双层石墨烯的情况比较复杂。除了由γ_0给出的能量尺度,存在一组描述邻近层原子内传输的矩阵元。为了简化,接下来只考虑上层内A原子间的传输和下层内B原子间的传输[13]。积分γ_1的大小约为0.35 eV。价带和导带分别裂分为两个亚能带(sub-bands),分别为π_1、π_2、π_1^*和π_2^*,其能级差为γ_1。在π_1和π_2的电子结构间不存在间隙,彼此在狄拉克点相交。能带在K点附近抛物线形分布[14]。

拉曼散射中电子-光子耦合是必然现象。本质上而言,该过程是电子吸收一个光子从费米能级以下的填充状态进入空能带。只有矩阵元为非零时这种迁移才能发生。基于群论(group theory)的分析可以准确地告诉我们电子态的对称性、光的偏振矢量及第一布里渊区内光吸收发生的位置[2]。对于双层石墨烯,跃迁发生在π和π^*之间。此外,在双层石墨烯内存在两种可能的跃迁,即 $\pi_1 \to \pi_1^*$ 和 $\pi_2 \to \pi_2^*$,对应每个亚能级对。电子跃迁必须保持动量守恒。由于光子的动量相对于布里渊区的大小而言很小,价带和导带的跃迁基本上是竖直的。

7.4.1 科恩异常效应(Kohn anomaly)

科恩异常现象为电子激发体系内屏蔽微扰的效应(an effect of screening a perturbation)。1959 年,Walter Kohn 提出,如果波矢 p 为费米能级处两能态的静态微扰,两能态能量为 ε_k 和 ε_{k+p},$\varepsilon_k - \varepsilon_{k+p} \to 0$,则电子会显著地屏蔽微扰[15-16]。同样的,也可以说电子从能态 k 激发到能态 $(k+p)$ 屏蔽了微扰。对于自由电子云,这种作用会一直存在,只要 $|p| < 2k_F$(费米球的直径)。当 $|p| > 2k_F$ 时,费米能级处不存在任何可以相连接的点。屏蔽效应大大减弱,表现为介电函数值的大幅度下降。Kohn 认为,如果扰动为声子,声子频率将在声子波矢越过 $2k_F$ 时出现突变。

对于石墨烯,存在相似的现象。由于声子的能量 $\omega(q)$ 较小,则科恩异常现象的条件变为 $\varepsilon_k - \varepsilon_{k+p} + \hbar\omega(q) \to 0$[17]。当需要一个能量为两个电子能级差的能量 $\hbar\omega(q)$ 的声子时,需要用到石墨烯的线性电子能带。发生从价带到导带的垂直跃迁,达到科恩异常发生的条件。因此,Γ 点($q=0$)处具有拉曼活性的声子被电子屏蔽,其频率减小[18]。另一个具有科恩异常现象的声子为价带在 K 点附近导带在 K' 点附近。该声子波矢为 $|K|$,由于科恩异常现象声子色散在 K 点出现异常下降。

科恩异常是强电子-声子耦合的直接表现。声子衰变成为一个电子-空穴对(e-h)。显然,当初始电子态为满,最终电子态为空时这个过程是可能的。否则,e-h 激发由于 Pauli 不相容原理不能发生。因此,科恩异常现象对于费米能级的位置非常敏感,也就是说,受石墨烯的掺杂的影响。声子波矢 $|q|$ 的大小在拉曼散射过程中由于动量守恒是固定的。唯一可能的改变 $|q|$ 使其与 $2k_F$ 相当的方式为,通过掺杂改变 $2k_F$ 的大小。图 7.2 为根据计算结果给出的 G 带位置和峰宽与自由电荷(电子或空穴)浓度 n 的对应关系[17]。当费米能级 ε_F 与 $\hbar\omega_\Gamma/2$ 相等时,低温下掺杂能级表现为对数发散(logarithmic divergences)。其原因是石墨烯内电子弛豫时间比 $1/\omega_G \sim 3$ fs(G 带晶格振动的时间尺度 $\omega_G \approx 1\,580$ cm^{-1})要大。在这种情况下,根据 BO 近似的假设,声子不能作为电子系统内的一个静态扰动。当 $\varepsilon_F > \hbar\omega_G/2$ 时,电子能量增加,科恩异常被抑制。对于电子和空穴掺杂,背离的位置对于零掺杂能级对称,且在常温下表现不出来。除了非绝热对于

声子频率的贡献,掺杂引起的晶格膨胀、远离费米能级能态间的跃迁等绝热效果也被计算在内。这些因素对电子和空穴掺杂的影响不同,导致频率和掺杂间出现不对称。声子的线宽对于低掺杂能级是最大的,这可由声子衰变为电子-空穴对而使声子寿命(与线宽呈反比)减小进行理解。对于$\varepsilon_F>\hbar\omega_G/2$,电子-空穴对由于不相容原理不能产生,声子寿命增加[17]。

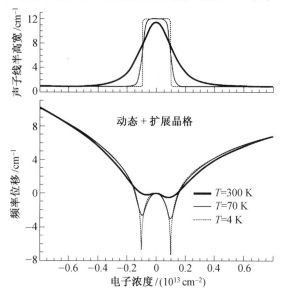

图7.2 不同温度下计算所得的单层石墨烯G带线半峰宽(上)和动态频率(下)与电子浓度的关系计算(包含非绝热效应)[17]

7.5 石墨烯拉曼光谱

单层石墨烯的典型拉曼光谱如图7.3所示。光谱内唯一的一个声子带为1583 cm^{-1}处的G带。其他谱线来源于二阶过程,包括单声子缺陷诱导或双声子散射。接下来,以单层和双层石墨烯为例描述其谱带。

7.5.1 一阶拉曼散射

(1)单层石墨烯。1583 cm^{-1}附近的G带为单层石墨烯内唯一的具有拉曼活性的一阶散射。正如第7.3节中所讨论的,其谱线位置和宽度受掺杂状态的影响。这种影响在图7.4中进行描述[19]。电子掺杂样品当采用正栅极电压时,G带上移(upshifts)。随着掺杂的进行线宽增加。即使对于

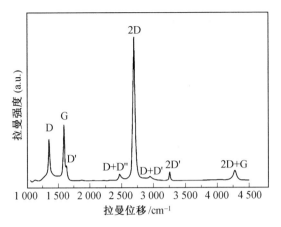

图 7.3　单层石墨烯的拉曼谱图及其归属

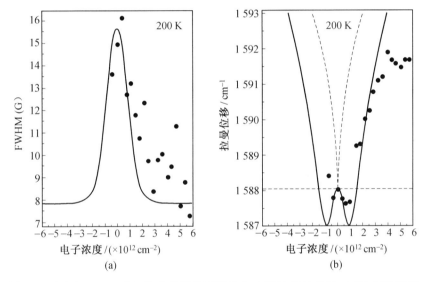

图 7.4　(a) 在 200 K 时 G 带半高宽(FWHM)与电子浓度的关系：(点)实验结果；(线)有限温度非绝热计算所得 FWHM。(b) 在 200 K 时，G 带位置与电子浓度的关系：(点)实验结果；(虚线)Born-Oppenheimer 绝热计算；(实线)有限温度非绝热计算[19]

很少量的空穴掺杂，也具有类似的规律。这两个样品图中实线为考虑非绝热效应的计算结果。可见尤其是谱线位置，计算结果与实验结果吻合。由于局部温度影响和石墨烯表面掺杂的非均匀性导致实验结果存在发散，与 BO 近似不相符的情况在图 7.4 中以点画线表示。这是在不考虑动力学效应的基础上计算的结果,与实验结果显然不符。

通过电化学顶栅方法(electrochemical top gating method)可以实现高掺杂[20]。G 带的位置和半高宽(FWHM)如图 7.5 所示。掺杂范围比 Pisana 等人的测量值高一个数量级[19]。结果与实验结果基本吻合。

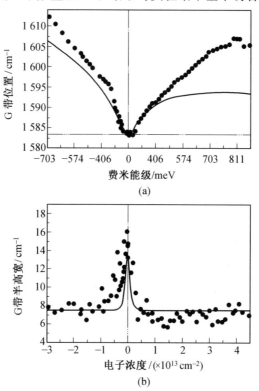

图 7.5 石墨烯 G 带位置(a)和 FWHM(b)与电子空穴掺杂的关系。实线为非绝热趋势的预测[19-20]

在较宽范围内,Lazzeri 和 Mauri 的理论与实验趋势相符[17],例如电子和空穴不对称性。但是理论结果与实验结果在高掺杂浓度时不符。对于掺杂引起的电子和空穴不对称性方面,前者所得偏移数值较小。掺杂增加或减少系统内的电子,引起 C—C 键强变化[21]。空穴掺杂增强键强,因此增强 G 带频率。电子掺杂恰好相反。

这些影响均以电子-声子耦合表现为频率变化。对于空穴掺杂,这两种机制的作用是累加的,而对于电子掺杂,这两种机制作用相反。因此,电子掺杂引起的频率偏移小于空穴掺杂所引起的偏移。

G 带的强度受掺杂的影响较为复杂。Kalbac 等发现对于高空穴掺杂[22],G 带强度显著增加,而对于电子掺杂却不具备该趋势。在零电压时强度下降,这是由于电子-声子耦合引起的线宽增加。G 带强度随掺杂基本保持不变后,G 带强度在低激光能量时迅速增加。Basko 等认为这一现象的原因为电子-声子矩阵元引起 G 峰在费米能级接近激光能量 1/2 时达到最大值[23]。激光能量越低,所需要达到 G 带强度突然上升所需的掺杂越低。

G 带在 Γ 点的二重简并可以采用单轴应变解除[24]。如图 7.6 所示,随着应变的增加,G 带裂分为两个峰,并线性地红移。然而,两条谱线的红移速度并不相同。原因是导致这两个拉曼分支的声子特征向量在应变下不同。与应力垂直的特征向量所引起的拉曼分支受应变影响较小,而与应变平行的特征向量所引起的分支受应变影响较大。拉曼光谱相对强度随入射激光的变化为研究应变方向与样品晶体学取向提供可能。可得到适用于石墨烯和碳纳米管的 G 带偏移与应变相互关系。这可被用于构建拉曼位移与应变传感器[25]。

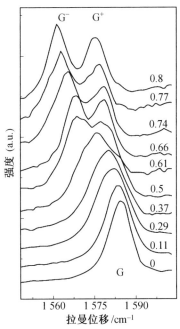

图 7.6 G 带位置与单轴应变的关系。测量时入射光沿应变方向并采集散射光[14]

(2)双层石墨烯。对于双层石墨烯,其电子结构与单层石墨烯不同。

导带和价带分裂为两部分,彼此间距离约 0.35 eV,费米能级附近的能带为抛物线[14]。双层石墨烯内也存在科恩异常,Yan 等认为其反映了 G 带的掺杂行为[26]。对比单层石墨烯,双层石墨烯内电子和空穴掺杂所表现出的谱线位置屏蔽效应均可被清晰地观察到,如图 7.7 所示。双层石墨烯内屏蔽效应存在几个可能的原因。双层石墨烯内两层原子间耦合作用,导致费米能级附近的态密度相对较大。此外,对于双层石墨烯,其 E_F 在低掺杂范围受电荷密度的影响较小。电荷分布的不均匀性导致 E_F 的变化较单层石墨烯较小。在后者中,当完全去除声子软化时 δE_F 大约为 100 meV。

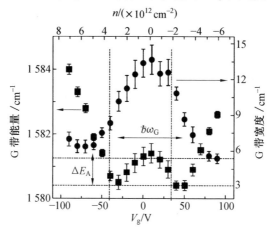

图 7.7 双层石墨烯 G 带线宽与栅极电压(电子和空穴浓度)的关系。声子能量演化为两个异常声子(方形)[26]

Malard 等研究了 G 带裂分为两部分的现象[27]。与单层石墨烯相比,裂分后的两部分对应邻近石墨烯层内的对称(S)和不对称(AS)振动。对于未受干扰的样品而言,只有前者具有拉曼活性。当层与层出现不等价时(由于非均匀掺杂),对称振动降低,而出现具有拉曼活性的非对称振动,如图 7.8 所示[27]。当费米能级接近狄拉克点(40 V)时,S 和 AS 间的能量区别可以忽略不计,该能量区别随负偏压增加而增大。其原因是,除了带间跃迁,双层石墨烯内存在带内跃迁,如图 7.9 所示。

对称模式涉及带间和带内的电子-空穴对,而不对称振动仅与带内跃迁有关。对于零掺杂,只有对称振动模式可见,这是由于带内跃迁被 Pauli 原理禁止。对于掺杂的情况,电子-空穴对的带内跃迁可以发生。(π_1-π_2)能级差 0.35 eV 大于 G 带的声子能量 0.2 eV。这就意味着对于弱掺杂,由于 π_1 能带的最终电子态被封闭(图 7.9),电子-空穴对是虚拟的,AS

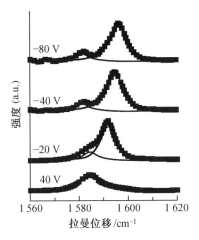

图7.8 不同栅极电压下(-80 V、-40 V、-20 V 和 40 V)双层石墨烯拉曼 G 带。对于电压为-80 V、-40 V、-20 V 时,G 带需要两个洛伦兹曲线拟合[27]

模式受费米能级位置影响较弱。采用理论和实验相结合可获得更详细的比较[28-29]。通过该过程,零栅压时可以估算由于衬底或电解液引起的样品额外载流子的浓度,这使得拉曼光谱成为探索石墨烯静电环境的有效手段。

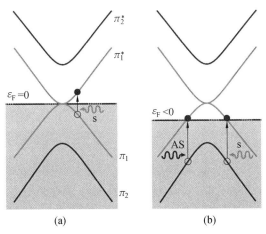

图7.9 双层石墨烯 K 点附近抛物线能带结构。垂直箭头为两种可能的迁移,分别为(a) $\varepsilon_F=0$ 的对称($q=0$)声子带间产生电子-空穴对,(b) $\varepsilon_F<0$ 的反对称($q=0$)声子带内产生电子-空穴对[27]

除了 G 带,群理论(group theory)预测了双层石墨烯存在另一个具有

拉曼活性的一阶 E_{2g} 模式[30]。它对应两层石墨烯相对的剪切运动。其频率与层数有直接关系。对于双层石墨烯，其谱线位置在 31 cm^{-1}，且随着石墨烯层数的增加发生蓝移，直至达到石墨所对应的约 43 cm^{-1} 处[31]。

7.5.2 二阶拉曼散射

双层石墨烯除了 G 带和低能带外，还存在其他具有拉曼活性的二阶过程。这意味着要解释这些拉曼谱线涉及更多的散射过程。这些谱线所涉及具体的散射过程，一般被分为两个主要谱带，即 D 带和 2D 带。

（1）单层石墨烯。对于所有碳材料，D 带位于 ~ 1 300 cm^{-1} 处，而 2D 带位于 2 680 cm^{-1} 处。对于声子散射，D 带归因于布里渊区 K 点附近的 iTO 声子，2D 带为其谐波。这些谱带频率由激发光能量决定，表现出色散性质。D 带以 50 cm^{-1}/eV^{-1} 的斜率发生上移（upshift），而 2D 带斜率是其 2 倍。这种行为本质上被认为是双共振过程的结果[32]。首先，处于 K 点附近的初始状态 i 的能量为 k 的电子，吸收能量为 E_1 的光子后激发进入导带（图 7.10[6]）。然后，电子发生几种不同的散射行为。对于 D 带，电子释放出波矢 q 和能量 ω_{ph} 的声子，进入布里渊区 K' 点附近的 b 能态，其波矢为 ($k + q$)。之后，电子散射回缺陷态 c。该背散射过程（backscattering）引起电子动量 $-q$ 的变化。电子与能态 i 处的空穴进行重组，释放出能量为 E_s 的光子。

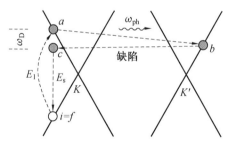

图 7.10　石墨烯内 D 带的二阶拉曼过程[6]

这个过程包含两个散射：声子发射引起的非弹性散射和缺陷引起的弹性散射。这两种散射过程在其他阶拉曼散射中都可能发生。显然，样品中必须存在无序才能在拉曼光谱中出现该谱线。对于 2D 带，散射过程为两个 K 点附近的声子的非弹性散射。如图 7.11 所示，该谱线不需要无序和缺陷作为条件，因此总会存在[6]。

电子在 K 和 K' 间的散射被称为谷间散射。声子矢量 q 的大小服从

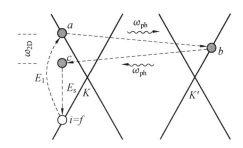

图 7.11 拉曼 2D 带的二阶拉曼过程。两个声子具有相反的波矢确保散射过程中动量守恒[6]

$q \sim 2k$,k 为电子的波矢。该条件是由动量守恒和两波矢均在 K 点决定。具有特定波矢 k 的电子能态可以由拉曼实验过程的激光能量决定。通过改变激光的能量 E,声子的能量发生变化,同时影响 K 点附近的 iTO 分支。由于分支在 K 点具有局部最小值,D 和 2D 带的位置往往随激光能量的增加而增加。

双共振过程存在三个中间态,对应微扰理论(公式 7.6)中的第四阶。对照图 7.10 和图 7.11,这三个状态为 a、b 和 c。三者中两个为真实的电子能态。K' 点附近的 b 能态为真实能态,而初始或最终能态(a 或 c)中只有一个为真实能态。

这三个散射均会共振产生 2D 带,散射过程涉及两个声子。为此,电子和空穴的散射过程都必须考虑。如图 7.11 所示,电子从初始态通过能态 a 散射到 K' 附近的 b 能态[6]。然后,K 点附近价带中的一个空穴通过发射声子散射到 K' 点附近价带能态。电子和空穴具有相同的波矢($k+q$),最终它们重组共振。这个所谓的三共振过程[33]可以很好地解释为什么单层石墨烯的拉曼图谱中 2D 带是其主要谱带[2]。

讨论至此,在散射过程中,传递动量 q 在 K 和 K' 点连接狄拉克锥的外部。此外,散射过程也可连接狄拉克锥的内部[34]。其原理如图 7.12 所示[34]。

这两个过程都有助于形成石墨烯的 D 带和 2D 带。该外部过程是由于 K-M 方向的声子,而内部散射过程为 Γ-K。由于在这两个方向上声子分支一般是不同的,即使不存在应变,2D 带也包含两部分。当存在单轴应变时该裂分变得更加明显。在布里渊区内,石墨烯晶格和电子能带的对称性沿不同方向发生不同变化。这取决于第一布里渊区内电子沿哪个方向散射,不同矢量 q 的声子参与双共振过程。

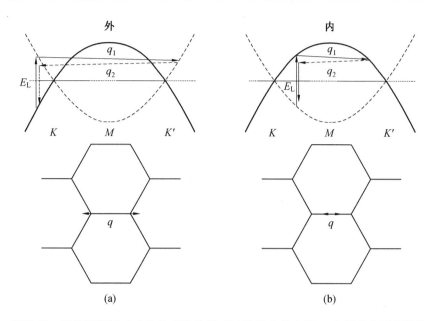

图 7.12 双共振机制:(a)狄拉克锥外部,(b)狄拉克锥内部。上部为在石墨烯能带结构 KMK' 内示意图,下部为布里渊区内机制示意图[34]

到这里,我们已经讨论了谷间散射的机理,然而谷内散射也是可能发生的。电子在相同 K 点的不同能级间散射。该过程的动量转移很小,所对应的声子来源于 Γ 点附近的 LO 分支。由于该分支的过分弯曲,图 7.13 中声子-缺陷散射导致拉曼图谱中出现能量高于 G 带的谱带[6]。该谱带称为 D′带,一般位于 1 620 cm^{-1} 附近。然而,其谱带强度明显低于 G 带。其谐波出现在 3 240 cm^{-1} 附近,为双 D′声子参与的散射过程。

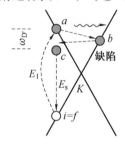

图 7.13 D′带谷内散射的二阶拉曼过程[6]

对于 2D 带的位置和强度与掺杂的关系,Das 等[20]和 Kalbac 等[22]进行了研究。在低掺杂水平下,对于电子掺杂所引起的位置变化不大。在高掺杂水平下,电子掺杂引起谱线位置发生约 20 cm^{-1} 的红移,而对于空穴掺杂表现出相同数值的蓝移。采用电化学掺杂样品的 2D 带也具有相似的现象[22]。掺杂对谱线位置存在两种主要影响:①它改变了平衡晶格常数——电子掺杂导致晶格膨胀,引起 2D 模式软化,而空穴掺杂引起晶格收缩和声子硬化;②由于科恩异常 K 点附近的声子色散发生变化。然而,后者所引起的 2D 带位置变化较小,这归因于 $q \sim 2k$ 条件导致相应的声子远离 K 点[35]。

2D 带强度受掺杂的影响明显区别于 G 带受掺杂的影响,表现为电子和空穴掺杂引起强度降低,且速率与激光能量无关[22]。2D 带和 G 带的强度比在表现出很强的掺杂依赖性,尤其在低电子浓度时,如图 7.14 所示[20]。

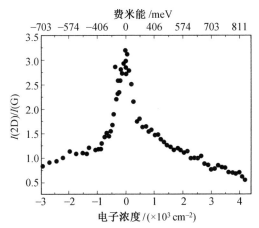

图 7.14 单层石墨烯内 2D 带和 G 带强度比与电子和空穴掺杂的关系[20]

高的 2D/G 强度比被认为是单层石墨烯的典型特点。然而,如图 7.14 所示,测试过程影响了该参数的可信度,这是由于石墨烯样品很容易受环境影响产生非人为掺杂。Basko 等从理论上研究了 2D 带强度的依赖性[36]。结果表明,2D 带强度的依赖性可以解释为电子-电子相互作用影响光生电子和空穴的散射概率。

(2)双层石墨烯。对于 AB 堆垛的双层石墨烯,其 2D 带由四部分组成,如图 7.15 所示[37]。该四部分归因于双层石墨烯内电子能带分裂或声子分支分裂。然而声子分支分裂仅为 1 cm^{-1} 量级[37]。因此,2D 带的分裂和峰数量是双层石墨烯内电子结构引起的。图 7.16 给出了 K 点和 K' 点附近四种可能的电子谷间散射过程[38]。

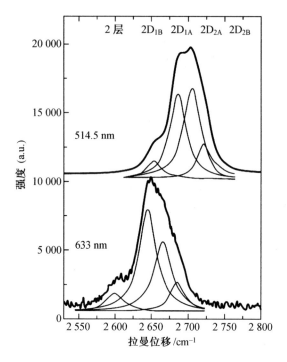

图 7.15 双层石墨烯 2D 带在 514 nm 和 633 nm 激发下可分为四个组成部分[37]

具体的过程取决于电子能带的对称性及所涉及的声子。双层石墨烯的 π 和 π^* 电子能带属于不同的对称群。在具有相同对称性的能带间跃迁为对称声子参与的过程(图 7.16 中的 P_{11} 和 P_{22} 跃迁),而非对称振动对应不对称性的带间跃迁(P_{12} 和 P_{21})。P_{11} 所对应声子波矢较大,而 P_{22} 所对应声子波矢较小。因为 iTO 频率随着 q 的增加而增加,2D 谱带中最强的裂分对应 P_{11} 散射机制,而最弱部分对应 P_{22} 散射机制。如果价带和导带为彼此的镜像,那么拉曼光谱中 P_{12} 和 P_{21} 过程所对应峰将很难区分。图 7.17 给出了采用洛伦兹拟合所获得的 2D 峰的四个组成部分[38],所采用 FWHM 为 24 cm^{-1},该值为单层石墨烯的 2D 峰所对应的数值。

对于层数 $N > 2$ 时其 2D 谱带峰数量,可采用与双层石墨烯相同的方法。对于三层石墨烯,在 K 和 K' 点间由双 iTO 声子参与的电子散射存在 15 种可能[39]。实际上不需要将 2D 峰拟合为 15 个单独的峰,因为它们间频率差别很小,且它们中的一些表现为一个典型特征。然而,石墨烯与石墨的 2D 谱带存在显著的不同,如图 7.18 所示[37]。

少层石墨烯(few-layer graphene (FLG))的 2D 谱带当 $N = 5$ 或更高时

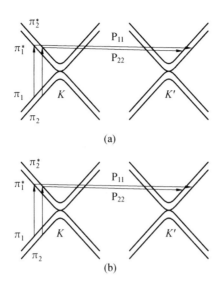

图 7.16 （a）P_{11} 和 P_{22} 双共振拉曼过程涉及对称声子,（b）P_{12} 和 P_{21} 双共振拉曼过程涉及反对称声子[38]

与石墨几乎相同。结果证明[40]石墨的 2D 带是由无数个介于 P_{22} 和 P_{11} 频率的峰组成的(图 7.17)。不同于双层石墨烯,石墨内存在一个垂直于石墨烯片层的波矢 q_z。π_2 和 π_2^* 间的散射受 q_z 影响,而 π_1 和 π_1^* 间的散射不受影响。当考虑到所有从 0 到 π/c(c 为点阵矢量)的 q_z 时,P_{12}、P_{21} 和 P_{22} 过程导致石墨 2D 谱带在低能量侧出现宽肩峰。另一方面与 P_{11} 过程(唯一的一个不包含 π_2 和 π_2^* 能带的散射)有关的声子频率对于不同 q_z 时为固定值。考虑到所有 q_z 所引起的变化,使 2D 谱带在高能量一侧出现一个主峰。

上述讨论给出了关于 AB 堆垛石墨烯层数的鉴定方法。对于非 AB 堆垛的石墨烯,其 2D 谱带很容易与单层石墨烯的单一对称谱带混淆。与 AB 堆垛石墨烯相比,两层随机结合的石墨烯层表现出单一的 2D 谱带[41]。在铜衬底上 CVD 法制备的石墨烯也表现出类似的情况[42]。拉曼图谱中表现出单一的 2D 谱带,但是峰位置和缝宽取决于石墨烯的层数,表明层与层杂乱排布。与单层石墨烯比较,其谱线位置发生 20 cm^{-1} 的上移(对于 514 nm 的激光),且 2D 谱带变宽(为 40 ~ 45 cm^{-1})[42]。少层石墨烯之所以表现出与单层石墨烯类似的 2D 谱带,是由于层与层间缺乏层间相互作用。少层石墨烯的电子能带未受影响,保持了其线性(狄拉克)特征[43]。因此,单一的 2D 并不能成为单层石墨烯的明确证据。

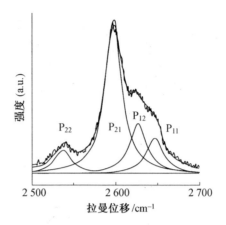

图7.17 由1.57 eV能量下获得的双层石墨烯的2D拉曼谱带可以按照FWHM为24 cm^{-1}拟合为四个洛伦兹峰[38]

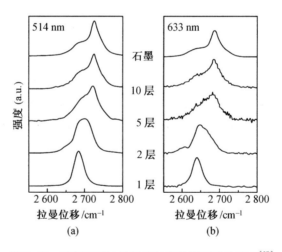

图7.18 采用不同波长激光时谱线随层数的演变[37]

2D谱带的形状可以用来确定少层石墨烯的堆垛方式。AB堆垛(bernal)和ABC堆垛(rhombohedral)的少层石墨烯的拉曼光谱存在显著的区别[44]。ABC堆垛三层石墨烯比ABA堆垛三层石墨烯具有更宽的2D谱带,且形状不对称。对于非堆垛的双层石墨烯,两层间的扭转角度导致电子结构发生变化,进而反映出2D谱带形状、位置和线宽的变化[45]。与单层石墨烯比较,价带导带能级的电子态密度出现范霍夫奇异点(van Hove singularities)。奇异点的能量区别由取向扭转角决定。针对特定的激光能量E_{laser},通过旋转角度可获得强的"共振"效应。

Gupta 等在研究石墨烯自身折叠后所获得的双层结构时发现了有趣的现象[46]。虽然单层石墨烯不产生 D 带,而折叠的石墨烯出现双峰的特征。两层之间存在随机的扭转角。双峰中的一个峰为无色散,另一个为随激发激光能量而变化。有别于从无序状态获得的静电势激发 D 带,其中一层的周期性势场可作为另一层的微扰。该扰动的电势修正了傅里叶分量,该分量随激光能量不发生变化,因此所获得拉曼谱带无色散。

石墨烯内其他二阶谱带

讨论至此,已经介绍了 D 和 D′谱带及其倍频(2D 和 2D′)。有时候,在拉曼谱图中高拉曼位移处也会出现一些谱线。这些谱线归因于各种振动模式间的组合。

图 7.3 中约 2 450 cm^{-1} 处的谱带是 D 带和 K 点的 LA 声子分支组合获得的。后者的谱带对于存在缺陷的高度取向的热解石墨(HOPG)样品时,位于 1 084 cm^{-1},被称为 T 带(石墨烯内称为 D″)[47]。拉曼图谱中约 4 290 cm^{-1} 的谱带为 2D 和 G 模式的叠加[2]。

图 7.3 中约 2 950 cm^{-1} 的谱带相对于上述谱带比较特殊。本质上而言,D 带和 D′带代表了缺陷辅助双声子过程。按照第 7.2 节中所述的扰动理论,该过程是五阶过程(有一阶未包含在公式(7.6)中),即两个电子–声子矩阵元、一个电子–缺陷矩阵元(不包含两个电子–声子矩阵元)和分母中的额外因子[7]。

7.6 结 论

石墨烯的拉曼图谱相对而言比较简单,然而从该图谱可以获得大量信息。一阶声子带反映出石墨烯晶格的对称性,该谱带随着外界扰动的变化而变化,如掺杂应变和多层石墨烯的层数。二阶谱带反映出电子和声子的各种散射机制。由于基本选择条件 $q=0$ 的放宽,通过光散射可以检测到布里渊区内的声子,该现象在拉曼光谱中很少见。二阶谱带的谱线位置、线宽和强度可以给出有关石墨烯层数和层间扭转角的信息。由于声子和电子间强耦合作用,还可获得有关石墨烯电子系统的丰富信息。通过掺杂改变了费米能级的位置进而影响了拉曼图谱中谱线的位置和形状,也就反映出了石墨烯电子系统的变化情况。

7.7 参考文献

[1] GEIM A K, NOVOSELOV K S. The rise of graphene[J]. Nature Materials, 2007, 6:183.

[2] MALARD L M, PIMENTA M S, DRESSELHAUS G, et al. Raman spectroscopy in graphene[J]. Physics Reports, 2009, 473:51.

[3] RAO C N R, SOOD A K, SUBRAHMANYAM K S, et al. Graphene: the new two-dimensional material[J]. Angewandte Chemie International Edition, 2009, 48:7752.

[4] DRESSELHAUS M S, JORIO A, SAITO R. Characterizing graphene, graphite, and carbon nanotubes by Raman spectroscopy[J]. Annual Review of Condensed Matter Physics, 2010, 1:89.

[5] DRESSELHAUS M S, JORIO A, SOUZA FILHO A G, et al. Defect characterisation in graphene and carbon nanotubes using Raman spectroscopy[J]. Phiosophical Transactions of the Royal Society A, 2010, 368:5355.

[6] DAS A, CHAKRABORTY B, SOOD A K. Probing single and bilayer graphene field effect transistors by Raman spectroscopy[J]. Modern Physics Letter B, 2011, 25:511.

[7] FERRARI A C, BASKO D M. Raman spectroscopy as a versatile tool for studying the properties of graphene[J]. Nature Nanotechnology, 2013, 8:235.

[8] JORIO A, DRESSELHAUS M S, SAITO R, et al. Raman spectroscopy in graphene related systems [M]. Weinheim: Wiley-VCH Verlag GmbH, 2011.

[9] HAYES W, LOUDON R. Scattering of light by crystals[M]. New York: John Wiley and Sons, 1978.

[10] WALLIS R F, BALKANSKI M. Many-body aspects of solid state spectroscopy[M]. Amsterdam: North-Holland Physics, 1986.

[11] YU P Y, CARDONA M. Fundamentals of semiconductors[M]. Berlin: Springer Verlag, 1996.

[12] CARDONA M. Light scattering in solids[M]. Berlin: Springer Verlag, 1983.

[13] MALARD L M. Probing the electronic structure of bilayer graphene by

Raman scattering[J]. Physical Review B,2007,76:201401(R).
[14] MCCANN E,ABERGEL D S L,FAL'KO V I. Electrons in bilayer graphene[J]. Solid State Communications,2007,143:110.
[15] KOHN W. Image of the Fermi surface in the vibration spectrum of a metal[J]. Physics Review Letters,1959,2:393.
[16] WOLL JR E J,KOHN W. Images of the Fermi surface in phonon spectra of metals[J]. Physics Review,1962,126:1693.
[17] LAZZERI M,MAURI F. Nonadiabatic Kohn anomaly in a doped graphene monolayer[J]. Physics Review Letters,2006,97:266407.
[18] PISCANEC S,LAZZERI M,MAURI F,et al. Kohn anomalies and electron-phonon interactions in graphite[J]. Physics Review Letters,2004,93:185503.
[19] PISANA S. Breakdown of the adiabatic Born-Oppenheimer approximation in graphene[J]. Nature Materials,2007,6:198.
[20] DAS A. Monitoring dopants by Raman scattering in an electrochemically top-gated graphene transistor[J]. Nature Nanotechnology,2008,3:210.
[21] YAN J,ZHANG Y,KIM P,et al. Electric field effect tuning of electron-phonon coupling in graphene[J]. Physics Review Letters,2007,98:166802.
[22] KALBAC M,REINA-CECCO A,FARHAT H,et al. The influence of strong electron and hole doping on the Raman intensity of chemical vapor-deposition graphene[J]. ACS Nano,2010,10:6055.
[23] BASKO D M. Calculation of the Raman G peak intensity in monolayer graphene:role of Ward identities[J]. New Journal of Physics,2009,11:095011.
[24] MOHIUDDIN T M G. Uniaxial strain in graphene by Raman spectroscopy:G peak splitting, Gruneisen parameters, and sample orientation[J]. Physical Review B,2009,79:205433.
[25] FRANK O. Development of a universal stress sensor for graphene and carbon fibres[J]. Nature Communications,2011,2:255.
[26] YAN J,HENRIKSEN E A,KIM P,et al. Observation of anomalous phonon softening in bilayer graphene[J]. Physics Review Letters,2008,101:136804.
[27] MALARD L M,ELIAS D C,ALVES E S,et al. Observation of distinct e-

lectron-phonon couplings in gated bilayer graphene[J]. Physics Review Letters,2008,101:257401.

[28] GAVA P,LAZZERI M,SAITTA A M,et al. Probing the electrostatic environment of bilayer graphene using Raman spectra[J]. Physical Review B,2009,80:155422.

[29] MAFRA D L. Characterizing intrinsic charges in top gated bilayer graphene device by Raman spectroscopy[J]. Carbon,2012,50:3435.

[30] SAHA,S K,WAGHMARE U V,KRISHNAMURTHY H R,et al. Phonons in few-layer graphene and interplanar interaction:A first-principles study[J]. Physical Review B,2008,78:165421.

[31] TAN P H. The shear mode of multilayer graphene[J]. Nature Materials, 2012,11:294.

[32] THOMSEN C,REICH S. Double resonant raman scattering in graphite [J]. Physics Review Letters,2000,85:5214.

[33] KüRTI J,ZÓLYOMI V,GRÜNEIS A,et al. Double resonant Raman phenomena enhanced by van Hove singularities in single-wall carbon nanotubes[J]. Physical Review B,2002,65:165433.

[34] MOHR M,MAULTZSCH J,THOMSEN C. Splitting of the Raman 2D band of graphene subjected to strain[J]. Physical Review B,2010,82: 201409(R).

[35] SAHA S K,WAGHMARE U V,KRISHNAMURTHY H R,et al. Probing zone-boundary optical phonons in doped graphene[J]. Physical Review B,2007,76:201404(R).

[36] BASKO D M,PISCANEC S,FERRARI A C. Electron-electron interactions and doping dependence of the two-phonon Raman intensity in graphene[J]. Physical Review B,2009,80:165413.

[37] FERRARI A C. Raman spectrum of graphene and graphene layers[J]. Physics Review Letters,2006,97:187401.

[38] MAFRA D L. Observation of the Kohn anomaly near the K point of bilayer graphene[J]. Physical Review B,2009,80:241414(R).

[39] MAFRA D L. Determination of LA and TO phonon dispersion relations of graphene near the Dirac point by double resonance Raman scattering [J]. Physical Review B,2007,76:233407.

[40] CANÇADO L G,REINA A,KONG J,et al. Geometrical 2007,approach

for the study of G′ band in the Raman spectrum of monolayer graphene, bilayer graphene, and bulk graphite[J]. Physical Review B,2008,77: 245408.

[41] PONCHARAL P,AYARI A,MICHEL T,et al. Raman spectra of misoriented bilayer graphene[J]. Physical Review B,2008,78:113407.

[42] LENSKI D R,FUHRER M S. Raman and optical characterization of multilayer turbostratic graphene grown via chemical vapor deposition[J]. Journal of Applied Physics,2011,110:013720.

[43] LATIL S,MEUNIER V,HENRARD L. Massless fermions in multilayer graphitic systems with misoriented layers: ab initio calculations and experimental fingerprints[J]. Physical Review B,2007,76:201402(R).

[44] LUI C H,LI Z,CHEN Z,et al. Imaging stacking order in few-layer graphene[J]. Nano Letters,2011,11:164.

[45] KIM K. Raman spectroscopy study of rotated double-layer graphene: misorientation-angle dependence of electronic structure[J]. Physics Review Letters,2012,108:246103.

[46] GUPTA A K,TANG Y, CRESPI V H,et al. Nondispersive Raman D band activated by well-ordered interlayer interactions in rotationally stacked bilayer graphene[J]. Physical Review B, 2010, 82: 241406(R).

[47] TAN P H,DENG Y M,ZHAO Q. Temperature-dependent Raman spectra and anomalous Raman phenomenon of highly oriented pyrolitic graphite [J]. Physical Review B,1998,58:5435.

第8章 低维碳材料的光电子能谱

本章介绍了采用光电子能谱、X射线光电子能谱(XPS)和角分辨光电子能谱(ARPES)研究sp^2杂化碳材料体系的研究进展。解释了如何采用上述手段分析碳纳米管和石墨烯的价带电子结构和导带电子结构。讨论了体系性能与价带态密度的关系。研究了石墨烯内杂原子的取代功能化(如氮原子)。

8.1 引 言

光电子能谱是了解材料电子结构的重要分析手段。通过光电子能谱可以研究分子价带能态、结构及其表面的信息。由于该手段同时检测能量和动量,实验过程中的分辨率和灵敏度是非常关键的因素。在本章开始,介绍碳原子的电子结构是非常有必要的,因为它是构成其他大量结构的基本单元。总体而言,碳材料在诸多方面比较特殊。例如,碳材料具有多种形状,包括团簇、块状晶体和分子等,均表现出独特的性能。这就使得碳材料成为光电子技术研究的重要对象。碳材料的变化归因于碳原子配置方式变化形式的多样化。碳是Ⅳ主族的第一个元素,这意味着它有两个强结合的内层电子($1s^2$)和4个基态价电子($2s^2$和$2p^2$)。由于碳原子是其主族内唯一的没有内层p电子的原子,只有碳具有sp^1、sp^2和sp^3杂化方式。对于碳原子以下的两个元素(Si和Ge),由于它们价带轨道和内部p电子相互作用,更倾向于形成sp^3杂化方式。Si在有机化学中得到了广泛的研究[1],其类石墨烯性也吸引着研究者不断地进行探索[2-3]。

碳材料具有多种不同的结构,在这里将主要讨论低维碳材料,即富勒烯、碳纳米管(CNTs)、石墨烯和石墨,需要再次强调的是这些碳材料的电子性质和光学性质主要与杂化碳原子的排列组合结构形态有关。这些结构形态产生了有趣的现象,而这些现象可以通过光谱技术进行分析和表征。在这样的背景下,光电子能谱及其所包含的系列分析检测手段和工具,使我们可以研究键合环境、电荷转移及掺杂等本质效应,接下来会讨论这些内容。这些技术对于原子排布类型非常敏感,从而可以研究采用杂原子替换碳原子后所引起的电子性质、振动性质、化学性质和机械性质的变化[4-5]。

该章主要在光电子能谱(包括角分辨和价带光电子发射)研究石墨烯及碳纳米管方面提供了全面的综述。有关富勒烯的讨论仅作为一个参照,因为有关富勒烯的研究进行较为广泛,也出现了较为系统的论述[6]。第一部分将详细描述 C 1s 光电子能谱的形状、位置和精细结构。对不同体系的特点和性能与价带态密度的关系进行详细讨论。对于功能化石墨烯体系,将重点讨论杂原子掺杂,例如氮原子。还将讨论纳米管的管壁功能化及金属性分类(metallicity-sorting)。最后,将讨论采用共振光电子发射研究不同能态价带电子结构的方法及过程。

8.2 光电子发射光谱学

光电子发射光谱学(photoemission spectroscopy,PES)是研究原子、分子和固体电子结构的广泛有效方法。它可以实现电子能量和动量的同步测定。PES 测量谱函数的 k 分辨电子结构。PES 是研究下列内容的最直接方法:电子结构和可控修饰碳结构,例如,单个掺杂缺陷引起局部载流子能量和晶格振动的变化。此外,借助该技术的最新研究进展,还可实现电子-电子、电子-声子、电子-等离子体等关联效应的影响[7]。采用 PES 研究单壁碳纳米管、富勒烯、石墨、石墨烯的 C 1s 的光电子能谱,可以灵敏地获得其修饰的信息(如功能化、管壁掺杂)。此外,PES 使多种分离的技术实现了统一应用。

8.2.1 X 射线光电子能谱

总体而言,PES 技术是以光电效应为基础,也就是说物质吸收电磁辐射(如 X 射线)的能量后释放出光电子[8]。X 射线光电子能谱(XPS)也被称为电子光谱化学分析(electron spectroscopy for chemical analysis (ESCA)),该技术是 20 世纪 60 年代中期 Kai Siegbahn 发明的,并于 1981 年获得诺贝尔奖以表彰其在该技术从发明到实际应用所做的贡献。在基于实验室的 XPS 研究中,X 射线来源于一个固定的源,一般常常采用 Mg K_α(1 253.6 eV)或 Al K_α(1 486.6 eV)。虽然这些光子在穿透固体方面的能力有限(仅为 1~10 μm),但是在研究碳材料方面却非常有用。从这些源产生的光子主要与浅层的原子发生相互作用,从而使发射出的电子服从光电效应中通用的 Hüffner 曲线所描述的深度范围(对于 Al K$_\alpha$ 为 4～10 nm)。所发射的电子动量 E_k 如下所示:

$$E_k = h\nu - E_B - \Phi_s$$

其中,$h\nu$ 为光子的能量;E_B 为电子所属原子轨道结合能;Φ_s 为仪器功函数(spectrometer work function),如图 8.1 所示。光子的能量是已知的,因为系统采用的为单色光激发。功函数可以通过采用绝缘样品施加偏置电压的方法,测量二次电子截止(secondary electron cutoff)获得。在上述两个参数已知的情况下,光电子的动量只受样品内结合能影响。而结合能变化为释放光电子前后能态的变化。每种离子具有很多个可能的最终态,因此所释放的电子动量也不同。样品所释放出的不同动量的电子对应着样品内不同的电子态密度。费米能级对应结合能为 0,也就是说费米能级代表电子发射后所剩余离子的相对能量或电子结合能。由于 X 射线能量的可调性,XPS 也可以采用高能同步加速器方式。

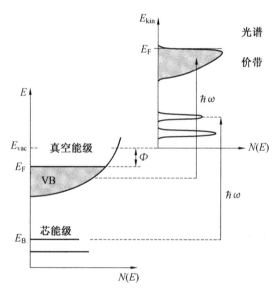

图 8.1　样品中电子态密度与动能分布之间的关系。一旦电子被能量高于真空能级的电子激发(电子能量 $\hbar$),其占有态即可被检测到[8]

除了光电子发射过程外,光电过程还存在俄歇电子发射,归因于光电发射所剩余离子的松弛。俄歇光谱也是一种非常有用的手段,但本章不进行描述。

谱线形状及数据解析

当考虑到设备分辨率和寿命效应等因素时,对于 XPS 的数据解析并不是一个简单的过程。在 PES 中碳核的信号会表现为一系列可能的谱线。采用简单的高斯-洛伦兹方程(Gaussian-Lorentzian (Voigtians) functions)即可确定碳原子的存在。然而,由于样品性质和设备参数的影响,谱图常表现出与理想谱图的偏离。此外,对于固体样品的 PES 谱图内往往存在由于非弹性光电子散射形成的背底。为了准确地研究谱图的形状和强度,首先需要正确识别所有可能的背底信号。存在多种背底形状模型。简单的线型背底可用于谱图的快速分析。然而,为了获得更准确的谱线形状和化学计量分析,需要采用更为复杂的背底类型。1979 年 Shirley 提出了一个基于峰值区域内存在一个恒定能量和恒定散射概率的光电子散射谱[9]。Shirley 最早研究了金价带的高分辨 PES 谱。通过去除系统内各种较小因素对强度影响,扣除费米能级所引起的强度变化,他发现谱图在高结合能和低结合能时均表现为恒定值。在高结合能一侧谱线强度比低结合能一侧要高。他假定这一差异完全是由于价带电子在离开样品前的非弹性散射引起的。

此外,光电子峰的形状受峰的类型和样品的绝缘或金属性质决定。由于样品中同一个元素存在多种不同的化学态,这些化学态的叠加效果也会反映到谱峰形状。在多数情况下,对于半导体和绝缘材料谱峰的形状可以由高斯-洛伦兹(Gauss-Lorentz)方法描述,而金属材料可以采用 Doniach-Sunjic 方法[10]。Voigt 函数是采用 XPS 方法进行定量分析的基础。然而,直接采用高斯和洛伦兹进行卷积解析是无法实现的。正是由于此原因,实际操作过程中采用两个近似的 Voigt 方程,即高斯-洛伦兹乘积形式和高斯-洛伦兹求和形式。对于 Doniach-Sunjic 方法完全不同。谱线形状是洛伦兹宽度组合的结果,是光电子峰的内在性状,这与寿命效应有关;而高斯宽度是仪器展宽(光电子谱仪和光子束)和电子-声子散射的卷积。这涉及一个描述芯电子空位电导所引起的能量依赖性屏蔽(energy-dependent screening)的奇异性指数。根据相互作用类型不同,如果样品为金属或半导体,将获得不同的谱线形状。例如对于碳纳米管的 C 1s 谱图,样品的固有特征发生改变将获得不同的信号,该过程将在第 8.3 节描述。

8.2.2 价带光电子能谱

原则上,除了 PES 芯能级外,紫外光电发射可以用于获得和理解材料的电学性质。前者提供高结合能的芯能级态信息,后者的目标是研究价带能态。在标准实验室条件下,采用 He 气体放电灯作为光电子源,可以获得紫外光电子能谱(UPS)或者价带光电子能谱(VB-PES)。VB-PES 所达到的分辨率和灵敏度提供了有关富勒烯定域分子轨道和 SWCNTs 中范霍夫奇异点(van Hove singularities,vHSs)的准确信息[11-12]。

8.2.3 角分辨光电子能谱

碳原子电子结构的明确是设计和优化新器件的关键因素。因此,基于角分辨光电子能谱(ARPES)可以研究材料表面光子所产生的电子的能量和动量,在研究准粒子能带结构中被广泛应用,例如石墨烯和其他二维关联电子系统。ARPES 已被用于测量能带色散关系[13-14]、狄拉克锥[15]以及基本关联效应,如准粒子动力学和电子–等离子和电子–声子间耦合强度[13,15]。在一个典型的 ARPES 扫描过程中,二维探测器同时获得结合能和晶体动量在二维方向的光电子强度。该光电子强度受光电子发射截面、光的偏振和最终状态有关的自能量修正影响。这就实现了研究准粒子能带结构的基本相互关系,例如插层石墨烯和石墨中光学声子的耦合[15-16]。因此,能量和动量分辨率技术的发展和快速检测的进步使 ARPES 成为研究石墨烯和其他关联体系(如高 T_c 的超导体和拓扑绝缘体)电子性质的理想手段[17]。ARPES 技术对偏振变化非常敏感,是研究电子波函数相位的有力工具。该技术已广泛应用于研究固态系统中分子气体电子轨道的杂化或电子相关的其他效应[18]。有关该话题的研究较多,也出现了一些有关 ARPES 研究石墨烯的综述和文章。因此,在这里我们不进行全面的论述,而是提供该技术可以进行哪些研究的概述。对该话题感兴趣的读者,可以参考相关文献,如参考文献[18]和[19]。

8.3 碳 sp^2 杂化体系的电子性质:C 1s 芯能级

纵观石墨烯、富勒烯或碳纳米管的 PES 谱图,其 C 1s 芯能级谱线的形状非常类似。然而当仔细研究 C 1s 芯能级时,可以获得大量有意义的信息。

对于碳材料体系的研究,PES 有关技术要求样品无杂质和吸附[11,20-22]。样品的纯度是至关重要的,然而该方面却没有引起足够的重视。第一步为样品的全谱扫描,且该步骤不能跳过。该步骤可以给出样品可能污染物的信息,而在这种情况下污染物多为制备过程的副产物。只有当样品的元素组成确定后,才能对某个特定的峰进行更全面细致的研究,进而获得有价值信息。碳芯能级的研究进行得较多,然而主要研究碳原子与其他元素的结合能状态(如样品中不可避免的氧原子),因此有关碳芯能级的研究还远远不够。如果没有出现任何有关所研究化合物的信号,那么这些碳芯能级信号谱峰可能与所研究化合物无关,而是来源于碳纳米管或石墨烯表面的水或化学、物理吸附的碳氧化合物。

原则上,1s 芯轨道不会影响到碳材料的固态性质,归因于 C 1s 芯能级与价带相比距离费米能级较远[23]。由于固体内相邻原子 1s 轨道的轻微重叠,C 1s 芯能级谱线(如碳纳米管)较尖锐,且芯能级能量与孤立的碳原子接近。富勒烯、石墨、碳纳米管、石墨烯的 C 1s 峰接近 284.5 eV (图 8.2),根据碳材料类型不同谱线中心出现轻微的偏移。该能量受碳原子间电荷转移特别敏感。因此,C 1s 芯能级的偏移反映出空间相对能量的变化,而该能量变化是由紧邻原子相互作用引起的。

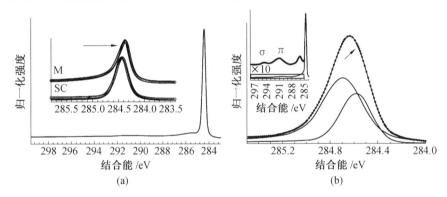

图 8.2 (a) C 1s 信号概况。碳分子系统的 C 1s 主峰信号位于 285 eV 附近。插图中给出了金属性 SWCNTs(metallicity separated SWCNTs)的谱图。可以看出二者存在显著区别。(b) 给出了混合金属性 SECNTs(metallicity mixed SWCNTs)的谱线。放大的矩形区域可以看出,C 1s 出现 σ 和 π 等离子体带间跃迁和激发

C 1s 谱的高结合能侧谱线响应存在波动,波动强度向高结合能侧逐渐降低。对于 CNTs,高结合能区域的放大可以发现存在两个小凸起,分别位于约 290 nm 和约 294.7 nm 处,对应等离子体激元振荡损失峰的卫星峰 (plasmon satellites)[24]。这些峰并非来源于 C 1s 能级附近的实际电子态,而是由于一些电子离开样品时的非弹性散射,因此所形成的卫星峰与主峰相比结合能比较高。这些等离子体激元振荡损失峰是传导电子的集体振荡带来的,尤其是带间跃迁及 σ 和 π 等离子体激发。另外,对于一些材料由于其表层区域内光电子和其他电子之间的相互作用,导致损失特定能量

的概率提高。能量损失现象使结合能母线信号上出现一个鲜明的尖锐信号。不同的固体介质表现不同,但对于金属材料该现象非常强。

例如,对于混合金属性单壁碳纳米管(metallicity-mixed SWCNTs),高分辨率的 PES 研究记录了 400 eV 激发的 PES 芯能级,结果表明 C 1s 实验谱图由两部分组成,分别以结合能 284.7 eV 和 284.6 eV 为中心,而且其相对强度与半导性部分和金属性部分的贡献比例直接相关(分别为 2/3 和 1/3),这是基于一束随机制备的 SWCNTs 进行的理论预测[25]。金属性 SWCNT 的结合能降低 0.1 eV 与两种因素有关。首先为金属性样品内增强的芯-空穴屏蔽效应(core-hole screening),其次为束内费米能级间的平衡。然而多年来,一直无法形成定论,这是由于以前采用扫描隧道谱研究纳米管束材料和单根的 SWCNTs 结果表明,带隙和 vHSs 在费米能级附近都不对称。因此,束内费米能级的平衡为最合理的解释。

随着纯化和分类技术的发展[20-22],可以获得纯度达到 99% 的新纳米管,而且该纯度考虑了催化剂剩余或其他形式的碳来源。在 0.5% 的检测极限下,未发现氧或其他催化颗粒的存在。此时,C 1s 的特点及纳米管的特性被很好地表达了出来,出现了期待已久的选择金属性单壁碳纳米管(metallicity selected SWCNTs)。Ayala 等[11]报道了 C 1s 的 PES 谱线,其中最高结合能为 283.48 eV 和 284.43 eV,清晰地观察到对应金属性(metallic)和半导体性(semiconducting)SWCNTs。也就出现了关于混合金属性(mixed metallicity)样品的初期假说,描述金属性和半导电性 SWCNTs 的叠加效应。比如混合金属性的管束中,这种下降可能是由于金属类样品内芯-空穴屏蔽(the core-hole screening),也可能是大量选择金属 SWCNTs 内不同的化学势。通过 PES 反应中 C 1s 谱线,金属性纳米管表现出 Doniach-Sunjic 线形,并以系数 $\alpha=0.11$ 进行 Gausssian 卷曲。这是描述电子导电引起的芯-空穴能量选择性筛选奇异性指数。这与之前论述的石墨非常吻合。

金属性纳米管的 C 1s 谱线峰的 FWHM 为 0.26 eV,该值小于石墨所对应的 0.32 eV,可归结为 SWCNTs 和石墨不同的金属性和耦合强度[26,27]。对于半导电 SWCNTs,表现出对称性的 Voigitian 型 C 1s 谱线。通过对谱线的形状进行分析,可得其 FWHM 值为 0.30 eV,该值与石墨的接近,但是低于 C_{60} 的数值 0.35 eV[28-29]。对于 C_{60} 富勒烯,宽度增加是由于其分子能带结构和曲率,其与单个碳原子的环境略微不同。需要特别指出的是,与石墨相比,无论金属性还是半导体性 SWCNTs 的 C 1s 均未表现出宽化。由此看来,平面石墨烯晶格与 SWCNTs 的 C 1s 间存在内在联系。在任何情况下,不同直径分布的 SWCNTs 内 Gaussian 传播都不会反映出 C 1s 线宽的变化。这是迄今为止碳材料中发现的 C 1s 谱线宽度最小值。

有时基于实验室的 XPS 系统不能满足该类型研究所需要的光谱仪分辨率,具有光子能量可调的同步加速器更适合于该类型研究。最近几年的研究为我们理解金属性和半导电性 SWCNTs 内掺杂对键合环境和电荷转移的影响提供了基础[11,12,30]。

8.4 化学态的识别:键合环境

众所周知,改变材料固态性能的一种有效方式是进行施主或受主掺杂。研究表明,SWCNTs 的费米能级可以通过电子或空穴掺杂发生移动。PES 是研究键合环境的最常用和有效的手段。这是由于元素结合能位移反映了化学势的变化。研究石墨烯时,ARPES 是最合适的方法。本章针对本方法进行论述,更详细的内容可以参考相关的文献和综述[18-19]。

对于碳纳米管,如果某些碳原子被杂原子取代(掺杂发生),根据芯能级的高分辨率和灵敏度变化,可以就掺杂水平和掺杂影响进行研究。特别是研究低掺杂浓度的 N 和 B 时,该手段具有极大的优势。对原始样品进行 XPS 全谱扫描,可以检测到低于 0.2% 的掺杂[31-33]。虽然该过程对于XPS 研究看起来是不太重要的步骤,但是该过程是非常有必要的。样品的质量直接决定了是否能有效地利用该技术研究样品的性质,更重要的是决定了该研究的可靠性。因此,对于 SWCNTs 的壁上掺杂和石墨烯而言,充分研究 C 1s 谱线是进行其他深层次分析的前提[34]。对于氮掺杂的纳米管,甚至可以检测出掺杂浓度在 0.1~0.2% 的 N 原子掺杂,通过 N 1s 谱图的研究可以获得 N 结合环境的相关信息。关于 B 掺杂的 SWCNTs 情况较为特殊。换句话说,想要研究石墨烯或纳米管内杂原子掺杂所引起的光电子响应,首先需要对 C 1s 谱线进行分析,然后进行杂原子芯能级信号的分析。除了碳原子和杂原子间可能的键合引起的影响,C 1s 信号给出了含碳材料的构成相关信息。例如,在 B 掺杂的情况下,C 1s 谱线的低结合能侧出现一个肩峰,对应 BC_4 的原子排列方式,而纯 SWCNTs 仅存在高结合能C 1s谱线[32,35]。该峰如果进一步突出和转移,则 C 1s 谱图将表现出两个显著的峰值,彼此间距约为 3 eV。只有对于纯 SWCNTs 样品才会出现没有位移的谱峰。

该类型的研究方法适用于功能化的纳米管、富勒烯及石墨烯。而对于石墨烯,需要涉及角分辨光电子能谱。

8.5　价带的电子结构

到目前为止,我们已经讨论了 C1s 谱线的光电子发射信号。而高分辨率的价带 PES(high-resolution valence-band (VB) PES)可以提供非常重要的信息,它可以检测样品占据态密度的矩阵元权重(the matrix element weighed density of occupied states)。通过对比理论和实验,可以了解 VB-PES 在纳米管研究中的优势。采用一阶紧束缚(tight-binding (TB))计算,对于已知的直径分布,考虑到金属性和半导体性纳米管功函数的不同,即可获得价带的态密度曲线。对于这两类材料,其 π 和 σ 态的整体特征可以反映出石墨烯的能带结构。此外,沿着谱图的黑色实线,很容易获得 vHSs 的位置和总体形状。这与上述的混合金属性 SWCNTs 的分析结构相吻合,例如对于半导电的纳米管样品,其 S1 和 S2 特征可以清楚地表达出来。此外,M_1 峰仅在金属性样品内可以检测到[10]。

8.6　结　　论

有关碳元素系统的光电子发射研究较多,也较为广泛,但如前所述仍然有几个需要解决的问题。大量有关 ARPES 的综述和研究性论文关注石墨烯的研究。本章主要针对碳元素系统的光电子发射及其相关研究技术进行了简短的介绍。旨在使研究者可以根据某种研究目的选择合适的研究手段和技术。

8.7　参考文献

[1] RAPPOPORT Z, APELOIG Y. The chemistry of organic silicon compounds [M]. London: Wiley, 2001.

[2] DE PADOVA P, QUARESIMA P, OTTAVIANI C, et al. Evidence of graphene-like electronic signature in silicene nanoribbons[J]. Applied Physics Letters, 2010, 96: 261905.

[3] AUFRAY B, KARA A, VIZZINI S, et al. Graphene-like silicon nanoribbons on Ag(110): a possible formation of silicene[J]. Applied Physics Letters, 2010, 96: 183102.

[4] AYALA P, ARENAL R, LOISEAU A, et al. The physical and chemical properties of heteronanotubes[J]. Review of Modern Physics, 2010, 42: 1843.

[5] AYALA P, RüMMELI M H, RUBIO A, et al. The doping of carbon nanotubes with nitrogen and their potential applications[J]. Carbon, 2010, 48: 575.

[6] PICHLER T. New Diamond and Frontier Carbon Technology. 2001, 11: 375.

[7] ZHANG Y, TAN Y-W, STORMER H L, et al. Experimental observation of quantum hall effect and Berry's[J]. Nature, 2005, 438: 201.

[8] HüFNER S. Very high resolution photoelectron spectroscopy[M]. Heidelberg: Springer Verlag, 2007.

[9] SHIRLEY D A. High-Resolution X-Ray Photoemission spectrum of the valence Bands of gold[J]. Physical Review B, 1972, 5: 4709.

[10] DONIACH S, SUNJIC M. Many-electron singularity in X-ray photoemission and X-ray line spectra from metals[J]. Journal of Physics C, 1970, 3: 285.

[11] AYALA P, DE BLAUWE K, SHIOZAWA H, et al. Disentanglement of the electronic properties of metallicity-selected single-walled carbon nanotubes[J]. Physical Review B, 2009, 80: 205427.

[12] ISHII H, KATAURA H, SHIOZAWA H, et al. Direct observation of Tomonaga-Luttinger-liquid state in carbon nanotubes at low temperatures [J]. Nature, 2003, 426: 540.

[13] BOSTWICK M A, OHTA T, SEYLLER T, et al. Quasiparticle dynamics in graphene[J]. Nature Physics, 2007, 3: 36.

[14] ZHOU S Y, GWEON G-H, GRAF J, et al. First direct observation of Dirac fermions in graphite[J]. Nature Physics, 2006, 2: 595.

[15] GRÜNEIS A, RUBIO A, VYALIKH D V, et al. Pichler angle-resolved photoemission study of the graphite intercalation compound KC8: a key to graphene[J]. Physical Review B, 2009, 80: 075431.

[16] VYALIKH D V, BALLARINI D, SANVITTO D, et al. Bloch observation of long-lived polariton states in semiconductor microcavities across the parametric threshold[J]. Physics Review Letters, 2008, 100: 056402.

[17] TASKIN A, ANDO Y. Berry phase of nonideal Dirac fermions in topolog-

ical insulators[J]. Physical Review B,2011,84:035301.
[18] LI H,PENG H L,LIU Z F. Two-dimensional nanostructures of topological insulators and their devices[J]. Acta Physico-Chimica Sinica,2012, 28:2423.
[19] LU D H,VISHIK I M,YI M C,et al. Angle-resolved photoemission studies of quantum materials[J]. Annual Review of Condensed Matter Physics,2012,3:129.
[20] MIYATA Y,YANAGI K,MANIWA Y, et al. Optical evaluation of the metal-to-semiconductor ratio of single-wall carbon nanotubes [J]. The Journal of Chemical Physics C,2008,112:13187.
[21] YANAGI K,UDOGUCHI H,SAGITANI S,et al. Transport mechanisms in metallic and semiconducting single-wall carbon nanotube networks [J]. ACS Nano,2010,4:4027.
[22] ARNOLD M S,GREEN A A,HULVAT J F,et al. Sorting carbon nanotubes by electronic structure using density differentiation [J]. Nature Nanotechnology,2006,1:2.
[23] SUZUKI S,BOWER C,KIYOKURA T,et al. Photoemission spectroscopy of single-walled carbon nanotube bundles[J]. Journal of Electron Spectroscopy and Related Phenomena,2001,114:225.
[24] KNUPFER M. Satellites in the photoemission spectra of a3c60 [J]. Physical Review B,1993,47:13944.
[25] KRAMBERGER C,RAUF H,SHIOZAWA H,et al. Unraveling van Hove singularities in X-ray absorption response of single-wall carbon nanotubes [J]. Physical Review B,2007,75:235437.
[26] BRUHWILER P A,KUIPER P,ERIKSSON O,et al. Core hole effects in resonant inelastic X-ray scattering of graphite[J]. Physics Review Letters,1996,76:1761.
[27] AHUJA R,BRUHWILER P A,WILL J M,et al. Theoretical and experimental study of the graphite 1s X-ray absorption edges[J]. Physical Review B,1996,54:14396.
[28] PRINCE K C,ULRYCH I,PELOI M,et al. Core-level photoemission from graphite[J]. Physical Review B,2000,62:6866.
[29] GOLDONI A,CEPEK K,LARCIPRETE R,et al. Core level photoemission evidence of frustrated surface molecules: a germ of disorder at the

(111) surface of C_{60} before the order-disorder surface phase transition [J]. Physics Review Letters,2002,88:196102.

[30] DE BLAUWE K,MOWBRAY D,MIYATA Y,et al. Combined experimental and ab initio study of the electronic structure of narrow-diameter single-wall carbon nanotubes with predominant (6,4),(6,5) chirality [J]. Physical Review B,2010,82:125444.

[31] AYALA P,GRÜNEIS A,KRAMBERGER C,et al. Effeet of the reation atmosphere composition on the synthesis of single and multiwalled nitrogen-doped nanotubes[J]. Journal of Chemistry Physics C,2007,127:184709.

[32] AYALA P,PLANK W,GRÜNEIS A,et al. A one step approach to B-doped single-walled carbon nanotubes[J]. Journal of Materials Chemistry,2008,18:5676.

[33] ELIAS A L,AYALA P,ZAMUDIO A,et al. Spectroscopic characterization of N-doped single-walled carbon nanotube strands:an X-ray photoelectron spectroscopy and Raman study[J]. Journal of Nanoscience and Nanotechnology,2010,10:3959.

[34] USACHOV D,VILKOV O,GRÜNEIS A,et al. Nitrogen-doped graphene:efficient growth,structure,and electronic properties[J]. Nano Letters,2011,11:5401.

[35] NAKANISHI R,KITAURA R,AYALA P,et al. Electronic structure of Eu atomic wires encapsulated inside single-wall carbon nanotubes[J]. Physical Review B,2012,86:115445.

第3部分　石墨烯及石墨烯器件的电子传输性能

第9章　石墨烯的电子传输:高迁移率

强烈的载流子散射影响了石墨烯中狄拉克费米子的固有反应,限制了石墨烯基器件的潜在应用。多重散射机理包括库仑散射、晶格无序散射和电子-声子散射。多重散射在石墨烯器件中具有一定作用。此外,散射机制的不同可用于确定石墨烯的类型和制备方法。本章论述了对于减少石墨烯中载流子散射的最新进展情况。本章首先讨论了不同参数,如载流子迁移率、平均自由程和散射时间,这些参数可以用来评价散射的强度。然后讨论了减少散射和提高载流子迁移率的方法。这些方法包括降低缺陷密度、使石墨烯悬浮、将石墨烯沉积到高品质衬底及用高 k 电介质覆盖石墨烯。最后,介绍了超净高迁移率石墨烯特定的物理现象及其器件应用。

9.1　引　言

石墨烯中载流子弱散射是该材料最显著的性能之一。散射强度通常由迁移率、载流子漂移速率和电场的比值来量化。探究石墨烯中电子传输的早期实验,报道了室温迁移率大约是 $10\ 000\ cm^2 \cdot V^{-1} \cdot s^{-1}$,比最常用的电子材料硅的典型迁移率高一个数量级[1]。早期研究把注意力放在了将石墨烯代替硅应用于电子领域。同时,在现有样品中载流子散射很强,足以掩盖石墨烯中狄拉克费米子之间相互影响的特性。

具有低载流子散射的石墨烯样品,对于实验探究狄拉克费米子的固有物理现象以及实现石墨烯的潜在应用是必要的。只有提高载流子迁移率,才有可能观察到石墨烯中丰富的现象,包括弹道传输(或无散射传输)、对称破坏量子霍尔效应、分数量子霍尔效应(FQHE)和Klein隧穿。由于石墨烯样品质量不够,影响了更多预测现象(包括玻色-爱因斯坦凝聚、超导、质量能带间隙的出现和 $\nu=5/2$ 分数量子霍尔态)的观察。高迁移率将有利于推进石墨烯的潜在应用,包括高频晶体管、传感器、光电调制器及透明导电电极。

本章论述了降低石墨烯中载流子散射的研究进程。首先,讨论了量化石墨烯中散射强度的不同参数。其次,讨论了石墨烯的不同类型和制备方法与不同的主要散射机理的对应关系。然后,讨论了石墨烯散射的不同机理,包括库仑散射、晶格无序散射和电子-声子散射。接下来,讨论了石墨烯中降低散射的不同方法,即降低缺陷密度、石墨烯悬浮、高品质基板上沉积石墨烯法及高 k 电介质覆盖法。最后,介绍了高迁移率石墨烯存在的物理现象以及这种材料的潜在应用。

9.2 散射强度的衡量参数

为了对不同研究团队石墨烯样品的散射强度进行比较,需要对样品质量有一个通用标准。最常用的参数是载流子迁移率以及载流子的漂移速率与电场的比值。尽管迁移率使用广泛,还是有几个可接受的定义,即使是同样的设备,通过这些定义法也会产生不同数值。此外,其他常用的参数有平均自由程、传输和量子散射时间及狄拉克峰的半峰宽。由于明确定义样品质量指标的重要性,本章简要讨论了测量石墨烯中散射的实验方法,并涉及了可能造成的假象,采用了 SiO_2 衬底上机械剥离的石墨烯作为对照标准。

载流子迁移率可通常定义为 $\mu \equiv v/E = \sigma/en$,其中,$v$ 是 Drude 载流体漂移速率;E 是施加电场(假定是低电场);σ 是电导率;n 是载流子密度。可采用霍尔条几何尺寸在低场磁电阻测试石墨烯场效应晶体管(FET)的载流子迁移率(图 9.1)。首先,施加板外磁场 B,记录霍尔电阻 R_{xy},得到载流子密度 $n=B/(eR_{xy})$ 和相应的栅极电压 V_G。然后确定纵向电阻 R_{xy}(低 B 时是常量)和电阻率 $\rho = R_{xx}(W/L)$,其中,W 和 L 是霍尔条的宽度和长度。由 $1/en\rho$ 计算获得霍尔迁移率。这种方法得到的迁移率被称为霍尔迁移率(μ_H)。

另外,通过测量石墨烯 FET 的电阻率 ρ 与栅极电压的关系,得到的场效应迁移率为

$$\mu_{FE} = C_g^{-1} d\rho^{-1}/dV_G = d\sigma/d(en)$$

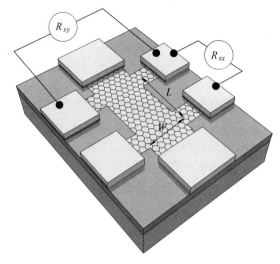

(a) 石墨烯器件的霍尔条几何结构

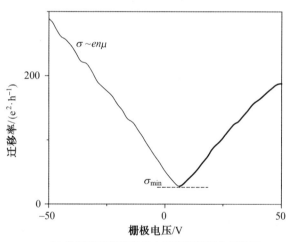

(b) 典型石墨烯器件的电导率和栅极电压关系

图9.1 石墨烯载流子迁移率的影响因素

其中，C_g 是石墨烯 FET 单位面积的栅极电容。可以通过测量两电极器件获得场效应迁移率，而无须施加磁场，且在假定 $\sigma \sim n$ 下（对于石墨烯体系而言是非常合理的近似），该场效应迁移率与霍尔迁移率具有相同的值。对于二氧化硅衬底上机械剥离的石墨烯，它的 μ_H 和 μ_{FE} 通常都是在 2 000 ~ 30 000 cm^2·V^{-1}·s^{-1} 变化。

我们对通过以上过程得到的迁移率数值持保留态度，原因如下：

(1) 虽然大多数石墨烯样品电导率和载流子密度都是成比例的,但并不确定。因此,迁移率是由载流子密度决定的。此外,因为石墨烯中不同散射机理对应不同的 n,迁移率在低载流子和高载流子密度下会受到不同散射机理的限制。因此,对比不同石墨烯器件的载流子迁移率时需要在相同载流子密度下进行(通常 n 约为 $10^{12} cm^{-2}$)。

(2) 场效应迁移率不受短程散射的影响,短程散射机制对 σ 产生的影响与载流子密度无关(详细论述见 9.3 节)。

(3) 采用每单元面积栅极电容 C_g 定义的场效应迁移率,通常采用平行板电容器近似 $C_g = k\varepsilon_0/d$,其中 k 是栅极材料的介电常数,d 是栅极厚度。只有当体系可近似为平行板时,也就是当石墨烯的尺寸远大于 d 时,这个公式才适用。当石墨烯器件的面积与 d 相近或小于 d 时,器件的载流子密度发生变化,迁移率的标准公式是不适用的。此外,平行板电容器公式没有考虑石墨烯的量子电容,而量子电容对于 d 值为纳米级的器件有着重要作用[2]。

(4) 对于霍尔条几何结构里的小样品而言,霍尔条"手臂"的宽度通常与器件的宽度和长度相近。这样,几何因素相关电阻和测量电阻之间就会出现误差。

(5) 迁移率的确定受随机掺杂分布造成的载流子密度空间不均匀性的影响。对于多探针的霍尔条样品而言,可以通过比较器件两边成对电极所测量的迁移率推测值来检验可能的载流子密度不均性。

载流子散射的其他几个指标将在下面章节有论述,这些指标通常与载流子迁移率有关。

9.2.1 平均自由程和传输散射时间

通常用散射事件的 Drude 平均时间 τ 和平均自由程 $l_{MFP} \equiv \nu_F \tau$ 描述散射强度和样品质量。对于石墨烯,散射长度(τ)与载流子迁移率关系可由 Drude 公式表示。

$$l_{MFP} = \mu(\hbar/e)\sqrt{\pi n}/\nu_F$$

其中,ν_F 是石墨烯的费米速度,ν_F 约为 $10^6\ ms^{-1}$。对石墨烯平均自由程可通过短于 l_{MFP} 的样品的弹道传输估算,也可以通过测量弯曲电阻估算[3]。

9.2.2 Shubnikov-de Hass 振荡开启场

在适度磁场下,2D 电子气体的电导率具有可重复振荡性,在高磁场下发展为量子霍尔平台。这些振荡本身就具有量子机械性,且当磁场强到载流子没有散射就可以完成回旋轨道时振荡就会出现。振荡开启场 B_c 与迁移率关系可通过半经典公式 $\mu \sim B_{c-1}$ 描述。虽然测试 Shubnikov-de Hass 振荡开启场仅提供了迁移率的定性估计,但该测试不会受到由几何因素 W/L 和栅极电容的不准确性所带来的影响。

9.2.3 最小电导率的温度依赖性

只有当 k_BT 大于特征能量 E_{puddle} ($E_{puddle} = \hbar v_F \sqrt{\pi \delta n}$) 时,温度才会显著地影响最小电导率,其中,$\delta n$ 是石墨烯载流子密度的不均匀性[4]。反过来,参数 δn 与载流体的迁移率有关[5]。对于二氧化硅上的石墨烯样本,δn 为 $10^{10} \sim 10^{11} cm^{-2}$,对应 $10 \sim 30$ meV 的能量[6]。这个能量范围决定了实验手段所能接近石墨烯的狄拉克点的程度[7]。

9.2.4 电中性点的电阻率峰的宽度和位置

电中性点的电阻率峰的半峰宽(Full Width at Half Maxima,FWHM)与电荷不均匀性程度有关,也与迁移率相关。几何因素不会影响半峰宽的测定。对于 SiO_2 上的石墨烯,典型的 FWHM 是 $10^{11} \sim 10^{12} cm^{-2}$。低质量样品的最小电导率点的位置通常是从零点开始偏移,只是除了载流子散射还存在其他集中无序导致的静电掺杂[5]。

9.3 石墨烯制备方法

多年来出现了很多针对石墨烯制备的实验方法。不同类型的石墨烯受不同的散射机理影响,降低散射也需要不同的方法。下面简单论述制备石墨烯最常用的方法:机械剥离法、化学气相沉积法、碳化硅上的外延生长法及化学法制备氧化石墨烯,并提出上述方法制备石墨烯的主要散射机理。

9.3.1 机械剥离法

从科学著作可知,用胶带或微机械剥离使高定向热解石墨晶体解离开,是由 Geim 和 Novoselov 的早期实验所使用的获得石墨烯的方法[1]。尽管该方法简单、产率低,但仍然能制出缺陷度很低的石墨烯,已出现大量关于微机械剥离法石墨烯迁移率的研究报道。二氧化硅衬底上剥离石墨烯的典型迁移率为 5 000~10 000 $cm^2 \cdot V^{-1} \cdot s^{-1}$。

9.3.2 金属衬底上化学气相沉积法

通过化学气相沉积(CVD)可以在镍、铜、锂、钌和其他金属上生长大面积的石墨烯[8]。在这种方法中,原料气体(通常是甲烷)在金属表面分解产生单层或多层具有相对低缺陷度的石墨烯。生长完毕后,石墨烯层可以转移到合适基板并进行表征。通常 CVD 生长的石墨烯的迁移率值是 1 000~10 000 $cm^2 \cdot V^{-1} \cdot s^{-1}$;主要的散射可能会出现在晶界、位错和其他与基板相关的特征。

9.3.3 还原氧化石墨法制备石墨烯

氧化石墨的化学还原制备单层石墨烯的图像要追溯到 20 世纪 60 年代早期[9]。在该方法中,大量石墨先进行化学氧化,然后通过超声分散在溶液中。所获得的氧化石墨片层再通过与还原剂(如联氨)进行化学反应被还原。虽然用这种方法可以制得大量石墨烯,但最终得到的石墨烯大部分具有无序性和缺陷特征。通常用化学方法得到的薄而多层石墨烯薄膜的迁移率变化范围为 0.1~1 $cm^2 \cdot V^{-1} \cdot s^{-1}$,而且可能受独立石墨烯片之间的载流子跳跃的限制[10]。

9.3.4 外延石墨烯

碳化硅衬底上的外延石墨烯是通过对碳化硅(SiC)的热降解得到的。在超高真空下,加热到高温(大于 1 000 ℃)会导致硅升华留下富碳层,然后是石墨化过程[11]。因为可以获得绝缘态的大块碳化硅,所以这种石墨烯不用转移到其他基板上就能够进行测量。通常这种方法得到的石墨烯的迁移率约为 1 000 $cm^2 \cdot V^{-1} \cdot s^{-1}$。通常外延石墨烯以其与下方衬底的强烈相互作用为特征。

9.4 石墨烯的散射来源

石墨烯中所有散射机制可以分为两大类。固有的散射机制(如缺陷、晶界或声子散射)都与石墨烯自身有关。相反的为外部散射机制,如石墨烯上面或下面的带电杂质发生的库仑散射,以及由于衬底上石墨烯的声子引起的散射,归因于石墨烯与其他紧邻材料的作用。本节主要讨论限制石墨烯器件中载流子迁移率的固有散射和外部散射。对于更详尽的探讨,可参考有关石墨烯的电子传输和缺陷的综述[12-13]。

9.4.1 库仑散射和其他长程机制

关于单层石墨烯中的电传输早期实验研究报道了单层石墨烯两个值的显著特点[14-15]：电导率 σ 与载流子密度 n（采用了几乎恒定的载流子迁移率 $\mu=\sigma/en$）大致的线性关系和电荷中性点处石墨烯的非零电导率 σ_{min}，如图 9.1(b)所示。与这种 $\sigma(n)$ 一致的最简单的是石墨烯中电荷载流子的库仑散射，由石墨烯下面的衬底（一般为二氧化硅栅极电介质）或石墨烯表面[5,16-18]的带电杂质产生。假定二氧化硅中带电杂质的密度，n_{imp} 约为 $50\times10^{10}\ cm^{-2}$，采用带有随机相近似的玻耳兹曼动力学方程理论模型[5]，重现了二氧化硅衬底上石墨烯场效晶体管的特征：载体子迁移率 μ 在为 $1\ 000\sim30\ 000\ cm^2\cdot V^{-1}\cdot s^{-1}$ 内 $\sigma(n)$ 与 μ 呈线性关系，非普适的最小电导率 σ_0 在 $2\sim12\ e^2/h$ 范围内变化，电荷中性点从 0 偏移至 $10\sim50$ V[19]。石墨烯中狄拉克特征的载流子导致库仑散射的作用相对而言不显著。首先，石墨烯由于不存在晶格无序具有赝自旋守恒，从而禁止了载流子的反向散射[20]。其次，在电荷中性点附近，载流子能够以 Klein 隧穿的形式穿过分割电子和空穴杂区域的势垒[21-22]。这些特征是由于二氧化硅和其他衬底上石墨烯具有较高的电荷掺杂浓度而表现出高的载流子迁移率[23]。

下面的实验为体现常见衬底上石墨烯的库仑散射的重要性，提供了有力的证据。第一，采用扫描探针技术可直接获得由 SiO_2 内杂质电荷的库仑场引起的石墨烯内局部载流子密度的随机波动[6]。这些波动的平均数量级 δn（约为 $10^{11}\ cm^{-2}$）与杂质密度的预测值一致。波动的结果导致，石墨烯中的电子气体分解为电中性点附近的定域化电子"坑"和空穴。第二，描述库仑散射的模型采用石墨烯上沉积钾离子进行了直接验证[24]。所观察到的载流子迁移率变化、电荷中性点位置以及最小电导率值与带电杂质散射的玻尔兹曼模型相一致（图 9.2）。第三，采用将冰层沉积到石墨烯上可降低散射，归因于库仑散射的电介质屏蔽[25]。

衬底相关的库仑散射是石墨烯散射中的一个未经证实的解释。而最近几项研究对这个看似简单的模型提出了重大难题：

(1) 不同衬底沉积的石墨烯样本具有相近的 μ 值[26]，退火之前悬浮的和有衬底的石墨烯之间的 μ 值相似[27]，其他一些证据也表明库仑散射可能与石墨烯表面的带电杂质或石墨烯和基板之间嵌入的杂质有关，而与 SiO_2 衬底内的电荷无关。

(2) 中间能隙态共振散射[28-29]或冻结脉冲散射[20]能够产生与载流子密度大致呈线性关系的电导，因此有可能被误认为是库仑散射。

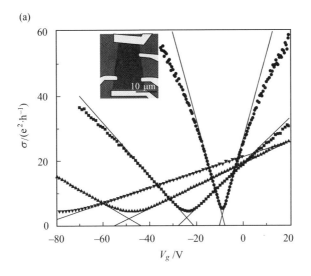

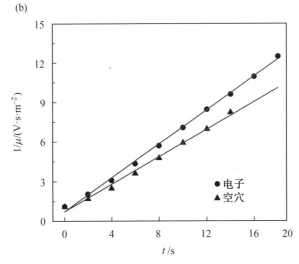

图 9.2 通过钾离子沉积到石墨烯来研究库仑散射[24]
(a) 四个逐渐增加锂浓度的石墨烯样品的电导率(σ)和栅极电压(V_g)的关系曲线。数据是在 20 K 超高真空环境下得到。直线适于描述库仑散射公式。(b) 电子迁移率的倒数和孔穴迁移率倒数与掺杂时间的关系

(3) $\sigma \sim n$ 的严格线性关系只在随机分布的库仑散射时才会出现;相反,聚集的带电杂质团簇会产生非线性 $\sigma(n)$[36]。这种非线性可以通过退火过程石墨烯上钾离子迁移获得[31]。

9.4.2 晶格无序散射

强的 sp^2 键合使石墨烯具有高的固有强度,且其晶格缺陷较少。然而,采用原子力显微镜、透射电镜等技术发现了石墨烯内部各种类型的缺陷,包括空位、拓扑缺陷(如 Stone-Wales 缺陷)、外来杂原子替换、化学键、晶界以及晶格的机械扭曲[32-36]。最早确定石墨烯内存在缺陷数量的方法是拉曼光谱[37]。由于区域边界声子引起所谓的 1 350 cm^{-1} 附近的 D 带,没有出现在平移不变的石墨烯中,只会在有缺陷的样本中出现。已出现了大量研究讨论不同类型的缺陷和无序性(包括晶界、位错和样本边缘)表现出的拉曼特征。通过比较 D 带的强度 I_D 和 G 带的强度(约 1 580 cm^{-1})I_G,平均缺陷密度可以通过以下经验公式进行估计:

$$n_d \approx 1.8 \times 10^{22} \lambda^{-4}(I_D/I_G)$$

其中,λ 是激发波长[38]。

与库仑散射相比,人们还没有彻底弄清楚石墨烯中其他类型无序对石墨烯电传输各种无序性的影响。最简单的模型把石墨烯的无序性看作是无关 δ 函数势,并预测一个由无序散射引起的与载流子密度无关的电导[16-18,25]。更多有关缺陷的实际模型(如空位[39]或吸附[29])预测了在电荷中性点或者附近的定域态的形成。这些中间带隙态使石墨烯中的载流子发生共振散射,并提高了与 n 大致呈线性关系的电导率,其作用与带电杂质类似。在研究石墨烯的晶界散射[23]以及由离子辐照和氢吸附产生的缺陷[28-29]时都发现了这样的电导率行为。

不同手段制备的石墨烯样品会表现出不同的缺陷类型:

(1)在机械剥离的石墨烯中,缺陷很少[33,40-42],并且通常被认为是电子或离子辐照的结果[33,35-36]。然而,高品质剥离的样本中出现的弱拉曼 D 峰和与密度无关的电阻,表明无序确实能引起散射,即使是在剥离器件中[28]。

(2)与衬底的共价键合对限制外延石墨烯迁移率可能会起到重要作用。在这些石墨烯中已发现密度相对较低的位错[43-44]。

(3)由化学气相沉积法制备的石墨烯存在更明显的缺陷。在这种石墨烯中缺陷往往以晶界的形式出现[23,44-48]。研究表明晶界会产生强烈散射而限制载流子迁移率[23,49-50]。

(4)通过化学法氧化-还原合成的石墨烯具有更多的缺陷,并含有大量的七边形-五边形缺陷和原子取代[51-52]。对于薄的还原的氧化石墨烯膜来讲,引起散射的主要原因可能是独立石墨烯片层之间的跳跃[10]。

9.4.3 电子-声子散射

许多电子材料固有的室温迁移率受声子对载流子散射的限制。例如,虽然 AlGaAs/GaAs 异质结中的 2D 电子气体的载流子迁移率在低温下能够达到大于 10^7 cm² · V⁻¹ · s⁻¹,室温时由于光学声子的散射会使这个体系的室温迁移率降到小于 2 000 cm² · V⁻¹ · s⁻¹[53]。相反,石墨烯中光学声子的能量相对高,纵向光学(LO)边界声子可达约 160 meV[54],接近室温和低于室温时,占主导地位的内在声子引发机制由于纵向声子(LA)而使散射相对较弱[39,55]。这种散射的传输特点是在 10~200 K 范围内与载流子密度无关的电阻与温度 T 呈线性关系,比例系数为 0.1 ~ 0.2 Ω · K⁻¹[4,56,57]。一直有争论的是,如果不考虑其他散射,这种声子引发散射导致石墨烯固有室温迁移率为 n 约为 10^{12} cm⁻² 时,极高限值 μ 为 200 000 cm² · V⁻¹ · s⁻¹[4,56]。这个值明显超过了室温半导体的最高迁移率(锑化铟的最高迁移率约为 7.7×10^4 cm² · V⁻¹ · s⁻¹)[58]。

在一些情况下,石墨烯中载流子可能与面外(弯曲的)声子相互作用。对于悬浮不受拉应力的石墨烯样本,弯曲声子散射占声学声子散射的主导地位,并把室温迁移率限制到小于 10 000 cm² · V⁻¹ · s⁻¹[59-60]。衬底支撑石墨烯样本的弯曲声子的作用还不清楚,仍存在争议。一些研究人员把温度 $T>100$ K 时,二氧化硅支撑的石墨烯电阻的快速增加归因于与衬底部分相连的石墨烯区域以及波纹内激发的弯曲声子引起的载流子散射[4]。

最后,石墨烯下面的极性衬底材料会产生使石墨烯内载流子发生散射的波动电场[60]。对二氧化硅上的石墨烯来说,这个远程界面声子(remote interfacial phonon,RIP)散射机制可以用来解释温度 $T>200$ K 时石墨烯电阻指数的变化[56,62]。一般认为 RIP 散射把二氧化硅上石墨烯的室温迁移率在 $n = 10^{12}$ cm⁻² 时限制到小于 40 000 cm² · V⁻¹ · s⁻¹。

9.4.4 多重散射的机制

在实际石墨烯器件中,多重散射同时起作用。例如,对于相对高品质的 SiO_2/Si 基板上剥离石墨烯器件,载流子密度 $n = 10^{12}$ cm⁻² 时,由于库仑散射或共振杂质散射使迁移率最大值 $\mu_{Coulomb}$ 约为 10 000 cm² · V⁻¹,SiO_2 的远程界面声子散射进一步引起迁移率 μ_{RIP} 约为 40 000 cm² · V⁻¹,石墨烯声子散射引起迁移率 $\mu_{phonons}$ 约为 2 000 000 cm² · V⁻¹。受几个不同散射机理影响的器件的迁移率可以通过 Matthiessen 规则获得:

$$\mu^{-1} = \mu_{Coulomb}^{-1} + \mu_{RIP}^{-1} + \mu_{phonons}^{-1}$$

最近普遍采用简单经验法则来区分多重散射机制的贡献。高载流子

密度时的电阻采用 $(ne\mu_L) - 1 + \rho_s$ 拟合,两个拟合参数为 μ_L 和 ρ_s。首先,与密度相关项被认为是远程散射机理的结果,如库仑散射、共振杂质散射及波纹散射。其次,与密度无关项 ρ_s 由近程机理引起,例如声子或无序散射。虽然方法简便,但也要谨慎使用。如上讨论,远程散射机理(如库仑和共振杂质散射)可以引起非线性 $\sigma(n)$,而近程散射有时也会存在密度依赖的 σ。

9.5 增加载流子迁移率的方法

提高石墨烯器件迁移率,需要同时抑制多重散射。首先,为了避免引起缺陷散射,高迁移率石墨烯器件不能包括多种缺陷或晶界。其次,应将石墨烯沉积到具有低电荷杂质密度且平整衬底上以避免石墨烯中褶皱散射和库仑散射。最后,理想的石墨烯衬底应只具有高能量的声子模式以避免远程界面声子散射。本节讨论减少石墨烯中散射和提高载流子迁移率可能用到的方法。

9.5.1 退火

经过转移或光刻过程制备的石墨烯的表面往往被严重污染。很难去除的杂质主要是电子束抗蚀剂的聚合物残留物,如聚甲基丙烯酸甲酯(PMMA)或光抗蚀剂。虽然普通溶剂(如丙酮)可以去掉杂质,但即使经过充分的化学清洗,石墨烯器件的表面通常还是会覆盖着纳米厚的不均匀残留物层,如图9.3(a)所示。这样的聚合物残留物层可能会引起石墨烯中的散射,比如这些残留物会成为电离杂质提供者或产生共振散射。

遗憾的是,去除聚合物残留薄膜最普通的方法(如氧等离子体蚀刻法)也会损坏石墨烯。相反,热退火方法可以去除部分杂质(但不是全部[63])并获得高品质石墨烯以及墨烯的原子分辨率图像[40],如图9.3(b)和(c)所示。所用方法的大体步骤都是由最初 Ishigami 等报道的方法演变而来的[40]:在氢或氩的环境下,400 ℃下退火约 1 h(图9.3)。也可采用在两个与石墨烯连接的终端之间通过一个大电流,再通过欧姆加热实现石墨烯的退火[64]。后一方法是在高真空低温容器里对石墨烯样本的杂质进行原位去除,因此避免了大气污染。

大多数研究报道都认为退火后支撑石墨烯的迁移率会得到提高。几乎所有研究具有高载流子迁移率的石墨烯器件都采用了某种形式的退火处理。对于 SiO_2 上的机械剥离石墨烯而言,退火前的迁移率通常是 2 000 ~ 30 000 $cm^2 \cdot V^{-1} \cdot s^{-1}$,退火后的迁移率是 10 000 ~

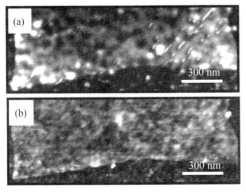
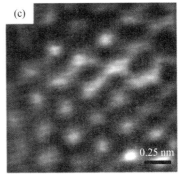

图9.3 石墨烯的原子力显微镜图像[40]

(a)退火之前。(b)退火之后。(c)退火后经扫描隧道显微镜获得石墨烯的原子分辨图像

$30\,000\ cm^2 \cdot V^{-1} \cdot s^{-1}$ [19,26]。对于化学气相沉积生长的石墨烯和氮化硼衬底沉积的石墨烯,通常迁移率从 $2\,000 \sim 5\,000\ cm^2 \cdot V^{-1} \cdot s^{-1}$ 增加到 $10\,000 \sim 30\,000\ cm^2 \cdot V^{-1} \cdot s^{-1}$ [65],而悬浮石墨烯器件的迁移率通过电流退火会提高到大于一个数量级[27]。

9.5.2 衬底工程

要使衬底引起的散射最低,理想的衬底材料为:①含有少量或没有电荷杂质;②保持平整以减少石墨烯变形引起的散射,并避免石墨烯和衬底之间出现杂质;③无界面极化声子模式。

虽然早期报道称衬底不会影响载流子迁移率[26],但后来研究发现传输品质可以通过衬底工程得到很大提高。例如,单晶 $Pb(Zr_{0.2}Ti_{0.8})O_3$ (PZT)[66]上获得的多层石墨烯的迁移率超过 $70\,000\ cm^2 \cdot V^{-1} \cdot s^{-1}$。用六甲基二硅胺(HDMS)薄的疏水层覆盖 SiO_2/Si 基板虽然不会提高衬底的迁移率,但会大大减少石墨烯上残余的助剂[67]。然而,直到 Dean 等研究结果出现,这一问题才得以突破,通过将石墨烯置于结晶六方氮化硼(hBN)衬底上,迁移率会持续提高到 $60\,000\ cm^2 \cdot V^{-1} \cdot s^{-1}$(图9.4)[68]。

大多观点认为氮化硼对于石墨烯器件来说是一种具有诱惑力的衬底材料[68]。第一,氮化硼的原子结构中的强键合力使它具有相对惰性并且不存在电荷陷阱;第二,氮化硼是一种具有较宽能带间隙(约 6 eV)的电介质;第三,氮化硼的晶格常数与石墨烯的晶格常数可以很好匹配(98.3%);最后,氮化硼的光学声子模式的能量相当于 SiO_2 的 2 倍。

为了制作 hBN 上石墨烯器件,Dean 等首先用成熟的剥离石墨烯的技

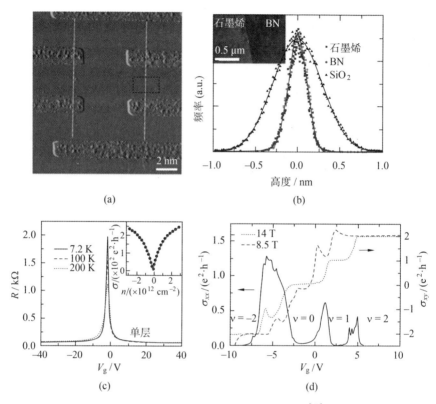

图 9.4 hBN 上高迁移率的剥离石墨烯[68]

(a) hBN 上石墨烯及电接触状态的 AFM 图像,白色虚线表示石墨烯边缘。(b) 高度分布直方图表明石墨烯粗糙度很难与 hBN 的粗糙度区分开。(c) 不同温度下 hBN 上单层石墨烯的电阻和所加栅极电压的关系。(d) hBN 上石墨烯在 $B=14$ T (虚线)和 $B=8.5$ T(短划线)的纵向霍尔电导率与栅极电压的关系

术剥离 hBN 层到 SiO_2/Si 衬底上[68]。石墨烯被剥离到分离聚合物衬底上并通过精确的光学定位机械转移到 hBN 上。为了消除产生的残留物,器件要在 Ar/H_2 气氛下退火。后续研究报道了转移技术的多方面改善,包括完全的干燥转移过程和利用上面另一层 hBN 对石墨烯的封装以防止石墨烯被环境污染[3,69]。许多研究团队所做的大量实验都证实了 hBN 上石墨烯器件的低散射。有报道称低温下场效应迁移率超过 300 000 $cm^2 \cdot V^{-1} \cdot s^{-1}$,室温下约为 100 000 $cm^2 \cdot V^{-1} \cdot s^{-1}$,甚至可能出现更高迁移率[3,68-69]。预计平均自由程 n 在 10^{12} cm^2 下会超过 3 μm。从透射电镜和扫描电镜测试可以推断,石墨烯的电荷密度量级可减少到 10^9 cm^{-2} [68,70]。在与温度有关的测试中没有发现 hBN 内有界面声子引起

的散射[68]。

值得注意的是,由化学气相沉积法生长的石墨烯的迁移率也可以通过将其转移到 hBN 上得到显著提高。该种结构器件的迁移率值可高达 60 000 cm^2·V^{-1}·s^{-1}[65,71],如图 9.5(d)~(f)所示。

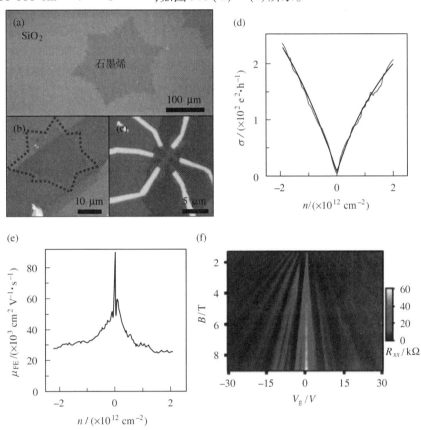

图 9.5 CVD 法生长的大尺寸石墨烯的高迁移率[71]

(a)转移到 SiO$_2$ 上的大尺寸 CVD 石墨烯的光学形貌。(b) 用干转移法转移到 hBN 片上的大尺寸 CVD 石墨烯晶体(为了清晰,用虚线画轮廓)。(c) hBN 上霍尔条形貌。(d)电导率图(黑色线为拟合曲线;灰色为数据)。(e)图(c)中石墨烯/hBN 器件在 1.6 K 下的载流子密度和场效应迁移率的函数关系。(f)器件的朗道扇形图

9.5.3 缺陷/晶界的消除

对于机械剥离的石墨烯,缺陷密度已相当低,通过热退火可以更

低[27]。然而主要的问题还是如何在器件制造和测试过程中避免缺陷的形成。通常在制造石墨烯器件的电子束光刻过程中,电子辐照会对石墨烯产生显著破坏[72,73]。石墨烯图案化后得到的粗糙边缘也可能造成散射。

对于化学气相沉积获得的石墨烯,无序散射起主要作用。最近的研究表明消除晶界会使载流子迁移率大大提高。用生长技术制作的大于几百微米的石墨烯晶片(图 9.5(a)),迁移率高达 45 000 $cm^2 \cdot V^{-1} \cdot s^{-1}$ [71]。这些器件中衬底引起的散射可以通过将 CVD 生长石墨烯转移到 hBN 衬底上而被抵消。外延生长的石墨烯也同样存在迁移率随晶粒大小而提高的现象[74]。

还原氧化石墨烯薄膜的电传输受相互连接的石墨烯晶片间的跳跃现象的影响。薄膜的迁移率可以通过提高晶片尺寸而提高到 5 $cm^2 \cdot V^{-1} \cdot s^{-1}$。通过增加还原氧化石墨烯膜的厚度,迁移率可以达到 100 $cm^2 \cdot V^{-1} \cdot s^{-1}$ [70]。

9.5.4 电介质工程

现在许多研究者认为库仑散射在石墨烯器件中即使不占主导地位也是非常重要的。可以通过把具有高介电常数 κ 的材料沉积到石墨烯上面或下面来减少这种散射。因为与石墨烯中载流子和石墨烯附近的带电杂质相连接的电场线路都是通过石墨烯周围的电介质传播的,使用高介电常数材料,可以"分开"这些线路,因此就减小了库仑作用的有效强度[75]。

事实上,有报道称石墨烯被冰覆盖,迁移率会增加约 30%。这个增加相当于用一薄层冰屏蔽了库仑散射[25]。后来的大量研究虽然没有直接研究不同 κ 值的影响,但是石墨烯和其他二维材料在高 κ 基板中都表现出迁移率极大的提高[66,76]。推断认为,对于完全埋入电介质材料中的石墨烯,屏蔽带来的作用应该是最强的。有报道讨论了屏蔽对迁移率的重大影响,在 κ 值可变的液体里悬浮的石墨烯,室温下当 κ 值很大时,迁移率能达到 60 000 $cm^2 \cdot V^{-1} \cdot s^{-1}$(图 9.6)。后来人们发现通过在石墨烯旁放置另一层石墨烯,可能会屏蔽掉库仑散射的影响。在这种体系中观察到的金属绝缘体转变是由屏蔽引起的电子-空穴坑密度降低而产生的[77]。

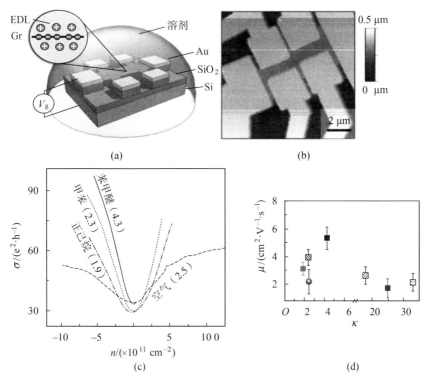

图9.6 悬浮在高介电常数液体里的石墨烯的高迁移率[81]
(a)悬浮在组成可控的液体内的多端石墨烯器件的示意图。(b)AFM形貌图。(c)不同介电常数的液体内石墨烯电导率和载流子密度的关系。(d)石墨烯器件的迁移率和石墨烯周围环境的介电常数的关系

9.5.5 悬浮器件

消除衬底引发的散射最根本的方法是使用悬浮石墨烯装置。悬浮石墨烯器件的最早电学测试表明迁移率比 SiO_2 中石墨烯器件的迁移率提高10倍[78-79]。有报道称多端悬浮器件在 $n=2\times10^{11}$ cm^{-2} 的迁移率数量级为 200 000 $cm^2 \cdot V^{-1} \cdot s^{-1}$,估计低温下二端悬浮石墨烯的迁移率可达百万数量级[7,80]。

用于制造高迁移率悬浮石墨烯的几种方法,都采用机械剥离石墨烯以减少无序性产生的影响。在第一种方法中,在 SiO_2/Si 基板上制得与金属电极相连的石墨烯 FET,如图 9.7(a)所示。SiO_2 基板通过化学蚀刻去除掉,通常用稀释后的氢氟酸。有趣的是,酸在处理石墨烯和基板时极其有

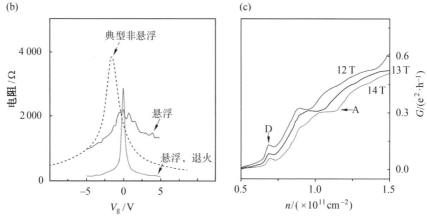

图 9.7 悬浮石墨烯的高迁移率[26]

(a)悬浮单层石墨烯的霍尔条几何结构。(b)欧姆退火前后图(a)中器件的电阻率和栅极电压的关系,退火的器件载流子迁移率大于 2 000 000 $cm^2 \cdot V^{-1} \cdot s^{-1}$。(c)高磁场下悬浮石墨烯的分数量子霍尔态 $\nu=1/3$(标记"A")

效,极短时间内蚀刻就会产生悬浮器件。最后样品干燥非常关键,因为干燥液的表面张力可能会引起石墨烯撕裂或坍塌到衬底上。在 CO_2 临界点干燥或在低表面张力的溶液里干燥,都会避免石墨烯的损坏[27,79]。

通常,新制作的悬浮石墨烯会含有残余物污染,传输质量与 SiO_2 上的石墨烯没有太大不同。幸运的是,残余物可以通过欧姆加热法有效除去[27]。通过对退火后悬浮石墨烯的机械共振频率测试表明几纳米厚的残余物可以通过退火方法完全除掉[82],如图 9.7(b)所示。

几种其他可选择的制作工艺也已经用于解决残余物污染问题。第一,石墨烯直接被剥离到预先设计并制造的电极上[83]。第二,可以通过荫罩进行金属蒸镀的方法沉积电极得到悬浮石墨烯,然后进行化学刻蚀衬底[84]。在这两种工艺中,石墨烯都保持了高迁移率,这是因为石墨烯没有暴露在聚合物抗蚀剂下。

多个指标都表明悬浮石墨烯器件具有极低的载流子散射。低温下悬浮石墨烯的霍尔效应迁移率超过 200 000 $cm^2 \cdot V^{-1} \cdot s^{-1}$ [27]。有报道称场效应迁移率和 Shubnikov-de Haas 振荡开启场迁移率可超过 $10^6 \ cm^2 \cdot V^{-1} \cdot s^{-1}$。高品质器件中的电荷密度不均匀性低到 $10^8 \ cm^{-2}$，因此平均自由程达到微米级别[1,80]。一些研究认为，室温下迁移率超过 100 000 $cm^2 \cdot V^{-1} \cdot s^{-1}$ [85]，而其他研究认为至少在不受力的器件中，迁移率由于弯曲声子的散射而被限制到小于 40 000 $cm^2 \cdot V^{-1} \cdot s^{-1}$ [59]。

低温下超净悬浮石墨烯器件中的电子传输明显不同于 SiO_2 上低迁移率的石墨烯的电子传输。双层石墨烯中的绝缘行为明显，费米速率通过电子-电子相互作用可以重新调整，在整数量子霍尔效应(IQHE)中可以看到对称性破缺，出现分数量子霍尔效应。下面将对这些现象加以讨论。

9.6 高迁移率石墨烯的物理现象

当石墨烯迁移率增加时，它的物理性能会发生显著变化。下面简要论述纯净石墨烯器件中出现的一系列新奇的物理现象。

弹道传输(也称冲猾导)在石墨烯的平均自由程 l_{MFP} 接近器件尺寸时对电传输起到重要作用。弹道传输的表征包括 $\sigma \sim \sqrt{n}$ 关系和 σ 以 2 e^2/h 为单位的量子化[79,86]。通过在 l_{MFP} 量级尖锐势垒可以观察到相对论狄拉克费米子的 Klein 隧穿[22,87]。

当 l_{MFP} 大于磁长度 $l_B = \sqrt{h/eB}$ 时，Shubnikov-de-Haas 效应和整数量子霍尔效应(IQHE)主导石墨烯的磁电导。四重简并和 Berry 相是石墨烯 IQHE 的显著特征[14-15]。有趣的是，随着样品质量的提高，在磁场低于 100 mT 下观察到了 IQHE。

对于更纯净的石墨烯样品，电子-电子之间相互作用产生的影响会很明显[88]。一个影响就是石墨烯朗道能级谱四重简并的破缺，以在 IQHE 出现 $\nu=1$ 态和 $\nu=0$ 绝缘态为标志。在更高品质的悬浮石墨烯样本和 hBN 上的石墨烯样本中，可以观察到更丰富的现象：

(1)由高激发能量和特殊层次结构为特征的分数量子霍尔态[78,89-92]，如图 9.7(c)所示。

(2)可能由电子-电子相互作用引起的双层石墨烯从金属态到绝缘态的相转变[93]。

(3)由电子-电子相互作用导致的费米速度的重正化。

(4)图案化石墨烯的电导率梯度是 $\sigma \sim 0.6 \times 2 \ e^2/h$。这个梯度与在一般二维电子气体(2DEGs)里所看到的"0.7 异常"类似,可能也是由于电子-电子的相互作用[86]。

大量预测的由电子-电子相互作用导致的现象,还没有被观察到,包括狄拉克费米的波色-爱因斯坦凝聚、超导性、FQHE 中非阿贝尔的 $\nu = 5/2$ 形态和石墨烯中质量间隙的出现[88]。

几乎所有石墨烯的应用都离不开高迁移率的石墨烯。在石墨烯场效应晶体管中,高载流子迁移率产生的高频和低散热特性以及石墨烯基传感器的敏感性,通常与迁移率成正比。对于基于石墨烯的透明导电电极,迁移率是决定石墨烯是否能胜过传统氧化铟锡薄膜的重要参数。石墨烯电化学调节器的转换率在某种程度上取决于载流子迁移率。此外,最近研究提出可将超净石墨烯中电子-电子之间的相互作用应用于电子领域[94]。

9.7 结 论

在首次测试分离的石墨烯电传输至今,通过各种先进技术已使石墨烯的载流子迁移率从 10^4 增加到 $10^6 \ cm^2 \cdot V^{-1} \cdot s^{-1}$。这些高迁移率和低散射石墨烯器件为研究狄拉克费米子之间丰富的物理相互作用提供了一个独特平台[88]。高迁移率石墨烯器件的应用可能很快出现。我们可以发现,石墨烯正循着它的近亲 GaAs/AlGaAs 异质结中的二维电子气体的步伐发展。20 年里,二维电子气体的迁移率被提高了 3 个数量级,最好器件的迁移率现在已超过 $3 \times 10^7 \ cm^2 \cdot V^{-1} \cdot s^{-1}$[95]。如果进一步减少石墨烯的散射而达到这些迁移率级别,许多预测的现象包括非阿贝尔的 FQHE 状态和狄拉克费米子的波色-爱因斯坦凝聚就可能出现[88]。

重要的是人们解决了石墨烯固有迁移率的问题,从而降低了散射。已经证实,由于声子散射作用室温下迁移率大于 $2 \times 10^5 \ cm^2 \cdot V^{-1} \cdot s^{-1}$,已经接近固有迁移率极限值 $2 \times 10^5 \ cm^2 \cdot V^{-1} \cdot s^{-1}$[3,56]。采用化学气相沉积批量生长的石墨烯也接近该值。低温时的情况还不清楚。有报道称一些低载流子密度的石墨烯器件 $\mu > 10^6 \ cm^2 \cdot V^{-1} \cdot s^{-1}$。此外石墨上方的石墨烯的迁移率超过 $10^7 \ cm^2 \cdot V^{-1} \cdot s^{-1}$,说明制作质量超过最好的 AlGaAs/GaAs 异质结的石墨烯基器件是可能的。

9.8 参考文献

[1] NOVOSELOV K S, FAL'KO V I, COLOMBO L, et al. A roadmap for graphene[J]. Nature, 2012, 490:192.

[2] XIA J L, CHEN F, LI J H, et al. Measurement of the quantum capacitance of graphene[J]. Nature Nanotechnology, 2009, 4:505.

[3] MAYOROV A S, GORBACHEV R V, MOROZOV S V, et al. Micrometer-scale ballistic transport in encapsulated graphene at room temperature[J]. Nano Letters, 2011, 11:2396.

[4] MOROZOV S V, NOVOSELOV K S, KATSNELSON M I, et al. Giant intrinsic carrier mobilities in graphene and its bilayer[J]. Physical Review Letters, 2008, 100:016602.

[5] ADAM S, HWANG E H, GALITSKI V M, et al. A self-consistent theory for graphene transport[J]. Proceedings of the National Academy of Sciences of the United States of America, 2007, 104:18392.

[6] MARTIN J, AKERMAN N, ULBRICHT G, et al. Observation of electron-hole puddles in graphene using a scanning single-electron transistor[J]. Nature Physics, 2008, 4:144.

[7] MAYOROV A S, ELIAS D C, MUKHIN I S, et al. How close can one approach the dirac point in graphene experimentally?[J]. Nano Letters, 2012, 12:4629.

[8] AVOURIS P, DIMITRAKOPOULOS C. Graphene: synthesis and applications[J]. Materials Today, 2012, 15:86.

[9] DREYER D R, RUOFF R S, BIELAWSKI C W. From conception to realization: an historial account of graphene and some perspectives for its future[J]. Angewandte Chemie-International Edition, 2010, 49:9336.

[10] WANG S, ANG P K, WANG Z Q, et al. High mobility, printable, and solution-processed graphene electronics[J]. Nano Letters, 2010, 10:92.

[11] RIEDL C, COLETTI C, STARKE U. Structural and electronic properties of epitaxial graphene on SiC(0001): a review of growth, characterization, transfer doping and hydrogen intercalation[J]. Journal of Physics D Applied Physics, 2010, 43:374009.

[12] BANHART F, KOTAKOSKI J, KRASHENINNIKOV A V. Structural de-

fects in graphene[J]. ACS Nano,2011,5:26.
- [13] DAS SARMA S,ADAM S,HWANG E H,et al. Electronic transport in two-dimensional graphene[J]. Reviews of Modern Physics,2011,83: 407.
- [14] NOVOSELOV K S,GEIM A K,MOROZOV S V,et al. Two-dimensional gas of massless Dirac fermions in graphene[J]. Nature,2005,438:197.
- [15] ZHANG Y,TAN Y W,STORMER H L,et al. Experimental observation of the quantum Hall effect and Berry's phase in graphene[J]. Nature, 2005,438:201.
- [16] NOMURA K,MACDONALD A H. Quantum hall ferromagnetism in graphene[J]. Physical Review Letters,2006,96(25):256602.
- [17] ANDO T. Screening effect and impurity scattering in monolayer graphene [J]. Journal of the Physical Society of Japan,2006,75:4716.
- [18] HWANG E H,ADAM S,DAS SARMA S. Carrier transport in two-dimensional graphene layers[J]. Physical Review Letters,2007,98:186806.
- [19] TAN Y W,ZHANG Y,BOLOTIN K,et al. Measurement of scattering rate and minimum conductivity in graphene[J]. Physical Review Letters, 2007,99:246803.
- [20] MCEUEN P L,BOCKRATH M,COBDEN D H,et al. Disorder, pseudospins,and backscattering in carbon nanotubes[J]. Physical Review Letters,1999,83:5098.
- [21] KATSNELSON M I,GEIM A K. Electron scattering on microscopic corrugations in graphene[J]. Philosophical Transactions of the Royal Society A Mathematical Physical and Engineering Sciences,2008,366:195.
- [22] YOUNG A F,KIM P. Quantum interference and Klein tunnelling in graphene heterojunctions[J]. Nature Physics,2009,5:222.
- [23] HUANG P Y,RUIZ-VARGAS C S,VAN DER ZANDE A M,et al. Grains and grain boundaries in single-layer graphene atomic patchwork quilts[J]. Nature,2011,469:389.
- [24] CHEN J H,JANG C,ADAM S,et al. Charged-impurity scattering in graphene[J]. Nature Physics,2008,4:377.
- [25] JIANG C,ADAM S,CHEN J H,et al. Tuning the effective fine structure constant in graphene:opposing effects of dielectric screening on short-and long-range potential scattering[J]. Physical Review Letters,2008,101:

146805.

[26] PONOMARENKO L A, YANG R, MOHIUDDIN T M, et al. Effect of a high-κ environment on charge carrier mobility in graphene[J]. Physical Review Letters, 2009, 102:206603.

[27] BOLOTIN K I, SIKES K J, JIANG Z, et al. Ultrahigh electron mobility in suspended graphene[J]. Solid State Communications, 2008, 146:351.

[28] NI Z H, PONOMARENKO L A, NAIR R R, et al. On resonant scatterers as a factor limiting carrier mobility in graphene[J]. Nano Letters, 2010, 10:3868.

[29] WEHLING T O, YUAN S, LICHTENSTEIN A I, et al. Resonant scattering by realistic impurities in graphene[J]. Physical Review Letters, 2010, 105:056802.

[30] LI Q Z, HWANG E H, ROSSI E, et al. Theory of 2D transport in graphene for correlated disorder[J]. Physical Review Letters, 2011, 107:156601.

[31] YAN J, FUHRER M S. Correlated charged impurity scattering in graphene[J]. Physical Review Letters, 2011, 107:206601.

[32] HASHIMOTO A, SUENAGA K, GLOTER A, et al. Direct evidence for atomic defects in graphene layers[J]. Nature, 2004, 430:870.

[33] GASS M H, BANGERT U, BLELOCH A L, et al. Free-standing graphene at atomic resolution[J]. Nature Nanotechnology, 2008, 3:676.

[34] MEYER J C, KISIELOWSKI C, ERNI R, et al. Direct imaging of lattice atoms and topological defects in graphene membranes[J]. Nano Letters, 2008, 8:3582.

[35] TAPASZTO L, DOBRIK G, NEMES-INCZE P, et al. Tuning the electronic structure of graphene by ion irradiation[J]. Physical Review B, 2008, 78:233407.

[36] UGEDA M M, BRIHUEGA I, GUINEA F, et al. Missing atom as a source of carbon magnetism[J]. Physical Review Letters, 2010, 104:096804.

[37] DRESSELHAUS M S, JORIO A, SAITO R. Characterizing graphene, graphite, and carbon nanotubes by raman spectroscopy[J]. Annual Review of Condensed Matter Physics, 2010, 1(1):89-108.

[38] CANCADO L G, JORIO A, FERREIRA E H M, et al. Quantifying defects in graphene via Raman spectroscopy at different excitation energies[J].

Nano Letters,2011,11:3190.
[39] STAUBER T,PERES N M R,GUINEA F. Electronic transport in graphene: a semiclassical approach including midgap states[J]. Physical Review B,2007,76:205423.
[40] ISHIGAMI M,CHEN J H,CULLEN W G,et al. Atomic structure of graphene on SiO_2[J]. Nano Letters,2007,7:1643.
[41] MEYER J C,GEIM A K,KATSNELSON M I,et al. The structure of suspended graphene sheets[J]. Nature,2007,446:60.
[42] STOLYAROVA E,RIM K T,RYU S M,et al. High-resolution scanning tunneling microscopy imaging of mesoscopic graphene sheets on an insulating surface[J]. Proceedings of the National Academy of Sciences of the United States of America,2007,104:9209.
[43] RUTTER G M,CRAIN J N,GUISINGER N P,et al. Scattering and interference in epitaxial graphene[J]. Science,2007,317:219.
[44] SUTTER P W,FLEGE J I,SUTTER E A. Epitaxial graphene on ruthenium[J]. Nature Materials,2008,7:406.
[45] CORAUX J,N'DIAYE A T,BUSSE C,et al. Structural coherency of graphene on Ir(111)[J]. Nano Letters,2008,8:565.
[46] PARK H J,MEYER J,ROTH S,et al. Growth and properties of few-layer graphene prepared by chemical vapor deposition[J]. Carbon,2010,48:1088.
[47] KIM K,LEE Z,REGAN W,et al. Grain boundary mapping in polycrystalline graphene[J]. ACS Nano,2011,5:2142.
[48] ZHAO L,RIM K T,ZHOU H,et al. Influence of copper crystal surface on the CVD growth of large area monolayer graphene[J]. Solid State Communications,2011,151:509.
[49] JAUREGUI L A,CAO H L,WU W,et al. Electronic properties of grains and grain boundaries in graphene grown by chemical vapor deposition[J]. Solid State Communications,2011,151:1100.
[50] SONG H S,LI S L,MIYAZAKI H,et al. Origin of the relatively low transport mobility of graphene grown through chemical vapor deposition[J]. Scientific Reports,2012,2:337.
[51] MATTEVI C,EDA G,AGNOLI S,et al. Evolution of electrical,chemical,and structural properties of transparent and conducting chemically

derived graphene thin films[J]. Advanced Functional Materials, 2009, 19:2577.

[52] GOMEZ-NAVARRO C, MEYER J C, SUNDARAM R S, et al. Atomic structure of reduced graphene oxide[J]. Nano Letters, 2010, 10:1144.

[53] KAWAMURA T, DASSARMA S. Phonon-scattering-limited electron mobilities in $Al_xGa_{1-x}As/GaAs$ heterojunctions[J]. Physical Review B, 1992, 45:3612.

[54] YAO Z, KANE C L AND DEKKER C. High-field electrical transport in single-wall carbon nanotubes[J]. Physical Review Letters, 2000, 84:2941.

[55] HWANG E H, DAS SARMA S. Acoustic phonon scattering limited carrier mobility in two-dimensional extrinsic graphene[J]. Physical Review B, 2008, 77:115449.

[56] CHEN J H, JIANG C, XIAO S D, et al. Intrinsic and extrinsic performance limits of graphene devices on SiO_2[J]. Nature Nanotechnology, 2008, 3:206.

[57] EFETOV D K, KIM P. Controlling electron-phonon interactions in graphene at ultrahigh carrier densities[J]. Physical Review Letters, 2010, 105:256805.

[58] HROSTOWSKI H J, MORIN F J, GEBALLE T H, et al. Hall effect and conductivity of InSb[J]. Physical Review, 1955, 100:1672.

[59] CASTRO E V, OCHOA H, KATSNELSON M I, et al. Limits on charge carrier mobility in suspended graphene due to fl exural phonons[J]. Physical Review Letters, 2010, 105:266601.

[60] MARIANI E, VON OPPEN F. Temperature-dependent resistivity of suspended graphene[J]. Physical Review B, 2010, 82:195403.

[61] MOORE B T, FERRY D K. Remote polar phonon-scattering in Si inversionlayers[J]. Journal of Applied Physics, 1980, 51:2603.

[62] FRATINI S, GUINEA F. Substrate-limited electron dynamics in graphene [J]. Physical Review B, 2008, 77:195415.

[63] LIN Y C, LU C C, YEH C H, et al. Graphene annealing: how clean can it be? [J]. Nano Letters, 2012, 12:414.

[64] MOSER J, BARREIRO A, BACHTOLD A. Current-induced cleaning of graphene[J]. Applied Physics Letters, 2007, 91:165513.

[65] GANNETT W, REGAN W, WATANABE K, et al. Boron nitride substrates for high mobility chemical vapor deposited graphene[J]. Applied Physics Letters,2011,98:242105.

[66] HONG X,POSADAS A,ZOU K,et al. High-mobility few-layer graphene field effect transistors fabricated on epitaxial ferroelectric gate oxides[J]. Physical Review Letters,2009,102:136808.

[67] LAFKIOTI M,KRAUSS B,LOHMANN T,et al. Graphene on a hydrophobic substrate:doping reduction and hysteresis suppression under ambient conditions[J]. Nano Letters,2010,10:1149.

[68] DEAN C R,YOUNG A F,MERIC I,et al. Boron nitride substrates for high-quality graphene electronics[J]. Nature Nanotechnology,2010,5:722.

[69] ZOMER P J,DASH S P,TOMBROS N,et al. A transfer technique for high mobility graphene devices on commercially available hexagonal boron nitride[J]. Applied Physics Letters,2011,99:232104.

[70] XUE J M,SANCHEZ-YAMAGISHI J,BULMASH D,et al. Scanning tunneling microscopy and spectroscopy of ultra-flat graphene on hexagonal boron nitride[J]. Nature Materials,2011,10:282.

[71] PETRONE N,DEAN C R,MERIC I,et al. Chemical vapor deposition-derived graphene with electrical performance of exfoliated graphene[J]. Nano Letters,2012,12:2751.

[72] RYU S,HAN M Y,MAULTZSCH J,et al. Reversible basal plane hydrogenation of graphene[J]. Nano Letters,2008,8:4597.

[73] TEWELDEBRHAN D,BALANDIN A A. Modification of graphene properties due to electron-beam irradiation[J]. Applied Physics Letters, 2009,94(1):013101.

[74] EMTSEV K V,BOSTWICK A,HORN K,et al. Towards wafer-size graphene layers by atmospheric pressure graphitization of silicon carbide [J]. Nature Materials,2009,8:203.

[75] JENA D,KONAR A. Enhancement of carrier mobility in semiconductor nanostructures by dielectric engineering[J]. Physical Review Letters, 2007,98:136805.

[76] RADISAVLJEVIC B,RADENOVIC A,BRIVIO J,et al. Singlelayer MoS_2 transistors[J]. Nature Nanotechnology,2011,6:147.

[77] PONOMARENKO L A, GEIM A K, ZHUKOV A A, et al. Tunable metal-insulator transition in double-layer graphene heterostructures[J]. Nature Physics, 2011, 7: 958.

[78] BOLOTIN K I, GHAHARI F, SHULMAN M D, et al. Observation of the fractional quantum Hall effect in graphene[J]. Nature, 2009, 462: 196.

[79] DU X, SKACHKO I, BARKER A, et al. Approaching ballistic transport in suspended graphene[J]. Nature Nanotechnology, 2008, 3: 491.

[80] TOMBROS N, VELIGURA A, JUNESCH J, et al. Large yield production of high mobility freely suspended graphene electronic devices on a polydimethylglutarimide based organic polymer[J]. Journal of Applied Physics, 2011, 109: 093702.

[81] NEWAZ A K M, PUZYREV Y S, WANG B, et al. Probing charge scattering mechanisms in suspended graphene by varying its dielectric environment[J]. Nature Communications, 2012, 3: 734.

[82] CHEN C Y, ROSENBLATT S, BOLOTIN K I, et al. Performance of monolayer graphene nanomechanical resonators with electrical readout[J]. Nature Nanotechnology, 2009, 4: 861.

[83] BUNCH J S, VAN DER ZANDE A M, VERBRIDGE S S, et al. Electromechanical resonators from graphene sheets[J]. Science, 2007, 315: 490.

[84] BAO W Z, LIU G, ZHAO Z, et al. Lithography-free fabrication of high quality substrate-supported and freestanding graphene devices[J]. Nano Research, 2010, 3: 98.

[85] BOLOTIN K I, SIKES K J, HONE J, et al. Temperaturedependent transport in suspended graphene[J]. Physical Review Letters, 2008, 101: 096802.

[86] TOMBROS N, VELIGURA A, JUNESCH J, et al. Quantized conductance of a suspended graphene nanoconstriction[J]. Nature Physics, 2011, 7: 697.

[87] KATSNELSON M I, NOVOSELOV K S, GEIM A K. Chiral tunnelling and the Klein paradox in graphene[J]. Nature Physics, 2006, 2: 620.

[88] KOTOV V N, UCHOA B, PEREIRA V M, et al. Electron-electron interactions in graphene: current status and perspectives[J]. Reviews of Modern Physics, 2012, 84: 1067.

[89] DU X, SKACHKO I, DUERR F, et al. Fractional quantum hall effect and insulating phase of Dirac electrons in graphene[J]. Nature, 2009, 462: 192.

[90] DEAN C R, YOUNG A F, CADDEN-ZIMANSKY P, et al. Multicomponent fractional quantum Hall effect in graphene[J]. Nature Physics, 2011, 7:693.

[91] FELDMAN B E, KRAUSS B, SMET J H, et al. Unconventional sequence of fractional quantum Hall states in suspended graphene[J]. Science, 2012, 337:1196.

[92] LEE D S, SKAKALOVA V, WEITZ R T, et al. Transconductance fluctuations as a probe for interaction-induced quantum Hall states in graphene [J]. Physical Review Letters, 2012, 109:056602.

[93] BAO W, VELASCO J J, ZHANG F, et al. Evidence for a spontaneous gapped state in ultraclean bilayer graphene[J]. Proceedings of the National Academy of Sciences of the United States of America, 2012, 109: 10802.

[94] BANERJEE S K, REGISTER L F, TUTUC E, et al. Bilayer Pseudospin field-effect transistor (BiSFET): a proposed new logic device[J]. IEEE Electron Device Letters, 2009, 30:158.

[95] STORMER H L. Nobel lecture: the fractional quantum hall effect[J]. Reviews of Odern Physics, 1999, 71:875.

第10章 双层石墨烯的电子传输

本章研究了双层石墨烯的电子传输,并讨论了相关的基本物理现象和概念;描述了模型汉密顿体系和产生能带间隙的方法,讨论了传输特性,包括 p-n 结的电导、自洽波恩相近和在偏置双层石墨烯中 RKKY(Ruderman-Kittel-Kasuya-Yosida)的相互作用;讨论了关于悬浮双层石墨烯和 SiC 上新一代双层石墨烯样本的研究情况,并探究了多体效应在这些体系中的作用,讨论了对称和不对称的电荷密度通道中的集体模式,检测了作为准粒子理论的基本量的有效质量,深层研究了双层石墨烯中电荷的可压缩性。

10.1 引　言

近几年,石墨烯的科学研究具有飞速有时近乎是革命性的发展。毋庸置疑,这些成功大部分源于实验中突破性的进展。石墨烯是一种单原子二维晶体,具有一系列特殊的性质。双层石墨烯由两个单层的石墨烯彼此分开小段距离构成,一般采用将薄石墨的机械剥离或碳化硅热解而得到,由于其具有许多独特的电子特性已经引起了人们的广泛关注。在低能量下,双层石墨烯准粒子表现为大的手性费米子,展现出许多有趣的特性,包括双层石墨烯悬浮减少无序性引起的极弱磁场下的破缺对称态以及量子霍尔机制下的反常激子凝聚。

本章内容是双层石墨烯的电子传输,重点从理论和实验的角度介绍相关的基础物理学以及概念性的问题。没有深入阐述能带结构特性,只在介绍传输现象中有所提及。声子散射、应变和褶皱的影响以及双层石墨烯中的光电导率也没有涉及,而是详细论述了石墨烯的电学性能[1-4]。现在我们详细阐述一下本章的组成。在讨论双层石墨烯之前,我们引入哈密顿模型,尤其是紧束缚法和它在量子力学领域的"特征值"。然后讨论通过调整外部垂直电场打开能带间隙的方法,并计算体系的电荷不平衡性。本章主要阐述了几乎无相互作用的双层石墨烯的传输性能,重点强调了双层石墨烯的导电性及其显著效果。

等离子体振荡是常见的电子液体高频集合密度振荡,常出现在金属和半导体中。人们已经认识到它涉及基础物理和应用物理等各领域的重要性。比如,它在等离子体光子学及场效应晶体管远红外探测领域的研究中就起到了重要作用。作为集体激发,由于电子之间的库仑作用,等离子体模式是电子相关的直接结果。等离子体模式的实验检测也已实现,并可用来确定石墨烯中电子的动力学行为。因此,了解双层石墨烯中电子气体的相关性与屏蔽性能是非常有必要的。10.4节中,讨论了四带模型中的非相关响应函数并采用随机相位近似得到等离子体模式。

因为热力学量与状态方程密切相关,所以它可以很好地描述相关电子体系中的多体效应,如电子可压缩性。在10.4节中,我们计算了四带连续模型下双层石墨烯的零度电子可压性。通过使用随机相位近似,包含了Hartree-Fock相关贡献外的基态能量。因为双层石墨烯由两个以小距离分离的单层石墨烯组成,所以层间的电子-电子相互作用对于体系的物理性质是至关重要的。因此,双层石墨烯中的多体效应已成为多项研究的共同课题。关于接近电荷中性点的相互作用效应的研究也已备受关注。因为在电荷中性点的双层石墨烯具有许多有趣的不稳定性,包括子晶格赝自旋铁磁性,一种可引发自发反演对称性破缺的轨道次序。我们重点介绍描述双层石墨烯体系的这些不稳定性的最新的实验性传输测试。

10.2 双层石墨烯的历史发展

近几年,随着由碳原子紧密堆叠而成的二维蜂窝晶格结构的单层石墨烯[5],以及由两个按照伯纳尔(AB)堆叠排列的石墨烯层构成的双层石墨烯[6-8]成功制备,石墨烯纳米带的传输特性引起了人们极大关注,尤其是低能载流子激发行为。[9-11]理想的单层石墨烯是在狄拉克点具有零态密度的无间隙半金属。在狄拉克点附近的低能量电子激发呈线性分散,可由有效无质量的狄拉克哈密顿来描述。另外双层石墨烯中的低能量电子具有二次离散关系。对于单层和双层石墨烯来讲,波函数都是由A和B两个子格组成,引起电荷载流子的手性性质。因此,石墨烯中的载流子具有手性特征,这使得石墨烯具有很多有趣特性,比如Berry相位效应,单层石墨烯中是π而双层石墨烯是2π。虽然双层石墨烯本来是零间隙半金属,但当施加栅极电压时,它具有非常有趣的特性,可以使双层石墨烯成为可调能带间隙的半导体。[12-13]能带间隙决定了场效应晶体管及二极管的阈值电压和开关比。因此,双层石墨烯在纳米电子领域的应用比单层石墨烯更有优势[14-17]。

10.2.1 双层石墨烯的鉴定

原子力显微镜(AFM)是用来测量衬底上方石墨烯相对高度的常用方法,从而确定样本的层数。系统的 AFM 测试研究[18]表明少层石墨烯的层间距大约是 0.35 nm,是由所测直方图峰值的高度获得的。隧道电子显微镜是另一个用于测试层数多达六层的样品中堆叠缺陷的方法。拉曼光谱是一种与样本中声子模式有关的测试手段,用于区分石墨烯中层数的可靠手段。石墨烯和其他石墨材料的拉曼光谱中三个最明显的峰是在 1 580 cm^{-1} 处的 G 带、2 680 cm^{-1} 处的 2D 带和 1 350 cm^{-1} 处由无序引发的 D 带。2D 峰的谱线形状以及它的强度与 G 峰强度之比,可以用来表征石墨烯的层数。单层石墨烯具有以下特征,极其尖锐的对称洛仑兹 2D 峰,且 2D 峰强度比 G 峰的 2 倍还强。随着层数的增加,2D 峰变得越来越宽,对称性下降,强度降低。所测得的拉曼光谱[19-21]表明单层石墨烯中双重简并的 2D 峰在双层石墨烯里分裂成四个非简并模式。这个分裂导致 2D 峰变宽并向高频率轻微偏移。

在单层和双层石墨烯中都有一个奇怪的现象,就是出现反常量子。[5,22-24] 单层石墨烯中无质量手性狄拉克载流子引发了霍尔平台特性,描述为 $\sigma_{xy} = \pm\sigma_0(N+1/2)$,其中,$N$ 是朗道能级系数,$\sigma_0 = 4\ e^2/h$。因数"4"源于谷和自旋简并。在无掺杂的双层石墨烯中,霍尔平台表现为 $\sigma_{xy} = \pm\sigma_0 N$。第一个 $N=0$ 的平台消失,表明双层石墨烯在中性点具有金属性,而双层石墨烯中标准的量子霍尔效应可以通过施加阈值电压得到恢复。Feldman 等通过实验已经观察到磁场存在下完全量子化的量子霍尔态,以及中间填充因数(如 0、±1、±2 和 ±3)时的破缺对称态[25]。

10.2.2 双层石墨烯的哈密顿模型

这里我们介绍一下双层石墨烯的紧束模型。图 10.1 所示是双层石墨烯晶体结构图,在 xy 平面内两个单层晶格彼此分离,四个原子以上面 A 子晶格位于下面 B 子晶格正上方的形式存在于单胞中,这些原子对间形成层间二体键合。另外两个原子在另一层没有配对原子。我们假定碳原子的 sp^2 杂化电子是惰性的,考虑来自 π 键的 2p$_z$ 电子。波函数可以写成四部分旋量 ψ_{A_1}、ψ_{B_1}、ψ_{A_2} 和 ψ_{B_2}。因此,双层石墨烯的传递积分矩阵就是一个 4×4 的矩阵:

$$H = \begin{pmatrix} \varepsilon_{A_1} & -\gamma_0 f(k) & \gamma_4 f(k) & -\gamma_3 f^*(k) \\ -\gamma_0 f(k) & \varepsilon_{B_1} & \gamma_1 & \gamma_4 f(k) \\ \gamma_4 f^*(k) & \gamma_1 & \varepsilon_{A_2} & -\gamma_0 f(k) \\ -\gamma_3 f(k) & \gamma_4 f^*(k) & -\gamma_0 f^*(k) & \varepsilon_{B_2} \end{pmatrix} \quad (10.1)$$

其中,$f(k) = \sum_{i=1}^{3} e^{ik\sigma_i}$,$\delta_i$是相对于A原子最近的三个B原子的位置。层内最近邻跳跃能量是$\gamma_0 = 3.16$ eV,层间由上层顶部次晶格A和不同层的B之间的跳跃能量$\gamma_1 = 0.39$ eV 和 $\gamma_3 = 0.315$ eV,表示两层间的非顶部次晶格A和B之间的跳跃能量[13]。

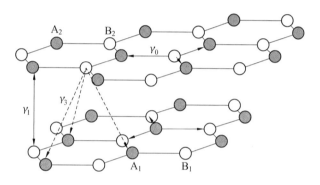

图 10.1　包含所有耦合能量的双层石墨烯的晶体结构 3D 示意图

底层原子 A_1 和 B_1,顶层 A_2 和 B_2 分别如图中灰色和白色圆所示。原胞矢量可定义为$(1+\sqrt{3})b/2$,其中 b 是邻近晶胞之间的距离,$b = 0.246$ nm

最近邻层之间的另一个跳跃能量 $\gamma_4 = 0.04$ eV,相比于 γ_0 是非常小的,可以忽略。在普遍情况下,四个原子位置上的原位能量 ε_{A_l} 和 ε_{B_l},$l = 1$ 或2,不再相等。原位能量由描述层间不对称的独立参数、每层内两原子之间的能量差异以及二体处和非二体之间的能量差异组成。

为了描述接近狄拉克点的电子特性,需要扩大狄拉克点的动量 K_+ 和 K_-。为了达到这个目的,引入 $p = \hbar k - K_\pm$ 并采用 p 扩展函数 $f(k)$,得到 $f(k) = f_p \approx -\sqrt{3}a(sp_x - ip_y)/2\hbar$,其中 $s = \pm$ 表示谷标记。这个表达式只有在接近狄拉克点时才有效。因此,可将等式(10.1)简化,得到四个谷简并带色散关系。只考虑原位能量下由垂直电场产生的层间非对称 $\Delta = \varepsilon_{A_1} = \varepsilon_{B_1} = -\varepsilon_{A_2} = -\varepsilon_{B_2}$,色散关系可写成 $E = \pm E_l(p,s)$,其中 $l = 1$ 或2。

$$E_1(p,s) = \sqrt{\frac{\gamma_1^2}{2} + \frac{\Delta^2}{4} + (\gamma_0^2 + \gamma_3^2/2)f_p^2 + (-1)^l \sqrt{g(p)}}$$

$$g(p) = [\gamma_1^2 + \gamma_3^2 f_p^2 + 4\Delta^2]f_p^2 + 2s\gamma_1\gamma_3 f_p^3 \cos(3\varphi) + \frac{(\gamma_1^2 - \gamma_3^2 f_p^2)^2}{4}$$
(10.2)

其中,$\phi = \arctan(p_y/p_x)$。将等式(10.2)扩展到动量的领头阶,并假定 $\Delta \ll \gamma_0$,得

$$E_l(p,s) = \left(\Delta - \frac{\hbar^2 p^2}{m} + \frac{\hbar^4 p^4}{2m^2\Delta}\right)$$
(10.3)

其中,双层有效质量 $m = \gamma_0/2\nu_F$ 或者大约 $0.03 m_e$,这与很小的有效质量相一致。非相互作用的石墨烯的狄拉克费米速率定义为 $\nu_F = 3\gamma_0 a/2$。

10.2.3 偏置双层石墨烯的能带间隙

最近邻耦合可能对低能量极限产生影响。二体键合里不包含原子位置之间的直接耦合,导致低能量色散关系变成各向异性,又叫作三角翘曲。该现象在角分辨光电子能谱中可观测到,并对双层石墨烯弱局域化有着重大影响[26,27]。

偏置双层石墨烯在 $p = \sqrt{2}\Delta/(\hbar\nu_F)$ 能带间隙是 2Δ,对于 $\Delta = 0$ 的情况,双层石墨烯是无间隙半导体,p 较小时具有抛物线的色散关系(图10.2)。抛物线色散关系仅应用于 p 值较小并满足 $\hbar\nu_F p \ll \gamma_0$。然而却存在一种单层石墨烯相类似线性色散关系。

双层石墨烯由于A/B晶格对称性表现为手性系统[22-24]。

值得注意的是,两层的破缺反演对称可能会引起非零能带间隙。有趣的是,通过对样本施加垂直电场会引起的连续可调能带间隙势垒。[28-30] 栅极化双层石墨烯的电传输测试表明,只有温度低于1 K时才有绝缘行为。栅极化双层石墨烯中具有可调的电子能带间隙已由实验得到证实。[31] 运用双栅极双层石墨烯场效晶体管红外光谱,可以得到一个栅极控制的连续可调的能带间隙,宽度可达250 meV。可调能带间隙的证据表现在从零到中红外的光谱区域。此外,该静电能带间隙的可控表明石墨烯在纳米电子和纳米光子器件的应用潜力。能带间隙可以采用光电子发射、磁传输、红外光谱和扫描隧道谱进行研究。

在 $\Delta < 250$ meV 整个能带间隙范围,研究了双层石墨烯能带对施加的位移场参数依赖性。[31] 不同理论模型和实验结论之间的对比说明石墨烯自屏蔽至关重要。这种情况也存在于自洽紧束缚模型[32]和密度泛函理论计算[33]。密度泛函计算预测的能带间隙稍小于紧束缚模型。通过局部密度泛函计算所获得能带间隙偏低是很普遍的。

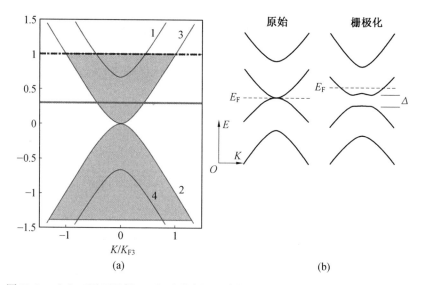

图10.2 (a)双层石墨烯$2p_z$电子形成的连续模型能带结构,根据掺杂水平,二维电子气体具有一个或两个导带费米表面。(b)由两层反演对称性的打破引起的非零带空隙散射关系

双层石墨烯的有效低能哈密顿可以由$1/\gamma_1$来表示。换句话说,有效排除二体键合所涉及的原子位置的是有效哈密顿。低能有效哈密顿可以表示为

$$H_{\text{eff}} = -\frac{1}{2m}(\boldsymbol{\sigma p})\sigma_x(\boldsymbol{\sigma p}) + s\Delta\sigma_z \qquad (10.4)$$

其中,$\boldsymbol{\sigma}$是波利矩阵。在双层石墨烯中,改变栅极电压,进而控制电子密度n和层间非对称的不同势能Δ。换句话说,能量的渐变ΔV与层密度n_1和n_2有关,$n = n_1 + n_2$,并且层密度最终取决于Δ。使用简化的双带有效哈密顿H_{eff},计算特征值和本征函数,层密度可以通过圆形费米面的积分来确定:

$$n_l = 4\int \frac{\mathrm{d}p}{(2\pi\hbar)^2} |\psi_l(p)|^2 \qquad (10.5)$$

其中,l代表层系数。用高斯法则和静电法则,独立层密度为[28,34]

$$n_l = \frac{n}{2} \mp \frac{n_0\Delta(n)}{2\gamma_1}\ln\left(\frac{n}{2n_0} + \frac{1}{2}\sqrt{\frac{n^2}{n_0^2} + \frac{\Delta^2(n)}{\gamma_1^2}}\right) \qquad (10.6)$$

非对称参数与密度的依赖关系为

$$\Delta(n) = -\frac{ed\varepsilon}{L\varepsilon_b}V_t + \frac{e^2 d}{\varepsilon_b}n_2 \qquad (10.7)$$

其中,V_t是距离双层石墨烯为L处、介电常数为ε的区域的外部栅极电压

双层石墨烯可以看作是被介电常数为 ε_b、距离为 d 的区域所分离的两个平行导电板。我们引入参数 $n_0 = \gamma_1^2/(\pi\hbar^2 v_F^2)$。层密度和能带间隙可以通过数值自洽性来获得需要的准确度。

在绝缘碳化硅上沉积可合成双层石墨烯薄膜,用角分辨光电子能谱研究其电子能带结构。[35] 通过选择性调整每层的载流子浓度,库仑电位的变化可以控制价带和导带之间能带间隙。最重要的是 K 点出现明显能带间隙变化。采用紧束模型计算能带间隙的变化。远离 K 点,能带间隙通常比预期的小,因为向 π 和 π^* 之间的能带间隙延伸的角比向模型延伸的角要尖锐得多。测量的双层石墨烯中的 π^* 态与紧束型带不同,低于 E 大约 200 meV,在该处带中可以观察到轻微扭结。

对于偏置体系,两个层具有不同的静电电位,相关的能量差异用 eV 表示。Ohta 等进行了理论计算[33]并将双层石墨烯中由电子密度 n 决定的电位(eV)与角分辨光电子能谱测量结果进行了比较。[35] 通过钾原子沉积到真空侧引发的电子密度 n,可用来改变整个密度。值得注意的是,在未屏蔽的情况下得到的结果不符合实验数据。而采用自洽过程屏蔽的结果与实验数据非常吻合。

10.3 双层石墨烯体系中的传输性能

10.3.1 Klein 隧穿

石墨烯具有一些非同寻常的特性,如 Klein 隧穿描述的不规则隧道效应。Klein 隧穿是一种通过 p-n 结符合手性能带态的隧穿[36-37],它和能量动量线性色散关系在石墨烯片中已经得到证实[38-41]。Klein 隧道预测了手性无质量载流子以一个概率穿过一个具有高静电电位的势垒,而无须考虑势垒的高度和宽度,而对于传统的非相对论质量载流子隧穿,其隧穿概率随着势垒高度的增加呈指数衰减,并受势垒特征的影响。[42-45] 当我们认识到狄拉克形式同时适用正能态和负能态时,Klein 隧穿就比较容易理解了。

双层石墨烯中的载流子具有抛物线形的能量谱,在零能量处具有有限态密度,这与传统的非相对论电子相似。另外,这些准粒子也是手性的,可由旋量波函数来描述。最令人惊奇的是,势垒外部电子进入势垒内空穴,反之亦然。这种传输与单层石墨烯的情形明显不同。一些入射角存在明显的传输共振,并且 T 也接近统一。与正常入射的狄拉克费米子的完美传输不同,质量手型费米子总是在角度接近 $\phi = 0°$ 的情况发生完全反射。完全反射可以看成是 Klein 悖论的一个体现,因为它来自于电荷共轭对称。对于单层石墨烯来说,势垒界面的电子波函数与具有赝自旋同向的空穴相关波函数一致。而对于双层石墨烯,电荷共轭需要一个波矢为 k 的电子进入波矢为 ik(而不是 $-k$)空穴,这是势垒内部的一个短暂波。换句话说,当粒子能量 E 远小于势垒势能 V_0 时,双层石墨烯中的传输可通过 $T(\phi) = E/V_0 \sin^2(2\phi)$ 进行简化。图 10.3 为质量狄拉克粒子隧穿势垒的散射情况。正如所解释的,允许一定值的入射角 ϕ 完全通过。

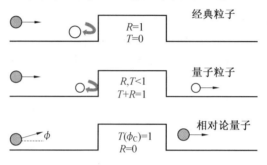

图 10.3　质量狄拉克粒子隧穿势垒的散射情况

一个正常入射经典粒子,当势垒高度大于粒子能量时,它的传输完全被禁止,因此量子粒子的概率是有限的,传输随着势垒的高度和宽度呈指数下降。对于质量狄拉克粒子,特定入射角 ϕ 的传输是必然发生的,与势垒的几何结构无关。这一点可以通过赝自旋守恒来理解

10.3.2　p - n 连接点的电导

在单层和多层石墨烯中所观察到的奇怪现象之一是反常量子霍尔效应。[5,22-24] 单层石墨烯中的无质量手性狄拉克电荷载流子形成了霍尔平台,表示为 $\sigma_{xy} = \pm \sigma_0 (N + 1/2)$,其中,$N$ 是朗道能级系数;$\sigma_0 = 4\,e^2/h$。因数 4 来自谷和自旋简并。在无掺杂的双层石墨烯中,霍尔平台表示为 $\sigma_{xy} = \pm \sigma_0 N$。$N = 0$ 的首个霍尔平台消失了,表明双层石墨烯在中性点处具有金属性,而双层石墨烯中的标准量子霍尔效应可通过施加栅极电压而恢复。Feldman 等研究发现,由于磁场的存在而完全量子化的量子霍尔态,以及中间填充因子(如 0、±1、±2 和 ±3)下的破缺对称态[25]。

在完美纳米带中,由于电子态横向约束的存在,通过子带的电子传输显示了以 $G_0 = 2e^2/h$ 为单位的电导的量子化[46]。最近,通过计算得到了不

存在无序的单层和无偏置双层石墨烯纳米带在恒定的垂直磁场下的零温度电导[47]。对于锯齿形边缘单层石墨烯纳米带电导为 $2(n+1/2)G_0$,扶手椅形边缘时电导为 nG_0。另外,对于双层石墨烯纳米带,锯齿形边缘的电导可量子化为 $2(n+1)G_0$,扶手椅边缘的电导为 nG_0,其中 n 是整数。

多个团队已经通过理论和实验探究了在双极性(p-n)和单极性(n-n 或 p-p)下的石墨烯结的量子霍尔效应和量化传输情况[48-50]。Long 等[48]用 Landauer-Buttiker 形式表明,在磁场下原位无序引起单层石墨烯 p-n 结传输性能的提高。另外,对于 n-n 结,在较宽的原位无序强度范围内,存在最低平台。研究还表明在无序强度的特定范围下,会出现新的平台(如 $G=3e^2/h$ 和 e^2/h)[48,51],有些情况通过实验也可以观察到。

对高品质双层石墨烯 pnp 结的传输进行了测试,其中无能带间隙体系的电子迁移率达到 10 000 cm^2·V^{-1}·s^{-1},能带间隙体系的开关比达到 20 000。[52] 此外,在不同掺杂区域之间发现了由量子霍尔边缘态平衡引起的分数量子霍尔平台。[52] 于是,界面的边缘态混合导致电导出现平台。

具有伯纳尔堆垛(AB)的双层石墨烯带,作为导体与左、右两侧的导线相连接。伯纳尔堆垛的双层石墨烯使导线结构化。哈密顿模型可表示为

$$H = H_{\text{center}} + H_L + H_R \tag{10.8}$$

其中, H_{center}、H_L 和 H_R 分别是中心区和左边线及右边线的哈密顿。左、右两条带被认为是完美的半无限双层石墨烯纳米带。我们认为最近邻紧束哈密顿在晶格每个位置都有一个 π 轨道。有磁场时双层石墨烯的有效单体哈密顿函数表示如下:

$$H = -\gamma_0 \sum_{l\langle i,j\rangle}(e^{i\phi_{i,j}}a^+_{l,i}b_{l,j} + h.c) - \gamma_1 \sum_i (a^+_{1,i}b_{2,i} + h.c) - \\ \gamma_3 \sum_{\langle i,j\rangle}(e^{i\phi_{i,j}}b^+_{1,j}a_{2,j} + h.c) + \sum_{l,i} v_l(a^+_{l,i}a_{l,i} + b^+_{l,i}b_{l,i}) + \\ \sum_{l,i}[w_i + (-1)^l\Delta](a^+_{l,i}a_{l,i} + b^+_{l,i}b_{l,i}) \tag{10.9}$$

其中, $a^+_{l,i}$、$a_{l,i}$($b^+_{l,i}$ 和 $b_{l,i}$)分别是层数 $l=1$ 或 2 上第 i 个位置的子晶格 A(B) 的产生和湮灭运算符。在外部垂直磁场 B 下,跳跃积分获得 Peierls 相因子 $e^{i\varphi_{i,j}}$,其中 $\varphi_{i,j} = \int_i^j \mathbf{A}\cdot\mathrm{d}(l/\varphi_0)$,磁通量子 $\varphi_0=\hbar/e$。采用朗道规范使 $\mathbf{A}=(-By,0,0)$。只在中心区域施加磁场,左(右)导线上的 $v_{l,i}$ 降低到偏压 $E_L(E_R)$,并通过栅极电压来控制。静电电位从右带变到左带,而且是线性变化 $v_l = k(E_R - E_L)/(M+1) + E_L$, $k=1,2,\cdots,M$,其中 M 是中心区域

的长度。中心区域的尺寸(例如导体)表示为 $4N \times M$ 个原子。

因为与霍尔平台的形成有关,所以有必要考虑度大于磁性长度的纳米带宽,$l_B = \sqrt{\hbar/eB}$。我们假定 $\varphi = 0.01$ 与 $l_B \approx 1.5$ nm 对应,该值小于所讨论的纳米带宽度 $L_y(N=45) \approx 10$ nm 和长度 $L_x(M=21) \approx 5.5$ nm。

无偏置的干净 zBGNR 的电导与 E_R 在不同带长、没有磁场($\phi=0$)和有磁场($\phi=0.01$)情况下的函数关系,如图 10.4 所示。没有磁场时,n-n 区域($E_R < 0$)的电导由于纳米带的横向几何限制而被量化,可描述为 $G = 2(n+1)G_0$,zBGNR 的最小电导为 $2G_0$[47]。此外,电导在低 E_R 值时与带长无关。在 $\phi=0$ 时 n-n 区域中看到,因为二次色散关系,平台之间的能量差异是不等的(而在单层石墨烯里是相等的)。电导台阶(step)宽度与能量谱的连续模式之间的能量尺度有关。因此电导对于 E_L 值敏感,且容易受平台的数量增加的影响,而平台数量随着偏置电压的增加而增加,如 $|E_L - E_R|$。对于 $E_R < E_L$ 的情况,不存在与单层石墨烯相同的平台[48]。在 n-p 区域,$E_R > 0$,电导来源于 n 和 p 区域之间的手性电荷载流子隧穿,且电导所具有的平台数少于 n-n 区域。该区域中,电导随带长度 M 的增加而降低,这是因为色散中心数量增加。

10.3.3 电导率的自洽玻恩近似

正如之前讨论的,在双层石墨烯中能量散射还包括 k 线性三角翘曲效应,该效应用 γ_3 参数表示,能引起低能量光谱各向异性。现在我们采用自洽玻恩近似来计算无磁场情况的对角线电导。狄拉克点附近的传输性能和电导率已有实验研究[54-57],并预测最小电导率[58-60]为 $8e^2/(\pi^2\hbar)$,是单层石墨烯中的 2 倍,当考虑三角翘曲时[59-60],电导率为 $24e^2/(\pi^2\hbar)$。有效哈密顿函数为

$$H = \frac{\hbar^2}{2m}\begin{pmatrix} 0 & \pi_-^2 \\ \pi_+^2 & 0 \end{pmatrix} - \frac{\hbar^2 k_0}{2m}\begin{pmatrix} 0 & \pi_+ \\ \pi_- & 0 \end{pmatrix} \quad (10.10)$$

其中,$\pi_\pm = k_x \pm k_y$,$\boldsymbol{k}$ 是狄拉克点的波矢;$k_0 = 2\sqrt{3}\gamma_3\gamma_1/(3a\gamma_0^2)$。很容易得到能量本征值及其对应的特征向量。[58] 在自洽玻恩近似中,无序平均 Green 函数的自能量 $\langle G_{\alpha,\alpha'} \rangle$ 可表示为

$$\Sigma_{\alpha\alpha'}(\varepsilon) = \sum_{\alpha_1\alpha'_1} \langle U_{\alpha,\alpha_1} U_{\alpha'_1,\alpha'} \rangle \langle G_{\alpha_1,\alpha'_1} \rangle \quad (10.11)$$

其中,$\alpha = (jks)$;U 是电子和杂质之间的相互作用;$\langle \cdots \rangle$ 表示杂质结构的平均情况。电导率可以由 Kubo 公式计算得到:

$$\sigma(\varepsilon) = \frac{\hbar e^2}{2\pi} \Re e Tr[v_x \langle G^R \rangle v_x^{RA} \langle G^A \rangle - v_x \langle G^R \rangle v^{RR} \langle G^R \rangle] \qquad 10.12$$

其中,$v_x^{RA(RR)} = v(\varepsilon + i0, \varepsilon \mp i0)$,满足 $v(\varepsilon, \varepsilon') = v_x + \langle UG(\varepsilon) v_x G(\varepsilon') U \rangle$,推迟和超前的 Green 公式可以由 $G^{R(A)}(\varepsilon) = (\varepsilon - H \pm i0)^{-1}$ 来定义,H 是包含无序电位的哈密顿。最终采用自洽方式计算方程组得到电导率。

对于短程散射来说,高能量 $|\varepsilon| > \varepsilon_0 (\varepsilon_0 = \hbar^2 k_0/(2m))$ 的电导率为

$$\sigma(\varepsilon) = \frac{4e^2}{\pi^2 \hbar W} \left(\frac{|\varepsilon|}{\varepsilon_0} + 1 \right) \qquad (10.13)$$

在低能量 $|\varepsilon| < \varepsilon_0$ 时,电导为

$$\sigma(\varepsilon) = \frac{8e^2}{\pi^2 \hbar} \left(1 + \frac{1}{2W} \right) \qquad (10.14)$$

说明电导率是通用的,在大范围无序的限制下也不会消失。其中,W 是无序强度。

通过这种方法计算本征能量,可以容易看出,在高能区 $\varepsilon > 0.25\varepsilon_0$,存在一个单一的三角扭曲费米线,而在低能区 $\varepsilon < 0.25\varepsilon_0$,费米线分裂为对应每个狄拉克点的四个分离凹陷。因此,电导率变成式(10.13)和式(10.14)中得到的值的四倍。

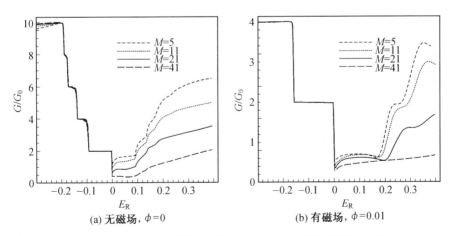

图 10.4 $E_L = -0.2$、$\Delta = 0$ 时不同长度纯净的锯齿形双层石墨烯纳米带的电导与 E_R 的函数关系[53]

10.3.4 双层石墨烯的 Ruderman-Kittel-Kasuya-Yosida (RKKY) 相互作用

石墨烯研究中的基本问题之一是非极性材料上两个局部磁矩之间的间接相互作用。载流子介导的交互作用被认为是 Ruderman-Kittel-Kasuya-Yosida (RKKY) 相互作用[61-62]，RKKY 在许多电子体系(自旋玻璃和合金)的磁性有序中起重要作用。相互作用的长程行为的两个主要特征，可以由电子气体的交换积分 J 测得，电子气体随着力矩之间距离 R 而发生摆动(出现符号和大小变化)，表现为铁磁性(FM)或反铁磁性(AFM)有序，也随着 R 发生衰减。两个主要特征根据主体材料尺寸和能量色散表现为不同的函数形式。

对于单层石墨烯，RKKY 相互作用已得到广泛研究。[63-70] 石墨烯的两个独有特点是狄拉克锥和引起 RKKY 相互作用的线性。对于不掺杂的石墨烯的两个主要特征观点是一致的。第一，不同于普通二维金属长程限制的 R^{-2} 衰减，在无掺杂石墨烯中长程有序按 R^{-3} 衰减，表现出"$1+\cos[(K-K')R]$"类型的振荡，该振荡存在一个受方向 R 影响的附加相位因子，相同次晶格的第二力矩表现出 FM 相互作用和相对次晶格的 AFM 相互作用，这是为了符合粒子空穴对称性。掺杂石墨烯的 RKKY 相互作用表现出类似于具有另一个狄拉克锥振荡因子的普通二维电子气长程行为。

双层石墨烯的 RKKY 相互作用在一些研究中已经得到解决[69,71,72]。Killi 等利用 Anderson 杂质模型的均场理论研究了双层石墨烯吸附原子局域磁矩的形成。研究发现局域力矩之间的 RKKY 相互作用可以通过调整化学电位或电场而发生改变，以至于引起双层石墨烯的能带结构变化。近期报道了关于半填充的双向晶格上的 RKKY 相互作用对称性的讨论[69]，发现未掺杂双层石墨烯内 RKKY 相互作用强度受距离影响。此外，Jiang 等[72]研究了多层石墨烯体系的 RKKY 相互作用，发现多层的厚度会对相互作用产生复杂的影响，在长程机制的双层石墨烯中相互作用的耦合随 R^{-2} 下降。这些研究只考虑了半填充的双层石墨烯的 RKKY 相互作用。掺杂情况的费米能量不再是微乎其微，垂直电场的影响以及能带间隙均会影响系统中的 RKKY 相互作用。

因为 RKKY 相互作用是以大部分系统内的巡游电子作为媒介，在偏置和掺杂的双层石墨烯中的电子直接影响了 RKKY 相互作用。[73] 为了描述狄拉克点附近的电子特性，采用垂直电场下的双带连续模型，哈密顿为

H_0。在线性反应理论中,RKKY 相互作用强度 J 由两步组成:第一步,由于位于原点的一阶力矩 S_1,人们运用非扰动态 $|\Psi^0\rangle$ 的周电子气(主体材料)的 Lippmann-Schwinger 方程 $|\Psi\rangle = |\Psi^0\rangle + G^0V|\Psi\rangle$,计算扰动态 $|\Psi\rangle$,第二步,位于晶格 R 处的二阶矩 S_2 存在时,出现自旋极化气体的能量的一阶修正,也就是 $E(R) = \langle\Psi|V(R)|\Psi\rangle$。这里,$G^0(E) = (E + i\eta - H_0)^{-1}$ 是无扰推迟格林函数。因此,相互作用的能量可以表示为

$$E(R) = J(0,\boldsymbol{R})S_1S_2 \qquad (10.15)$$

其中,RKKY 相互作用 $J(0,\boldsymbol{R})$ 与静态磁化系数 $\chi(0,R)$ 成比例,即

$$J(0,\boldsymbol{R}) = \frac{\lambda^2\hbar^2}{4}\chi(0,\boldsymbol{R}) \qquad (10.16)$$

其中,静态磁化系数描述了扰动 δV 和密度 δn 变化之间的比例,$\chi(\boldsymbol{r},\boldsymbol{r}') = \delta n(\boldsymbol{r})/\delta V(\boldsymbol{r}')$,可以得出 $\chi(\boldsymbol{r},\boldsymbol{r}')$ 为:

$$\chi(\boldsymbol{r},\boldsymbol{r}') = -\frac{2}{\pi}\int_{-\infty}^{\varepsilon_F}dE\Im m[G^0(\boldsymbol{r},\boldsymbol{r}',E)G^0(\boldsymbol{r}',\boldsymbol{r},E)] \qquad (10.17)$$

其中,$G^0(\boldsymbol{r},\boldsymbol{r}',E) = \sum_\lambda \psi_\lambda(\boldsymbol{r})\psi_\lambda^*(\boldsymbol{r}')(E + i\eta - E_\lambda)^{-1}$,是单自旋通道时推迟格林函数的实空间矩阵元,用 λ 标记全套的 H_0 本征态。积分后面的因数 2 代表两个自旋轨道。公式(10.17)可由电荷密度和扰动 GF 之间的关系来确定:

$$n(\boldsymbol{r}) = \sum_\lambda^{OCC}|\psi_\lambda(\boldsymbol{r})|^2 = -\frac{2}{\pi}\int_{-\infty}^{\varepsilon_F}dE\text{Im}G(\boldsymbol{r},\boldsymbol{r},E)$$

通过近似 Dyson 方程 $G = G^0 + G^0VG^0$,可以获得由扰动 $\delta V_\beta(\boldsymbol{r}')$ 引起的电荷差异 $\delta n(\boldsymbol{r})$。

等式(10.17)中的磁化系数的表达式可以很容易拓展到几个晶格自由度的体系,如双层石墨烯。用类似于以自旋密度泛函形式定义磁化系数的方式,密度的变化定义为

$$\delta n_{\alpha\beta}(\boldsymbol{r}) = n_{\alpha\beta}(\boldsymbol{r}) - n_{\alpha\beta}^0(\boldsymbol{r}) = \sum_{\alpha'\beta'}\int d\boldsymbol{r}'\chi_{\alpha\beta,\alpha'\beta'}(\boldsymbol{r},\boldsymbol{r}')V_{\alpha'\beta'}(\boldsymbol{r}') \qquad (10.18)$$

其中,α、β 代表晶格指数(双层石墨烯的 A1、B1、A2 和 B2),满足封闭关系 $\sum_\xi\int|\boldsymbol{r},\alpha\rangle\langle\boldsymbol{r},\alpha|d\boldsymbol{r} = 1, \xi = \alpha,\beta,\cdots$ 扰动势可定义为 $V_{\alpha\beta}(\boldsymbol{r},\boldsymbol{r}') = V_{\alpha\beta}(\boldsymbol{r})\delta(\boldsymbol{r} - \boldsymbol{r}')$。按照类似步骤,我们可以定义衍生的磁化系数为

$$\chi_{\alpha\beta,\alpha'\beta'}(\boldsymbol{r},\boldsymbol{r}') = -\frac{2}{\pi}\int_{-\infty}^{\varepsilon_F}dE\Im m[G_{\alpha\alpha'}^0(\boldsymbol{r},\boldsymbol{r}',E)G_{\beta\beta'}^0(\boldsymbol{r}',\boldsymbol{r},E)]$$

$$(10.19)$$

如果只限定对对角线外部势的响应,由对角密度矩阵所得的磁化系数为 $\chi_{\alpha\beta}(\boldsymbol{r},\boldsymbol{r}') = \delta n_\alpha(\boldsymbol{r})/\delta n_\beta(\boldsymbol{r}')$,公式(10.19)变为

$$\chi_{\alpha\beta}(\boldsymbol{r},\boldsymbol{r}') = -\frac{2}{\pi}\int_{-\infty}^{\varepsilon_F} \mathrm{d}E \Im m [G^0_{\alpha\beta}(\boldsymbol{r},\boldsymbol{r}',E) G^0_{\beta\alpha}(\boldsymbol{r}',\boldsymbol{r},E)] \quad (10.20)$$

由公式(10.20)、公式(10.16)的交换积分的次晶格分量可重新表达为

$$J_{\alpha\beta}(\boldsymbol{R}) = \frac{\lambda^2 \hbar^2}{4}\chi_{\alpha\beta}(0,\boldsymbol{R}) \quad (10.21)$$

因此,公式(10.20)和公式(10.21)是计算双层石墨烯中RKKY相互作用的不同晶格分量的主要公式。

如上所述,推迟格林函数是计算RKKY相互作用的重要量化方式。推迟格林函数可以通过哈密顿函数定义为 $G^0_{(q,E)} = (E - H + i0^+)^{-1}$,由此有

$$G^0 = \frac{1}{\Delta_2}\begin{pmatrix} E + \Delta & -\dfrac{q^2 \mathrm{e}^{-2is\theta_q}}{2m} \\ -\dfrac{q^2 \mathrm{e}^{2is\theta_q}}{2m} & E - \Delta \end{pmatrix} \quad (10.22)$$

其中,$\Delta_2 = E^2 - \Delta^2 - q^4/4m^2$。偏置双层石墨烯的能带发散可以通过函数 Δ_2 的零值得到,可表示为 $E = \pm\sqrt{\Delta^2 + \dfrac{q^4}{4m^2}}$。

推迟格林函数,如等式(10.22)所示,因为 $G^0_{B_2B_1}(q,E) = G^{0*}_{B_1B_2}(q,E)$,所以两个独立项为 $G^0_{B_1B_1}(q,E)$ 和 $G^0_{B_1B_2}(q,E)$。此外,$G^0_{B_2B_2}(q,E)$ 可以通过用 $-\Delta$ 代替 Δ 从 $G^0_{B_1B_1}(q,E)$ 得到。推迟格林函数的傅里叶变换表达式为

$$G^0_{\alpha\beta}(\boldsymbol{R},0,E) = \frac{1}{\Omega_{BZ}}\int \mathrm{d}\boldsymbol{q}\mathrm{e}^{i\boldsymbol{q}\cdot\boldsymbol{R}}[\mathrm{e}^{i\boldsymbol{K}\cdot\boldsymbol{R}}G^0_{\alpha\beta}(\boldsymbol{q}+\boldsymbol{K}+E) + \mathrm{e}^{i\boldsymbol{K}'\cdot\boldsymbol{R}}G^0_{\alpha\beta}(\boldsymbol{q}+\boldsymbol{K}'+E)]$$

$$(10.23)$$

其中,$\boldsymbol{R}$ 表示杂质的距离;Ω_{BZ} 是第一布里渊区的面积。

首先考虑杂质在同一层的情况,系统内能产生RKKY相互作用。动量空间中的推迟格林函数可以写为

$$G^0_{B_1B_1} = m^2(2E + 2\Delta)\sum_{i=1,2}\xi^{-1}[\xi + (-1)^i q^2]^{-1}$$

其中,$\xi = \sqrt{4m^2E^2 - 4m^2\Delta^2}$,因此真空间的格林函数是存在的[66],并且满足

$$G^0_{B_1B_1}(0,R,E) = \frac{2\pi m^2}{\Omega_{BZ}}\frac{2E + 2\Delta}{\xi}\Phi_{B_1B_1} \times [K_0(\sqrt{\xi}R) - K_0(i\sqrt{\xi}R)]$$

$$(10.24)$$

其中,$\Phi_{B_1B_1} = (e^{-iK\cdot R} + e^{-iK'\cdot R})$ 和 $K_0(x)$ 是第二类的改进 Bessel 函数。因为是讨论推迟格林函数,E 为 $(E + i0^+)$。采用同样的方式可获得 $G^0_{B_1B_1}(R,0,E)$,磁化系数为

$$\chi_{B_1,B_1}(0,R) = \frac{-16\pi^2 m^2}{2mR^2\Omega_{BZ}^2}(\{1 + \cos[(K-K')R]\}I_1(\Delta,R,\varepsilon_F)$$

(10.25)

我们引入 $y = mR^2E, I_1(\Delta,R,\varepsilon_F)$ 后,变为:

$$I_1(\Delta,R,\varepsilon_F) = \int_{-\infty}^{2m\varepsilon_F R^2} dy \Im m \left\{ \frac{y + 2m\Delta R^2}{y - muR^2} [K_0(-i\sqrt{\sqrt{y^2 - 4m^2\Delta^2 R^4}}) - K_0(\sqrt{\sqrt{y^2 - 4m^2\Delta^2 R^4}})]^2 \right\}$$

(10.26)

很显然,当 $I_1(\Delta,R,\varepsilon_F)$ 趋于常数时,长程机制中 RKKY 相互作用表现为 R^{-2},这样就很好理解 $I_1(\Delta,R)$ 积分。有趣的是,当 $\varepsilon_F = 0$ 时,$I_1(\Delta \to 0, R, \varepsilon_F)$ 趋于 π,因此 RKKY 相互作用按照无偏置和未掺杂的双层石墨烯的 R^{-2} 进行衰减[69]。

接下来要实现系统内 RKKY 相互作用,需要考虑不同层的杂质情况。动量空间中的推迟格林函数可以表示为 $G^0_{B_1B_2}(q,E)$ 是 $me^{2is\Theta_q} \cdot \sum_{i=1,2}(-1)^i \cdot [\xi + (-1)^i q^2]^{-1}$,其中,$\Theta_q$ 是 q 和 x 轴(x 轴沿着锯齿形方向)之间的夹角。真实空间的推迟格林函数由以下计算得到:

$$G^0_{B_1B_2}(0,R,E) = \frac{2\pi m}{\Omega_{BZ}}\Phi_{B_1B_2}[K_2(-i\sqrt{\xi}R) + K_2(\sqrt{\xi}R)] \quad (10.27)$$

其中,$\Phi_{B_1B_2} = e^{-iKR-2i\Theta_R} + e^{-iK'R+2i\Theta_R}$;$K_2(x)$ 是二阶改进的 Bessel 函数。最后,磁化系数 $\chi^0_{B_1B_2}(0,R)$ 表示为

$$\chi^0_{B_1B_2}(0,R) = \frac{16\pi m^2}{\Omega_{BZ}^2 2mR^2}\{1 + \cos[(K-K')R + 4\Theta_R]\}I_2(\Delta,R,\varepsilon_F)$$

(10.28)

其中

$$I_2(\Delta,R,\varepsilon_F) = -\int_{-\infty}^{2m\varepsilon_F R^2} dy \Im m [K_2(-i\sqrt{\sqrt{y^2 - 4m^2\Delta^2 R^4}}) + K_2(\sqrt{\sqrt{y^2 - 4m^2\Delta^2 r^4}})]^2$$

(10.29)

研究表明[73],掺杂双层石墨烯的 RKKY 相互作用明显不同于未掺杂的情况,而且当双层石墨烯在垂直电场下,还会表现出不同的性能。此外,当杂质在不同层时,在杂质短程区域内,采用双带连续模型和四带连续模

型获得的 RKKY 相互作用存在着差异。重要的是,存在偏置电压时,磁位置对 RKKY 相互作用的依赖性会发生明显变化。二维系统内的 RKKY 相互作用见表 10.1。

表 10.1 二维电子气体(2DEG)、单层石墨和双层石墨中 RKKY 相互作用的长程行为的分解。RKKY 相互作用与表中第三和第四栏中的数值成比例[73]

文献	体系	相同亚晶格的 RKKY 相互作用	不同亚晶格的 RKKY 相互作用
[76]	2DEG	$R^{-2}\sin(2k_F R)$	—
[77]	2DEG + 杂质	$R^{-2}\sin(2k_F R)\mathrm{e}^{-R/\delta}$	—
[67]	SLG($\varepsilon_F = 0$)	$-R^{-3}\Phi_{AA}$	$+3R^{-3}\Phi_{AB}$
[68]	SLG($\varepsilon_F \neq 0$)	$-R^{-2}\sin(2k_F R)\Phi_{AA}$	$+R^{-2}\sin(2k_F R)\Phi_{AB}$
[69,72]	SLG($\varepsilon_F = 0$, $u = 0$)	$-R^{-2}\Phi_{B_1B_1}$	$R^{-2}\Phi_{B_1B_2}$
[73]	BLG($\varepsilon_F \neq 0$, $\Delta \neq 0$)	$-R^{-2}\cos(k_F R)[\sqrt{2}\mathrm{e}^{-k_F R} + \sin(k_F R)]\Phi_{B_1B_1}$	$R^{-2}\cos(k_F R)[\sqrt{2}\mathrm{e}^{-k_F R} - \sin(k_F R)]\Phi_{B_1B_2}$
[73]	BLG($\varepsilon_F \neq 0$, $\Delta \neq 0$)	$-R^{-2}\sin(2k_U R)\Phi_{B_1B_1}$	$R^{-2}(\mathrm{e}^{-k_U R}\cos(k_U R) - \alpha\sin(2k_U R) - \alpha\sin(2k_U R) - \beta\frac{\cos(2k_U R)}{2k_U R})\Phi_{B_1B_2}$

注:①2DEG 是二维电子气体;SLG 为单层石墨;BLG 是双层石墨。
②δ 是 2DEG 中杂质引起的散射电子的有限平均自由程。α 和 β 是偏置双层石墨中由 E_F 和 Δ 控制的参数,$k_U \sqrt{2m}(\varepsilon_F^2 - \Delta^2)^{\frac{1}{4}}$。$\Phi_{AA} = 1 + \cos[(K-K')R]$,$\Phi_{BB} = \{1 + \cos[(K-K')R + \pi - 2\theta_R]\}$。

RKKY 相互作用与表 10.1 所示的第三和第四栏中所给的数值成一定的比例关系,δ 是 2DEG 中由杂质引起的散射电子的有限平均自由程。α 和 β 是由偏置双层石墨烯的 E_F 和 V 及参数 $l = k\nu R$ 共同控制的,其中 $k\nu = [(2mE_F)_2 - m_2V_2]_{1/4}$。对于 SLG,函数 Φ_{AA} 和 Φ_{AB} 分别由 $1 + \cos[(K-K_-)R]$ 和 $1 + \cos[(K-K_-)R + \pi - 2\theta_R]$ 给出,函数 $\Phi_{B_1B_1}$ 和 $\Phi_{B_1B_2}$ 分别由公式(10.18)和公式(10.24)定义。

10.4 双层石墨烯中传输性能的多体效应

关于库仑屏蔽和集体模式已有大量理论研究,外栅极下双层石墨烯中相互作用的重要性也受到关注。当存在磁场或低费米能量时相互作用显得尤为重要。本节将研究双层石墨烯多体效应的一些重要知识。

10.4.1 四频带连续模式的响应函数

双层石墨烯被认为是距离为 d(约为 0.3 nm)的两个隧道耦合的石墨烯单层。一般不考虑三角翘曲,只有在极低密度下,三角翘曲才显得非常重要(其他效应也很重要,如无序性),哈密顿动力学[78]表达式为

$$(\hbar = 1)\hat{T} = \sum_{k,\alpha,\beta} \hat{c}^+_{k,\alpha} T_{\alpha\beta}(k) \hat{c}_{k,\beta}$$

式中

$$T_{\alpha\beta}(k) = \begin{pmatrix} 0 & v_F k e^{-i\varphi k} & 0 & -\gamma_1 \\ v_F k e^{i\varphi k} & 0 & 0 & 0 \\ 0 & 0 & 0 & v_F k e^{-i\varphi k} \\ -\gamma_1 & 0 & v_F k e^{i\varphi k} & 0 \end{pmatrix} \quad (10.30)$$

其中,φ_k 是 k 和 x 轴之间的角度;α 和 β 约为 2×2 大小的顶部,左边模块指的是上层,底部右边模块指的是下层;对角线外模块代表层内跳跃,在此模型中,这种情况只发生在上层 A 晶格和位于上层正下方的 B 晶格之间。层中电子通过二维库仑电位 $V_s(q) = 2\pi e^2/(\varepsilon_0 q)$ 相互作用,ε_0 为平均介电常数,由双层石墨烯的介电环境决定。现在引入石墨烯耦合常数,表达式为 $\alpha_{ee} = e^2/(\hbar\varepsilon_0 v_F)$。不同层的电子以 $V_D(q) = V_s(q)\exp(-qd)$ 产生相互作用,从而使两层之间产生进一步耦合。

由于等式(10.30)中的非对角的层间隧穿项,层自由度并不是一个好的量子数:它本应使我们更加方便地获得呈对角线分布哈密顿动力学。这可以通过适当 k 依赖的单一转换 u_k 的来完成。[78] 人们发现四个杂化带(图 10.2):

$$\varepsilon_{1,2}(k) = \pm\sqrt{v_F^2 k^2 + \gamma_1^2/4} + \gamma_1/2 \text{ 和 } \varepsilon_{3,4}(k) = \pm\sqrt{v_F^2 k^2 + \gamma_1^2/4} - \gamma_1/2$$

在这个基础上,电子-电子相互作用通过下式来描述:

$$\hat{H}_{int} = (2S)^{-1} \sum_q [V_+(q)\hat{\rho}_q\hat{\rho}_{-q} + V_-(q)\hat{Y}_q\hat{Y}_{-q}]$$

其中，$V_{\pm}=(V_S\pm V_D)/2$，是层内和层间的对称和非对称的组合。

$$\hat{\rho}_q = \sum_{k,\lambda,\lambda'} \hat{c}^+_{k-q,\lambda}(D_{k-q,k})_{\lambda\lambda'}\hat{c}_{k,\lambda}$$

其中，$D_{k-q,k}=u^+_{k-q}u_k$，是密度波动算子。公式中和下文中的 Greek 指数 λ 和 λ' 指的是四频带 $\varepsilon_\lambda(k)$，$\lambda=1,\cdots,4$。横向算子 $\hat{Y}_q$ 定义为

$$\hat{Y}_q = \sum_{k,\lambda,\lambda'} \hat{c}^+_{k-q,\lambda}(S_{k-q,k})_{\lambda\lambda'}\hat{c}_{k,\lambda}$$

其中，$S_{k-q,k}=u^+_{k-q}\gamma^5 u_k$，这里 $\gamma^5 \equiv -i\gamma^0\gamma^1\gamma^2\gamma^3$，其中 γ^u 是手性表达式中的 4×4 狄拉克 γ 矩阵。[79] 对于 $q\to 0$，$S_{k-q,k}\to -\gamma^x\gamma^5$，只有反对角线才会有项，即 $q=0$ 时，只有满足对称性的 $\hat{Y}_q$ 描述的（垂直的）传输，是1和4（高能量）带之间的以及2和3（低能量）带之间。

从 $\hat{H}_{int}$ 可以看到，需要两个相应函数才能评价双层石墨烯的集体模式和基态性能：密度-密度相应函数 $\chi_{pp}(q,\omega)=\ll\hat{\rho}_q;\hat{\rho}_{-q}\gg_\omega/S$，和横向响应 $\chi_{YY}(q,\omega)=\ll\hat{Y}_q;\hat{Y}_{-q}\gg_\omega/S$，其中 $\ll\hat{A};\hat{B}\gg_\omega$ 是 Kubo 乘积（Kubo product）。[80] 很容易看出混合相应函数，如 $\chi_{p\gamma}(q,\omega)$ 和 $\chi_{\gamma p}(q,\omega)$ 对于每个 q 和 ω 都是0。这是一个确切结论，即使超出随机相近似也是正确的。其物理原因就是密度算子 $\hat{\rho}_q$ 和横向算子 $\hat{Y}_q$ 在倒置时以相反的方式改变（$\hat{\rho}_q$ 是不变的，而 $\hat{Y}_q$ 改变符号）。

在非相互作用下，以上介绍的线性响应函数可以很方便地用确切本征态表示：

$$\chi^{(0)}_{pp(YY)} = \sum_{\lambda,\lambda'}\int\frac{d^2k}{(2\pi)^2}\frac{n_{k,\lambda}-n_{k+q,\lambda'}}{\omega+\Delta_{\lambda\lambda'}(k,q)+i0^+}M_{\lambda\lambda'}(k,q) \quad (10.31)$$

其中，$n_{k,\lambda}$ 为零温度非相互作用带占位因子；$\Delta_{\lambda\lambda'}(k,q)$ 为带能量差，$\Delta_{\lambda\lambda'}(k,q)=\varepsilon_{k,\lambda}-\varepsilon_{k+q,\lambda'}$。在密度-密度通道 $M_{\lambda\lambda'}(k,q)=|(D_{k,k+q})_{\lambda\lambda'}|^2$，在横向通道 $M_{\lambda\lambda'}(k,q)=|(S_{k,k+q})_{\lambda\lambda'}|^2$。当带 $\varepsilon_4(k)$ 和 $\varepsilon_2(k)$ 都是满的而其他两个带是空的时，通过计算得到未掺杂双层石墨烯中 $\chi^{(0u)}_{pp(YY)}$ 的解析表达式[81]。这种情况用 $\chi^{(0u)}_{pp(YY)}$ 来标记。下面给出的结论只针对这些响应函数的虚数部分。从标准的 Kramers-Krönig 解析可得到实数部分的相应解析表达式，但非常烦琐。经过很复杂计算后，得到以下结论（每个自旋和每个谷）：

$$\Im m\chi^{(0u)}_{pp}(q,\omega) = \left\{\frac{1}{16v_F^2}\left[\frac{v_F^2 f^2(q,\omega)-2v_F^2 q^2}{\sqrt{g(q,\omega,\omega)}}+2\sqrt{g(q,\omega,\omega)}-\right.\right.$$

$$\frac{2}{\omega}\mid g(q,\omega,\omega_-)\mid\Big]\theta[g(q,\omega_-,\omega_+)]-$$

$$\frac{1}{8v_F^2\omega}[\omega\sqrt{g(q,\omega_-,\omega_-)}-\mid\sqrt{g(q,\omega,\omega_-)}\mid]\cdot$$

$$\theta[g(q,\omega,\omega_--\gamma_1)]\}+\{\cdots\}_{\gamma_1\to-\gamma_1} \tag{10.32}$$

和

$$\Im m\chi_{YY}^{(0u)}(q,\omega)=\Big\{\frac{1}{16v_F^2}\Big[\frac{v_F^2f^2(q,\omega_-)-2v_F^2q^2-2\gamma_1^2}{\sqrt{g(q,\omega_-,\omega_-)}}+2\sqrt{g(q,\omega_-,\omega_-)}-$$

$$\frac{2}{\omega}\mid g(q,\omega,\omega_-)\mid\Big]\theta[g(q,\omega_-,\omega_+)]-$$

$$\frac{1}{8v_F^2\omega}[\omega\sqrt{g(q,\omega,\omega)}-\mid\sqrt{g(q,\omega,\omega_-)}\mid]\cdot$$

$$\theta[g(q,\omega_-,\omega_+)]\}+\{\cdots\}_{\gamma_1\to-\gamma_1} \tag{10.33}$$

其中,$\omega_\pm=\omega\pm\gamma_1$;$g(q,\omega,\Omega)=\omega\Omega-v_F^2q^2$;$f(q,\omega)=q\sqrt{\dfrac{g(q,\omega_-,\omega_+)}{g(q,\omega,\omega)}}$;$\theta(x)$ 是普通阶跃函数。方程(10.32)和方程(10.33)是多体双层石墨烯中重要的量。很容易检验极限 $\gamma_1\to 0$ 时(其中"广义动量"$f(q,\omega)\to q$),$\Im m\chi_{\rho\rho}^{(0u)}(q,\omega)=\Im m\chi_{YY}^{(0,u)}(q,\omega)$,且它们减小到非掺杂单层石墨烯的响应函数虚部的 2 倍。

掺杂体系[81-82]的响应函数(对于任意费米能量 ε_F)可以写作

$$\chi_{\rho\rho}^{(0)}(q,\omega)=\chi_{\rho\rho}^{(0u)}(q,\omega)+\chi_{\rho\rho}^{(0)}(q,\omega)$$

和

$$\Im m\chi_{YY}^{(0)}(q,\omega)=\Im m\chi_{YY}^{(0u)}(q,\omega)+\Im m\chi_{YY}^{(0)}(q,\omega)$$

响应函数的实数部分可以由 Kramers-Krönig 关系得到。

静态响应函数 $\chi_{\rho\rho(YY)}^{(0)}(q,\omega=0)$ 的曲线如图 10.5(a) 和图 10.5(b) 所示。在低密度极限,$\chi_{\rho\rho}^{(0)}(q,\omega=0)$ 在 $q=2k_{F3}$ 时表现出非解析行为(科恩异常)[83],其中 k_{F3} 是 $\lambda=3$ 带的费米动量,如图 10.2(a) 所示。从曲线中可以看到,双带模型[83](图 10.5 中的短画线)里计算得到的响应函数 $\chi_{\rho\rho}^{(0)}(q,\omega=0)$,过高估计了非解析的强度并完全曲解了 q 很大时的表现。在高密度极限下,$\chi_{\rho\rho}^{(0)}(q,\omega=0)$ 形状与单层石墨烯的形状极为相似。在 $2k_{F1}$ 和 $2k_{F3}$ 中也会出现非解析。当分裂带 $\varepsilon_1(k)$ 被占用,Friedel 振荡(FOs)显示出一个从低密度到高密度的有趣交叉。在前者机制中,只有当带 $\varepsilon_3(k)$ 被占用时,在 $2k_{F3}$ 才有 FOs,而在高密度机制下有两个周期性,一个与 $2k_{F1}$ 有关,一个与 $2k_{F3}$ 有关,二者会冲撞。两个机制下的 FOs 的振幅具有不同幂律。

从图 10.5(b) 中可以清楚地看到双带模型对于横向通道更加不适用。双带模型 $\chi_{YY}^{(0)}(q,0)$ 是发散的,并且它取决于所使用紫外的截止 k_c 的精确值。

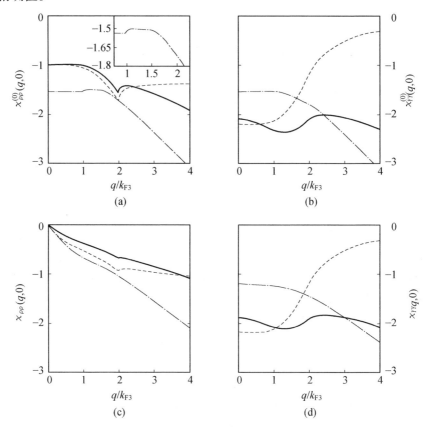

图 10.5 双层石墨烯静态响应与 q/k_{F3} 的函数关系,以 $\lambda = 3$ 带的费米能量态密度为单位,$i_{ev} = (\varepsilon_F + \gamma_1/2)/(2\pi v_F^2)$

(a) $\chi_{pp}^{(0)}(q,0)$,虚线是双带模型结果[83],而实线是四带模型结果。这些结果都是基于掺杂级别为 $n = 10^{12} \text{cm}^{-2}$。点划线是 $n = 5 \times 10^{13} \text{cm}^{-2}$,用四带模型计算结果。插图:主图中高密度小力矩的放大。(b) $\chi_{YY}^{(0)}(q,0)$。标识与(a) 中相同。二带模型所得 $\chi_{YY}^{(0)}(q,0)$ 采用截止(cut-off) $k_c = \gamma_1/V_F$。(c) 与(a) 相同,但采用了随机相近似屏蔽 $\alpha_{ee} = 0.5$。(d) 与(b) 相同,但采用了随机相近似屏蔽 $\alpha_{ee} = 0.5$[81]

相互作用的掺杂体系的随机相近似响应函数如下:

$$\chi_{\rho\rho(YY)} = \frac{\chi^{(0)}_{\rho\rho(YY)}}{1 - V_\pm \chi^{(0)}_{\rho\rho(YY)}} \equiv \frac{\chi^{(0)}_{\rho\rho(YY)}}{\varepsilon_{\rho\rho(YY)}} \quad (10.34)$$

相互作用体系的磁化系数完全取决于密度 n、层间距离 d(采用 $d = 0.335$ nm)和层间隧道 γ_1(采用 0.35 eV)。图 10.5(c) 和 (d) 所示是屏蔽静态响应 $\chi^{(0)}_{YY}(q,0)$ 的曲线图。从图 10.5(c) 和 (d) 清楚看到,屏蔽对纵向通道有极大影响,当 $q \to 0$ 时,有 $\chi^{(0)}_{YY}(q,0) \to 0$,并大大减小了科恩异常的尺寸,而在横向隧道中起的作用很小。

10.4.2 双层石墨烯的有效质量重正

为了研究单层和双层速度重正之间的相似和差异,对双层石墨烯运用了普遍采用的双组分模型[84],适用于能量低于层间隧道尺度时的情况,并解释双层的不寻常[85]量子霍尔效应。双层石墨烯的双组分模型也可用来计算双层的压缩性[86,87]和静态非相互作用的密度 – 密度线性响应函数[83]。

单层双带模型和双层双带模型之间的主要差异是:① 双层时的能带色散是有效质量 $m = \gamma_1/(2\nu_F^2)$ 的二次方;② 双层石墨烯的手性是 $J = 2$,而不是 $J = 1$;③ 双层的层内 $[V_k^{(S)} = \nu_k]$ 和层间 $[V_k^{(D)} = V_k^{(S)} \exp(-kd)]$ 的库仑相互作用是不同的。在讨论双层石墨烯时,采用了类托马斯 – 费米电位 $\nu_k = 2\pi/\varepsilon_0(\lambda + |k|)$,以避免在库仑体系中出现熟知的均场理论。双层时采用了截止 k_{max},虽然它的作用后来证明不太重要。计算双层时采用了跟单层一样的屏蔽(λ)和截止($k_{max}/k_F \equiv \Lambda$)参数,此外,决定于值约为 0.2 的无量纲层间距离参数 $\bar{d} = dk_{max}$。双层石墨烯的 Thomas-Fermi 屏蔽向量由 $q_{TF}/k_F = [\gamma_0/(2\nu_F k_{max})]\alpha_{ee}\vec{\Lambda} = t\alpha_{ee}\Lambda$ 给出。当 $d = 0.355$ nm 时,有

$$\frac{q_{TF}}{k_F} = \bar{t}\alpha_{ee}\Lambda \approx 0.38\alpha_{ee}\Lambda \quad (10.35)$$

双层石墨烯的双带模型均场理论计算精确,遵循与单层石墨烯相同的路线。均场 Hartree – Fock 哈密顿为

$$\hat{H}_{HF} = \sum_{k,\alpha,\beta} \hat{\Psi}^+_{k,\alpha}[\delta_{\alpha\beta}B_0(k) + \sigma_{\alpha\beta}B(k)]\hat{\Psi}_{k,\beta} \quad (10.36)$$

其中

$$B_0(k) = -\int \frac{d^2k'}{(2\pi)^2} V^{(S)}_{k-k'} f_+(k') \quad (10.37)$$

并且

$$B^{eq}(k) = \frac{k^2}{2m}u_2(k) - \int \frac{d^2k'}{(2\pi)^2} V_{k-k'}^{(D)} f_-(k') u_2(k') \qquad (10.38)$$

其中,$u_J(k) = [\cos J_{\varphi k}, \sin J_{\varphi k}]$指明能带结构的赝自旋空间中引起有效磁场的方向的$k$依赖性。单层情况时,手性$J$值为$J=1$,$u_1(k) = \hat{k}$。由此得出,对比单层和双层情况,只有Hartree-Fock自能量的赝自旋相关部分是不同的。

$$\Sigma(k) = -\int_0^{k_{\max}} \frac{dk'}{2\pi} k' f_-(k') V_2^{(D)}(k,k') \qquad (10.39)$$

在此情况下有

$$V_m^{(D)}(k,k') = \int_0^{2\pi} \frac{d\Theta}{2\pi} e^{-im\Theta} \nu(\sqrt{k^2 + k'^2 - 2kk'\cos\Theta}) \times$$

$$\exp(-d\sqrt{k^2 + k'^2 - 2kk'\cos\Theta}) \qquad (10.40)$$

这样,等式(10.39)可重新写为

$$\Sigma(k) = \frac{1}{4\pi} \int_{k_F}^{k_{\max}} dk' k' V_2^{(D)}(k,k') \qquad (10.41)$$

采用无量纲单位给出(用vk_F衡量能量,不是用费米能量)

$$\frac{\Sigma(x)}{vk_F} = \frac{1}{2}\alpha_{ee} \int_1^\Lambda dx' x' \overline{V}_2^{(D)}(x,x') \qquad (10.42)$$

其中,在单层情况下,所有波矢都用k_F重新衡量,也就是$x = k/k_F$,$x' = k'/k_F$。此处引入无量纲相互作用$\overline{V}_m^{(D)}$,$V_m^{(D)}$采用$2\pi e^2/(\varepsilon_0 k_F)$为单位衡量。因为双层中的准粒子有手性$J=2$,自能量表达式中的角积分选择了库仑相互作用的二阶矩$V_2(k,k')$,而不是一阶矩。

图10.6是$\alpha_{ee} = 0.5$和$\Lambda = k_{\max}/k_F = 10$时$\Sigma(k)$对$k/k_F$的依赖关系。从曲线看到,$\Sigma(k)$有两个不同屏蔽参数$\lambda$值:$\lambda = 10^{-2}$与基本非屏蔽库仑电位相关,$\lambda = 1$与Thomas-Fermi屏蔽电位相关。小$k$值行为受屏蔽影响不大,可以解析并理解。对于大的$\Lambda$值,可以以$d/\Lambda$指数的形式扩展式(10.40):

$$\exp\left(-\frac{\overline{d}}{\Lambda}\sqrt{x^2 + x'^2 - 2xx'\cos\Theta}\right) \to 1 - \frac{\overline{d}}{\Lambda}\sqrt{x^2 + x'^2 - 2xx'\cos\Theta} +$$

$$\frac{1}{2}\left(\frac{\overline{d}}{\Lambda}\right)^2 (x^2 + x'^2 - 2xx'\cos\Theta) + O[(\overline{d}/\Lambda)^3] \qquad (10.43)$$

对于非屏蔽库仑相互作用为

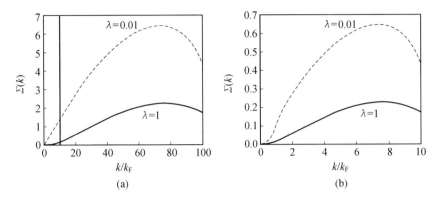

图 10.6 当 $\alpha_{ee} = 0.5$ 时,双层石墨烯中 Hartree-Fock 自洽能 $\Sigma(k)$(以 vk_F 为单位)的赝自旋相关部分与 k/k_F 的关系(波矢量直至截止)

(a)$\Lambda = 10$。垂直线表示点 $K = K_F$。(b)$\Lambda = 100$。虚线是 $\lambda = 10^{-2}$,而实线对应 $\lambda = 1^{[89]}$

$$\overline{V}_m^{(D)}(x,x') = \int_0^{2\pi} \frac{\mathrm{d}\Theta}{2\pi} \mathrm{e}^{-im\Theta} \frac{1}{\sqrt{x^2 + x'^2 - 2xx'\cos\Theta}} \cdot$$

$$\exp\left(-\frac{\overline{d}}{\Lambda}\sqrt{x^2 + x'^2 - 2xx'\cos\Theta}\right) \quad (10.44)$$

这就表示当 $m > 0$ 时,$\overline{V}_m^{(D)}(x,x')$ 在 $\overline{d}/\Lambda$ 二阶开始 $\overline{V}_m^{(D)}(x,x')$ 扩展:与 $\sqrt{x^2 + x'^2 - 2xx'\cos\Theta}$ 成比例的项,是 $\overline{d}/\Lambda$ 的一阶,在方程(10.44)中由于分母积分被消除。因此,方程(10.44)中该项的角平均只在 $m = 0$ 时有确定结果。

不考虑这项,当 $m > 0$,$\overline{V}_m^{(D)}(x,x')$ 可按照 $\overline{d}/\Lambda$ 扩展为

$$\overline{V}_m^{(D)}(x,x') \to \left[1 + \frac{1}{2}\left(\frac{\overline{d}}{\Lambda}\right)^2(x^2 + x'^2)\right]\overline{V}_m(x,x') -$$

$$\frac{1}{2}xx'\left(\frac{\overline{d}}{\Lambda}\right)^2 [\overline{V}_{m+1}(x,x') + \overline{V}_{m-1}(x,x')] \quad (10.45)$$

将该表达式用于方程式(10.42),得

$$\frac{\Sigma(k)}{vk_F} = \frac{1}{2}\alpha_{ee}\int_1^\Lambda \mathrm{d}x' x' V_2(x,x') + \frac{1}{4}\alpha_{ee}\left(\frac{\overline{d}}{\Lambda}\right)^2 x^2 \int_1^\Lambda \mathrm{d}x' x' V_2(x,x') +$$

$$\frac{1}{4}\alpha_{ee}\left(\frac{\overline{d}}{\Lambda}\right)^2 \int_1^\Lambda \mathrm{d}x' x'^3 V_2(x,x') -$$

$$\frac{1}{4}\alpha_{ee}\left(\frac{\bar{d}}{\Lambda}\right)^2 x \int_1^\Lambda dx' x'^2 [V_1(x,x') + V_3(x,x')] \qquad (10.46)$$

为了弄清极限 $x \to 0$ 时方程式(10.46)中双层的 Hartree-Fock 自能量,采用

$$\bar{V}_m(x \to 0, x') = \frac{x^m}{x'^{m+1}} \frac{1 \cdot 3 \cdot 5 \cdots (2m-1)}{2^m m!} \qquad (10.47)$$

在方程式(10.46)中插入这个表达式,计算 x' 积分,由于 $x \to 0$ 可忽略比 x^2 更快接近零的项,最终发现

$$\frac{\Sigma(x \to 0)}{v k_F} = \frac{3}{16}\alpha_{ee} x^2 \frac{\Lambda - 1}{\Lambda} - \frac{1}{32}\alpha_{ee} x^2 \bar{d}^2 \frac{\Lambda - 1}{\Lambda^2} + O(x^3) \qquad (10.48)$$

注意,在 $\Lambda \to \infty$ 时(零掺杂极限)第一项是确定的,而含有层分离依赖性的第二项趋于 0。在单层的情况下,对小波矢量的交换自能量的主要贡献来自对屏蔽不敏感的负能量能级的相互作用。

双层石墨烯的重正质量 m^* 可定义为

$$\frac{k_F}{m^*} \equiv \frac{\partial}{\partial k}\left[\frac{k^2}{2m} + B_0(k) + \Sigma(k)\right]\bigg|_{k=k_F} = \frac{k_F}{m} + \frac{1}{2}v\alpha_{ee}\frac{\partial \Gamma_\Lambda(x)}{\partial x}\bigg|_{x=1} \qquad (10.49)$$

恢复赝自旋独立的自能量贡献,这对于单层和双层是完全相同的,得

$$\Gamma_\Lambda(x) \equiv \int_0^\Lambda dx' x' V_2^{(D)}(x,x') - 2\int_0^1 dx' x' V_0(x,x') - \int_1^\Lambda dx' x' V_0(x,x') \qquad (10.50)$$

它遵循

$$\frac{m^*}{m} = \frac{1}{1 + \frac{1}{2}\alpha_{ee}\bar{t}\Lambda \frac{\partial \Gamma_\Lambda(x)}{\partial x}\bigg|_{x=1}} \qquad (10.51)$$

其中,$\bar{t}$ 是以 k_F 为单元进入 Thomas-Fermi 屏蔽矢量的相同数量。图 10.7 是 $\lambda = 10^{-2}$ 时 m^*/m 比值与 α_{ee} 的关系曲线,而图 10.8 中描绘的是 m^*/m 在不同屏蔽参数 λ 下与密度之间的函数关系。在图中,通过使 $q_{TF} \to \gamma q_{TF}$,也考虑了石墨烯的自旋和能谷简并因数 $g = 4$ 的情况。

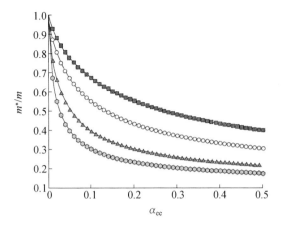

图 10.7　不同 Λ 下,由方程式(10.51)计算得到的双层石墨烯的 Hartree-Fock 重正质量 m^*/m 与 α_{ee} 的关系($\lambda = 10^{-2}$)

从顶部到底部,Λ = 10(方形)、20(圆形)、50(三角形)和 100(六边形),说明相互作用会抑制准粒子有效质量[89]

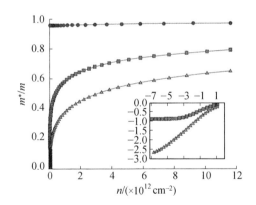

图 10.8　α_{ee} = 0.5 时,由方程式(10.51)计算得到的双层石墨烯的 Hartree-Fock 重正质量 m^*/m 与密度 n(以 $10^{12}\mathrm{cm}^{-2}$ 为单元)的函数关系

圆形代表 $\lambda = 1$,方形是 $\lambda = 10^{-2}$,三角形是 $\lambda = 10^{-4}$。插图是 $\lambda = 10^{-2}$ 和 $\lambda = 10^{-4}$ 时,$\log 10\,(m^*/m)$ 与 $\log 10\,(n)$ 的函数关系;清晰地表明了当 $n \to 0$,双层重正质量如何保持恒定值[89]

10.4.3　双层石墨烯的化学电位和电荷可压缩性

化学势简单来说就是 Hartree-Fock 在 $k = k_F$ 处的准粒子自能量[89],因

此

$$\mu = \frac{\hbar^2 k_F^2}{2m} + [B_0(k) + \sum(k)]|_{k=k_F} \quad (10.52)$$

显而易见，$B_0(k_F) = v_F k_F \alpha_{ee}$ 和电荷压缩性 $n^2 k = [\partial \mu / \partial n]^{-1}$，其中 n 是电子密度。图 10.9 是不同 α_{ee} 值下，双带 Hartree-Fock 逆热动力学态密度与掺杂的函数关系。

当电荷载子密度趋近于零时，计算逆压缩性很有意义。经过一些简单计算后，得

$$\partial \mu / \partial n \sim \frac{\pi \varepsilon_F}{k_F^2}\left[1 - 0.035 \frac{\alpha_{ee} \gamma_1}{\hbar v_F k_F}\right] \quad (10.53)$$

在 $n_c = \alpha_{ee}^2 \times 1.1 \times 10^{10}$ cm^{-2} 处符号改变。

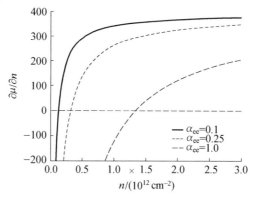

图 10.9　Hartree-Fock 逆热动力学态密度 $\partial \mu / \partial n$（采用双带模型计算，单位为 eV·Å^2）与掺杂 n（不同的 α_{ee} 值，单位为 10^{12} cm^{-2}）的关系

在 Hartree-Fock 近似下，因为对负的并与 $n^{1/2}$ 成比例的化学电位净贡献很小，所以可压缩性在极低密度时变成负值。对可压缩性的这一贡献可以想象成更大，但却是对出现在普通二维电子气中的化学电位的有关贡献。因为此情况下相互作用的相对强度和频带能量可以在长度变化范围内被吸附，$n^{1/2}$ 交换能量可以看作是以 e^2/k_F 幂值能量扩展中的主要阶项。对化学电位有相关贡献的主要阶项会与 n^0（上到对数因子）成比例而且不会在可压缩性中出现。因为连续极限哈密顿的层间跳跃和层内平面跳跃项不能以密度的相同方式进行缩放，所以简单的缩放属性不适于双层石墨烯。每个粒子的化学电位和能量分别采用 $n^{1/2}$ 和互动尺度 α_{ee} 进行扩展。

相互作用的双层石墨烯体系的整个基态能量 E 可以通过随机相近似

内的电荷 - 电荷响应函数[90]的积分进行计算。因此电荷可压缩性就可以通过 $k^{-1} = \frac{n^2}{S} \cdot \frac{\partial^2 E}{\partial^2 n}$ 得到,其中 S 是样品面积。Borghi 等[90]对基态能量进行了详细阐述,在此给出主要结论。

图 10.10 是反热动力学态密度 $\partial \mu / \partial n$,包括交换和随机相相关更正的计算:$\partial \mu / \partial n = (\partial^2 \varepsilon_k + \varepsilon_{xc}) / \partial^2 n$。定性看图 10.10 的结论与从 Hartree-Fock 得到的非常相似。而掺杂低于 $n_1 = 18 \times 10^{12}$ 时,可以清晰地看出修正的作用非常重要。为了便于比较,图 10.10 给出了悬浮的 $\alpha_{ee} = 2.2$ 单层石墨烯的 $\partial \mu / \partial n$。正如预想的,高度掺杂时双层和单层石墨烯可压缩性之间的差异非常小,尤其是在所有四个带都被占用的时候。而在低密度时,情况完全不同,因为在这个机制下,双层石墨烯光谱接近抛物线形式,与单层石墨烯的线性色散发生强烈偏离。特别是在单层情况下,当 $n \to 0$ 时,$\partial \mu / \partial n$ 会发生分离,而双层石墨烯时却是个确定值。这个显著差异的产生是由于双层石墨烯的准粒子有效质量在掺杂接近零时是有限

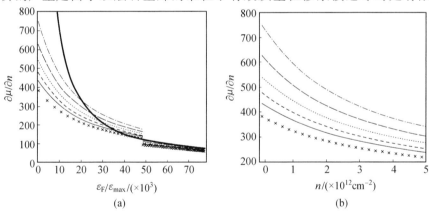

图 10.10 (a) 以 eV·Å 为单元的随机相近似逆热动力学态密度 $\partial \mu / \partial n$ 与费米能 ε_F(费米能量从 $\varepsilon_F = 7 \times 10^{-4}$ eV 到 $\varepsilon_F = 0.53$ eV)的函数关系。不同的 α_{ee} 的值(从底部到顶部):$\alpha_{ee} = 0.125$(实线)、0.25(短划线)、0.5(虚线)、1(长短划线)和 2.2(点划线)。交叉表明 $\alpha_{ee} = 0$,非相互作用石墨烯。加重实线为悬浮的 $\alpha_{ee} = 2.2$ 单层石墨烯的逆热动力学态密度。表明当 $n \to 0$ 时,双层石墨烯的可压缩性保持恒定,而单层石墨烯的可压缩性有分离。在 $n = n_1$、$\alpha_{ee} \neq 0$ 时,忽略了对 $\partial \mu / \partial n$ 负贡献的 δ 函数。
(b) 是(a)图低密度的放大图。x 轴表示以 $10^{12}\mathrm{cm}^{-2}$ 为单位的整体密度 n[90]

的。

最近对双层石墨烯的可压缩性进行了实验性的研究[91-93]。发现在平衡极限内,$\partial \mu / \partial n$ 在接近零载流子密度时存在峰,并随着载流子密度的增加而单调下降;双层石墨烯中未出现普通二维电子气体在低密度时的符号变化,这与上述讨论的一致。[90]理论计算表明这些实验结果受到相互作用的影响,因此在用带结构模型拟合可压缩性实验时要小心谨慎。

我们要强调的是当电屏蔽很弱,石墨烯层中相互作用很强时,相关性对于双层石墨烯的可压缩性很重要。对于悬浮双层石墨烯,忽视相关性作用会导致100%错误的出现。

10.4.4 双层石墨烯的不稳定性和对称性破缺

正如已提到的,相互作用效应在双层石墨烯中是非常重要的,尤其在低电子密度下。双层石墨烯具有由低能量费米有限态密度引发的二次能带。据预测接近半填充时,电子-电子的相互作用是不稳定的。由于双层石墨烯存在赝自旋,意味着层、能谷简并和电子自旋在不同对称破缺态下将表现出丰富的相图。相图取决于模型和微观细节。

通过弱耦合重整化群及强耦合扩张,可以对半填充蜂窝状双层的多体不稳定性进行研究[94-95],结果发现无自旋费米中有四个独立的四费米接触耦合。而占主导地位的不稳定性取决于耦合的微观值,对称破缺态通常是一个具有破缺反演对称性或破缺时间反演对称性的间隙绝缘子,具有量子化异常霍尔效应。排斥Hubbard模型通常最主要的弱耦合不稳定性是一种反铁磁性相。

悬浮双层石墨烯的实验证实了极低电子密度下相关态的存在。[96-98]在无辐射电阻中出现了电阻对电场非线性依赖的累计,可预测发生层能谷极化的异常量子霍尔或为无能隙向列相。此外,实验还观察到以绝缘态为标志的具有能隙的有序相形成。认为与垂直电场产生响应的电导与发生层自旋极化的层反铁磁态保持一致。

因此,双层石墨烯的电子特性是一个具有挑战性的问题,需要进行更多的实验和理论方面的研究。

10.5 结 论

本章讨论了基于低能有效哈密顿模型的双层石墨烯体系的一些性能。由于双层石墨烯独特的手性特征和与众不同的电荷载流子色散,使它的电学性质与传统的二维电子气体和单层石墨烯有所不同。这些独特性能对双层石墨烯的传输性有着重要影响,包括本章介绍的一些特性。

本章介绍了双层石墨烯中的电导由于 n-p 区域的不对称性而提高,还介绍了由于具有电子和空穴电荷载流子的不同区域里的量子霍尔边缘态平衡,而出现了霍尔平台。

此外,本章还讨论了双层石墨烯和偏置双层石墨烯的载流子介导的相互作用及掺杂体系的电子电导。对于非偏置和非掺杂体系,RKKY 相互作用与如粒子-空穴对称体系所预期的双向性相对应。还讨论了由双带模型计算得到的短程 RKKY 相互作用和从四带连续模型中得到的 RKKY 相互作用之间的差异。对于栅极双层石墨烯,长程行为由决定于费米能量和栅极电压的动量来衡量,并可通过栅极电压调整 RKKY 的相互作用。

从多体效应观点看,提高准粒子速度可以增加使化学电位远离狄拉克点的所需动能。提高速度也会抑制自发的自旋极化,而自发的自旋极化通常是低载流子密度电子气体系中最经常被考虑到的不稳定类型。

我们还发现电子-电子相互作用抑制了可压缩性,相互作用也起了重要作用。可压缩性降低与常规二维电子气体中的可压缩性大大提高,形成了鲜明对比,即使这两个体系具有相同的抛物线色散。本质原因是双层石墨烯中的导带载流子和狄拉克海负能量之间的相互影响。抑制可压缩性与准粒子速度的提高起因一样,这些现象最终都来自低能光谱的手性特征。可压缩性结论表明相互关系对双层石墨烯的相互作用的定量研究起到了至关重要的作用。

10.6 参考文献

[1] PERES N M. The transport properties of graphene: an introduction[J]. Review of Modern Physics,2010,82:2673.

[2] DAS SARMA S,ADAM S,HWANG E H,et al. Electronic transport in twodimensional graphene[J]. Review of Modern Physics,2011,83:407.

[3] ABERGEL D S L,APALKOV V,BERASHEVICH J,et al. Properties of graphene: a theoretical perspective[J]. Advances in Physics,2010,59:261.

[4] MCCANN E,KOSHINO M. The electronic properties of bilayer graphene [J]. Reports on Progress in Physics,2013,76:056503.

[5] NOVOSELOV K S,GEIM A K,MOROZOV S V,et al. Electric field effect

in atomically thin carbon films[J]. Science,2004,306:666.
[6] BERGER C,SONG Z,LI T,et al. Ultrathin epitaxial graphite:2d electron gas properties and a route toward graphene-based nanoelectronics[J]. The Journal of Chemical Physics B,2004,108:19912.
[7] NOVOSELOV K S,GEIM A K,MOROZOV S V,et al. Two-dimensional gas of massless Dirac fermions in graphene[J]. Nature,2005,438:197.
[8] ZHANG Y,TAN Y W,STORMER H L,et al. Experimental observation of the quantum Hall effect and Berry's phase in graphene[J]. Nature, 2005,438:201.
[9] CASTRO NETO A H,GUINEA F,PERES N M,et al. The electronic properties of graphene[J]. Review of Modern Physics,2009,81:109.
[10] CASTRO E V,NOVOSELOV K S,MOROZOV S V,et al. Electronic properties of a biased graphene bilayer[J]. Journal of Physics: Condensed Matter,2010,22:175503.
[11] NILSSON J,CASTRO NETO A H,GUINEA F,et al. Electronic properties of bilayer and multilayer graphene[J]. Physical Review B,2008, 78:045405.
[12] MAK K F,LUI C H,SHAN J,et al. Observation of an electric-field-induced band gap in bilayer graphene by infrared spectroscopy[J]. Physics Review Letters,2009,102:256405.
[13] KUZMENKO A B. Determination of the gate-tunable band gap and tight-binding parameters in bilayer graphene using infrared spectroscopy[J]. Physical Review B,2009,80:165406.
[14] OHTA T,BOSTWICK A,SEYLLER T,et al. Controlling the electronic structure of bilayer graphene[J]. Science,2006,313:951.
[15] CASTRO E V,NOVOSELOV K S,MOROZOV S V,et al. Biased bilayer graphene:Semiconductor with a gap tunable by the electric field effect [J]. Physics Review Letters,2007,99:216802.
[16] OOSTINGA J B,HEERSCHE H B,LIU X,et al. Gate-induced insulating state in bilayer graphene devices[J]. Nature Materials,2008,7:151.
[17] YAO W,XIAO D,NIU Q. Valley-dependent optoelectronics from inversion symmetry breaking[J]. Physical Review B,2008,77:235406.
[18] OBRAZTSOVA E A,OSADCHY A V,OBRAZTSOVA E D,et al. Statistical analysis of atomic force microscopy and Raman spectroscopy data

for estimation of graphene layer numbers[J]. Physica Status Solidi, 2008,245:2055.
[19] FERRARI A C, MEYER J C, SCARDACI V, et al. Raman spectrum of graphene and graphene layers[J]. Physics Review Letters, 2006, 97: 187401.
[20] GRAF D, MOLITOR F, ENSSLIN K, et al. Spatially Resolved Raman Spectroscopy of Single-and Few-Layer Graphene[J]. Nano Letters, 2007,7 (2):238-242.
[21] FRANK O, MOHR M, MAULTZSCH J, et al. Raman 2D-band splitting in graphene:theory and experiment[J]. ACS Nano,2011,5:2231.
[22] NOVOSELOV K S, MCCANN E, MOROZOV S V, et al. Unconventional quantum Hall effect and Berry's phase of 2π in bilayer graphene[J]. Nature Physics,2006,2:177.
[23] MCCANN E, FALKO V I. Landau level degeneracy and quantum hall effect in a graphite bilayer[J]. Physics Review Letters, 2006, 96: 086805.
[24] HSU Y-F, GUO G-Y. Anomalous integer quantum hall effect in AA-stacked bilayer graphene[J]. Physics Review Letters, 2010, 82: 165404.
[25] FELDMAN B E, MARTIN J, YACOBY A. Broken-symmetry states and divergent resistance in suspended bilayer graphene[J]. Nature Physics, 2009,5:889.
[26] GORBACHEV R V, TIKHONENKO F V, MAYOROV A S, et al. Weak localization in bilayer graphene[J]. Physics Review Letters,2007,98: 176805.
[27] KECHEDZHI K, FALKO V I, MACCANN E. Influence of trigonal warping on interference effects in bilayer graphene[J]. Physics Review Letters,2007,98:176806.
[28] MCCANN E. Asymmetry gap in the electronic band structure of bilayer graphene[J]. Physical Review B,2006,74:161403.
[29] MIN H K, SAHU B, BANERJEE S K, et al. Ab initio theory of gate induced gaps in graphene bilayers[J]. Physical Review B, 2007, 75: 155115.
[30] GUINEA F, NETO A H C, PERES N M. Electronic states and landau

levels in graphene stacks[J]. Physical Review B,2006,73:245426.
[31] ZHANG Y,TANG T-T,GIRIT C. Direct observation of a widely tunable bandgap in bilayer graphene[J]. Nature,2009,11:820.
[32] ZHANG L M. Determination of the electronic structure of bilayer graphene from infrared spectroscopy [J]. Physical Review B, 2008, 78: 235408.
[33] CASTRO E V, NOVOSELOV K S, MOROZOV S V. Biased bilayer graphene: Semiconductor with a gap tunable by the electric field effect[J]. Physical Review Letters,2007,99:216802.
[34] FOGLER M M, MCCANN E. Comment on "Screening in gated bilayer graphene"[J]. Physical Review B,2010,82:197401.
[35] OHTA T, BOSTWICK A, SEYLLER T, et al. Controlling the electronic structure of bilayer graphene[J]. Science,2006,313:951.
[36] KLEIN O Z. Die reflexion von elektronen an einem potentialsprung nach der relativistischen dynamik von Dirac[J]. Zeitschrift für Physik A Hadrons and Nuclei,1929,53(3):157.
[37] KATSNELSON M I, NOVOSELOV K S, GEIM A K. Chiral tunnelling and the klein paradox in graphene[J]. Nature Physics,2006,2:620.
[38] HUARD B. Transport measurements across a tunable potential barrier in graphene[J]. Physics Review Letters,2007,98:236803.
[39] GORBACHEV R V, MAYOROV A S, SAVCHENKO A K, et al. Conductance of p-n-p graphene structures with "air-bridge" top gates[J]. Nano Letters,2008,8:1995.
[40] STANDER N, HUARD B, GOLDHABER-GORDON D. Evidence for Klein tunneling in graphene p-n junctions[J]. Physics Review Letters, 2009,102:026807.
[41] YOUNG A F, KIM P. Quantum interference and klein tunnelling in graphene heterojunctions[J]. Nature Physics,2009,5:222.
[42] GREINER W, MUELLER B, RAFELSKI J. Quantum electrodynamics of strong fields[M]. Berlin:Springer,1985.
[43] SU R K, SIU G C, CHOU X. Barrier penetration and Klein paradox[J]. Journal of Physics A,1993,26:1001.
[44] DOMBEY N, CALOGERACOS A. Seventy years of the Klein paradox [J]. Physics Reports,1999,315:41.

[45] KREKORA P, SU Q, GROBE R. Klein paradox in spatial and temporal resolution[J]. Physics Review Letters, 2004, 92: 040406.

[46] IHNATSENKA S, KIRCZENOW G. Conductance quantization in strongly disordered graphene ribbons[J]. Physical Review B, 2009, 80: 201407(R).

[47] XU H, HEINZEL T, ZOZOULENKO I V. Edge disorder and localization regimes in bilayer graphene nanoribbons[J]. Physical Review B, 2009, 80: 045308.

[48] LONG W, SUN Q-F, WANG J. Disorder-induced enhancement of transport through graphene p-n junctions[J]. Physics Review Letters, 2008, 101: 166806.

[49] ABANIN D A, LEVITOV L S. Quantized transport in graphene p-n junctions in a magnetic field[J]. Science, 2007, 317: 641.

[50] WILLIAMS J R, DICARLO L, MARCUS C M. Quantum hall effect in a gate-controlled p-n junction of graphene[J]. Science, 2007, 317: 638.

[51] LI J, SHEN S-Q. Disorder effects in the quantum Hall effect of graphene p-n junctions[J]. Physical Review B, 2008, 78: 205308.

[52] JING L, VELASCO JR J, KRATZ P, et al. Quantum transport and field-induced insulating states in bilayer graphene pnp junctions[J]. Nano Letters, 2010, 10: 4000.

[53] HATAMI H, ABEDPOUR N, QAIUMZADEH A, et al. Conductance of a bilayer graphene in the presence of a magnetic field: effect of disorder[J]. Physical Review B, 2011, 83: 125433.

[54] OOSTINGA J B, HEERSCHE H B, LIU X, et al. Gate-induced insulating state in bilayer graphene devices[J]. Nature Materials, 2007, 7: 151.

[55] GORBACHEV R V, TIKHONENKO F V, MAYOROV A S, et al. Weak localization in bilayer graphene[J]. Physics Review Letters, 2007, 98: 176805.

[56] MOROZOV S V, NOVOSELOV K S, KATSNELSON M I. Giant intrinsic carrier mobilities in graphene and its bilayer[J]. Physics Review Letters, 2008, 100: 016602.

[57] FELDMAN B E, MARTIN J, YACOBY A. Broken-symmetry states and divergent resistance in suspended bilayer graphene[J]. Nature Physics, 2009, 5: 889.

[58] KOSHINO M, ANDO T. Transport in bilayer graphene: calculations within a self-consistent Born approximation[J]. Physical Review B, 2006,73:245403.

[59] CSERTI J. Minimal longitudinal dc conductivity of perfect bilayer graphene[J]. Physical Review B,2007,75:033405.

[60] CSERTI J,CSORDAS A,DAVID G. Role of the trigonal warping on the minimal conductivity of bilayer graphene[J]. Physics Review Letters, 2007,99:066802.

[61] RUDERMAN M A, KITTEL C. Indirect exchange coupling of nuclear magnetic moments by conduction electrons[J]. Physics Review,1954, 96:99.

[62] KASUYA T. A theory of metallic ferro-and antiferromagnetism on Zener's model[J]. Progress of Theoretical Physics 1956,16,45.

[63] VOZMEDIANO M A H, LóPEZ-SANCHO M P, STAUBER T, et al. Local defects and ferromagnetism in graphene layers[J]. Physics Review B, 2005,72:155121.

[64] DUGAEV V K, LITVINOV V I, BARNAS J. Exchange interaction of magnetic impurities in graphene[J]. Physical Review B,2006,74: 224438.

[65] BUNDER J E,LIN H-H. Ruderman-Kittel-Kasuya-Yosida interactions on a bipartite lattice[J]. Physical Review B,2009,80:153414.

[66] BLACK-SCHAFFER A M. RKKY coupling in graphene[J]. Physical Review B,2010,81:205416.

[67] SHERAFATI M,SATPATHY S. RKKY interaction in graphene from the lattice Green's function[J]. Physical Review B,2011,83:165425.

[68] SHERAFATI M,SATPATHY S. Analytical expression for the RKKY interaction in doped graphene[J]. Physical Review B,2011,84:125416.

[69] KOGAN E. RKKY interaction in graphene[J]. Physical Review B, 2011,84:115119.

[70] SHERAFATI M, SATPATHY S. On the ruderman-kittel-kasuya-yosida interaction in graphene[J]. AIP Conference Proceedings,2012,1461: 24.

[71] KILLI M,HEIDARIAN D,PARAMEKANTI A. Controlling local moment formation and local moment interactions in bilayer graphene[J]. New Journal of Physics,2011,13:053043.

[72] JIANG L,LI X,GAO W,et al. Journal of Physics C,2012,24:206003.

[73] PARHIZGAR F, SHERAFATI M, ASGARI R, et al. Ruderman-kittel-kasuya-yosida interaction in biased bilayer graphene[J]. Physical Review B, 2013, 87:165429.

[74] MOHN P. Magnetism in the solid state: an introduction[M]. Germany, Springer, 2005.

[75] KüBLER J. Theory of itinerant electron magnetism[M]. Oxford: Oxford University Press, 2009.

[76] FISCHER B, KLEIN M W. Magnetic and nonmagnetic impurities in two-dimensional metals[J]. Physical Review B, 1975, 11:2025.

[77] LARSEN U. Damping of the RKKY interaction in metals of arbitrary dimensions[J]. Journal of Physics F, 1985, 15:101.

[78] NILSSON J, CASTRO NETO A H, PERES N M R, et al. Electron-electron interactions and the phase diagram of a graphene bilayer[J]. Physical Review B, 2006, 73:214418.

[79] MAGGIORE M. A modern introduction to quantum field theory[M]. Oxford: Oxford University Press, 2005.

[80] GIULIANI G F, VIGNALE G. Quantum theory of the electron liquid [M]. Cambridge: Cambridge University Press, 2005.

[81] BORGHI G, POLINI M, ASGARI R, et al. Dynamical response functions and collective modes of bilayer graphene[J]. Physical Review B, 2009, 80:241402.

[82] GAMAYUN O V. Dynamical screening in bilayer graphene[J]. Physical Review B, 2011, 84:085112.

[83] HWANG E H, SARMA S D. Screening, kohn anomaly, friedel oscillation, and RKKY interaction in bilayer graphene[J]. Physics Review Letters, 2008, 101:156802.

[84] MCCANN E, FAL'KO V I. Landau-level degeneracy and quantum hall effect in a graphite bilayer [J]. Physics Review Letters, 2006, 96: 086805.

[85] NOVOSELOV K S, MCCANN E, MOROZOV S V, et al. Unconventional quantum hall effect and berry's phase of 2π in bilayer graphene[J]. Nature Physics, 2006, 2:177.

[86] KUSMINSKIY S V, NILSSON J, CAMPBELL D K, et al. Electronic compressibility of a graphene bilayer[J]. Physics Review Letters, 2008, 100:106805.

[87] KUSMINSKIY S V, CAMPBELL D K, CASTRO N A H. Electron-elec-

tron interactions in graphene bilayers[J]. Europhysics Letters, 2009, 85:58005.
[88] MIN H, BORGHI G, POLINI M, et al. Pseudospin magnetism in graphene [J]. Physical Review B, 2008, 77, 041407(R).
[89] BORGHI G, POLINI M, ASGARI R, et al. Fermi velocity enhancement in monolayer and bilayer graphene[J]. Solid State Communications, 2009, 149:1117.
[90] BORGHI G, POLINI M, ASGARI R, et al. Compressibility of the electron gas in bilayer graphene[J]. Physical Review B, 2010, 82:155403.
[91] HENRIKSEN E A, EISENSTEIN J P. Measurement of the electronic compressibility of bilayer graphene[J]. Physical Review B, 2010, 82: 041412(R).
[92] YOUNG A, DEAN C, MERIC I, et al. Electronic compressibility of layer-polarized bilayer graphene[J]. Physical Review B, 2012, 85:235458 (R).
[93] MARTIN J, FELDMAN B E, WEITZ R T, et al. Local compressiblility measurements of correlated states in suspended bilayer graphene [J]. Physics Review Letters, 2010, 105:256806.
[94] VAFEK O. Interacting fermions on the honeycomb bilayer: from weak to strong coupling[J]. Physical Review B, 2010, 82:205106.
[95] ZHANG F, MIN H, POLINI M, et al. Spontaneous inversion symmetry breaking in graphene bilayers [J]. Physical Review B, 2010, 81: 041402.
[96] WEITZ R T, ALLEN M T, FELDMAN B E, et al. Broken-symmetry states in doubly gated suspended bilayer graphene[J]. Science, 2010, 330: 812.
[97] MAYOROV A S, ELIAS D C, Mucha-Kruczynski M. Interaction-driven spectrum reconstruction in bilayer graphene[J]. Science, 2011, 333: 860.
[98] VELASCO JR J, JING L, BAO W, et al. Transport spectroscopy of symmetry-broken insulating states in bilayer graphene[J]. Nature Nanotechnology, 2012, 7:156.

第11章 吸附物对石墨烯电子传输的影响

通过讨论聚合物和金属吸附物对石墨烯电荷传输的影响,说明了表面污染一直是实现大面积超洁净平面石墨烯独特应用的巨大障碍,如用于电子显微镜研究分子的支撑膜、单分子分辨化学传感器和生物传感器以及超高速电子产品。本章介绍了采用光学方法来确定吸附物的浓度或覆盖面积,探究了去除聚合物残留物的方法。

11.1 引 言

石墨烯是由蜂窝晶格中的紧密结合碳原子组成的单原子膜。由于线性动能分散 $E=hv_F k$,石墨烯中的电荷载流子可具有极高迁移率,其中 k 是载流子波矢,v_F 是费米速度($10^6~\mathrm{ms^{-1}}$)。对于独立的石墨烯来说,费米能量与锥点重合,导带和价带在锥点相交,室温载流子迁移率达到 200 000 $\mathrm{cm^2 \cdot V^{-1} \cdot s^{-1}}$,是纯半导体中所发现的最高值[1-2]。当石墨烯置于衬底上时,高迁移率会受到很大影响[3]。任何可作为外部散射源的媒介与石墨烯接触,都会有效地减少电荷载流子的平均自由程,降低迁移率[4,5]。

这些外部散射源也会影响石墨烯场效应晶体管的最小电导率 $\sigma_{\min}$[6-8]。电导率不是呈线性下降,而是下降到接近 $4e^2/h$ 的一恒定值。这与早期基于自洽波恩近似的预测(Self-consistent Born Approximation)不一致[9-12],当存在无序性时,所得最小电导率 $\sigma_{\min}=4e^2/(\pi h)$。后来这个差异归因于中性点附近的电荷分布高度不均[13-15]。由于存在外部随机带电杂质,系统分解成空间不均匀的电子导电水坑和空穴液滴,在石墨烯中产生电位波动[14]。因此,经这样二维坑的传输简化为具有自旋或谷通道的一维随机网络,得到电导率(e^2/h)[13]。对于高载流子密度的均匀体系,散射的主要来源是变化的。高迁移率(干净的)样本的传输实验中,随栅极电压 V_g 线性增加的电导率曲线(σ-V_g)转变为与 V_g 具有次线性关系,说明短程散射起了更重要的作用。因此无论载流子密度大还是电荷杂质浓度低,点缺陷或分子吸附都显得尤为重要,并导致次线性电导率远离狄拉克点[6,13]。

本章首先简单讨论了石墨烯与金属和分子的相互作用。这些吸附对化学气相沉积法在金属衬底上生长的石墨烯的迁移率降低有很大关系。分析了聚合物和金属残留物的来源,介绍了足以表征独立石墨烯的表面清洁度的光学方法。然后介绍了这些吸附物在电子传输中的影响情况。最后,详细阐述了去除残留物的实验细节。

11.2 吸附物与石墨烯的相互作用

石墨烯的特殊性能和潜在应用激发了人们调控石墨烯电子性能的研究,包括与支撑衬底相互作用[16-18]、应变[19-20]或金属/分子吸附[21-22]。人们积极研究石墨烯与分子或金属吸附原子的相互作用,关于理论和实验的研究表明吸附过程对石墨烯电性能产生了显著影响,包括高密度掺杂、能带结构调整、工作函数调整和磁矩的产生。这些吸附是有意进行的,目的是得到石墨烯电性能的预期变化。其他的吸附认为是污染吸附。第一种无意吸附来自转移过程的聚合物[23-25]。旋涂于石墨烯上的聚合物薄膜用作支架,确保衬底金属蚀刻和移动过程中有足够的机械强度。虽然聚合物支架最终被有机溶剂溶解,一薄层聚合物残留物与石墨烯强烈作用,不能完全除去。第二种无意吸附是金属蚀刻后残留在石墨烯表面的金属纳米颗粒。对于生长在铜衬底上的 CVD 石墨烯,典型的金属残留物可能是铜或铁。当氯化铁作为铜的腐蚀剂时,在石墨烯上常发现有铁。

表 11.1 所示是在广义梯度近似[21]下用第一原理密度泛函理论(DFT)得到的 12 种不同吸附原子的电性能。吸附产生的电性能强烈依赖于碳和金属之间的离子和(或)键的共价性。因为碱金属强烈的离子键性质,所以碱金属是良好的电子供体,能有效增加石墨烯中电荷载流子的数量,并由于电荷转移而产生大的电偶极子[21,26]。不同于 I - Ⅲ 组元素,过渡金属 Ti、Fe 和 Pd 通过 d 态与石墨烯 p_z 态的杂化形成共价键,极大地改变了石墨烯的电子结构。在共价键中,吸附物和石墨烯间通过化学键共享电荷。共价键中电荷的重构和定位在原子上半芯态的极化都对电偶极子的产生也有一定作用。

表 11.1　12 种不同吸附原子的电性能[21]

原子	位置	$P(D)$	Φ/eV
Li	H	3.46	2.72
Na	H	2.90	2.21
K	H	4.48	1.49
Ca	H	0.85	3.18
Al	H	0.93	3.08
Ga	H	1.83	2.66
In	H	2.57	2.34
Sn	H	0.19	3.81
Ti	H	1.39	3.16
Fe	H	1.84	3.24
Pd	H	1.23	3.61
Au	H	-1.29	4.88

注：H、T 和 B 分别代表中空的、顶部和桥梁三个容易发生吸附的位置。列出的性能为每个吸附原子的电偶极矩(P)和功函数(Φ)

电偶极子与石墨烯中载流子迁移率降低最为相关。如 11.1 节所述，由于长程库仑散射作用，带电杂质对迁移率降低有很大影响。任何引起吸附物和石墨烯之间的强烈电荷传输的金属吸附原子或分子，都会产生大量可以通过长程库仑作用与流动电子和空穴进行相互作用的静电中心。除了晶界散射，它还解释了为什么剥离 HOPG 石墨烯中的电子迁移率总是比 CVD 石墨烯的高。

由于聚甲基丙烯酸甲酯（PMMA）处理和加工便捷，因而成为用于石墨烯转移的最为普遍的高分子。它也是电子束光刻器件制造过程最常用的分子抗蚀剂。用 PMMA 作为蚀刻掩模可制得大量石墨烯纳米结构，如纳米带和量子点[27]。然而，由于 PMMA 和石墨烯 p_z 轨道之间强偶极子作用导致石墨烯表面存有聚合物残留薄层，使用有机溶剂（如丙酮、氯仿或 1-甲基-2-吡咯烷酮）去除残留的长链分子基本不可能。众所周知，PMMA 薄层只能引起弱的 p 型掺杂[28]可用作石墨烯中电荷传输的无害吸附剂。PMMA 的主要影响发生在 150~350 ℃除水或 PMMA 残留物的热退火过程中。正如 11.4 节和 11.5 节所解释的，在 PMMA 随机断裂的缺陷位置发生碳从 sp^2 到 sp^3 的二次杂化，改变了局部能带结构并产生大量点散射源。

11.3 转移引发的金属和分子吸附

如第 2 章所述,运用 CVD 法可在各种过渡金属上制备高品质大面积的石墨烯薄膜[29-32]。在这些用于 CVD 生长的金属衬底中,最重要的是铜箔上大面积石墨烯层的生长[33-34]。在降低生长压力下由于碳在 Cu 里的溶解度可以忽略不计,CVD 过程可能是表面介导并具有某种自限性,主要生成由随机取向的区域组成的单层石墨烯。为了分离生长的石墨烯层,往往选用聚合物辅助的湿蚀刻法[23-24,35]。后来该技术发展为可实现任意衬底上[35]碳纳米管的更多可控安置,并用于转移石墨烯。如图 11.1 所示,转移过程中,有机溶剂溶解的聚合物先被旋涂到石墨烯上,形成薄而结实的支架。通过湿蚀刻法去除金属催化剂。然后将石墨烯转移到要求的基板上,使用同一种有机溶剂清除聚合物薄膜。

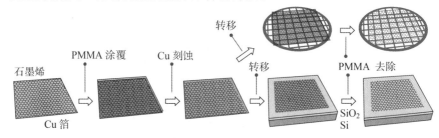

图 11.1 石墨烯转移到硅衬底或转移到 TEM 多孔栅格上悬浮的示意图

图 11.2 是经过 PMMA 辅助转移的石墨烯的透射电子显微镜(TEM)图像。在低倍率图像中,许多明显的带状表示 PMMA 残留物均匀分散于石墨烯表面。黑色点状物是以氧化物形式存在的金属残留物。金属残留物会影响长程库仑散射,导致与剥离 HOPG 石墨烯相比,CVD 石墨烯电子迁移率明显下降[15]。石墨烯上很难发现无 PMMA 残留物的清洁区,因此在当前状态下获得石墨烯的周期性六方晶格是不可能的,高分辨率的 TEM 图像如图 11.2(b)所示。先前使用扫描隧道显微镜的研究表明,PMMA 残留物的薄层厚度通常为 1 ~ 2 nm[36]。一般转移或光刻次数的增加会引起更多 PMMA 聚集在石墨烯上。

转移石墨烯的表面清洁度可以通过拉曼光谱表征。需要建立相同区域的高分辨率 TEM 图像和拉曼光谱图之间的联系。图 11.3(a)是样品的部分制备过程。在 CVD 石墨烯薄膜上制得形状可分辨的金属支架,将石墨烯转移到硅基上,以便通过标记来辨别独立石墨烯表面的相同位置。这

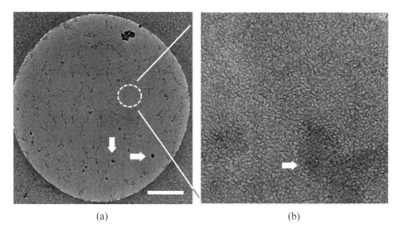

图 11.2 (a)转移 CVD 石墨烯的 TEM 图像,箭头指向的黑点是酸刻蚀后剩余的大量氧化铜。(b)高分辨率的转移 CVD 石墨烯的 TEM 图像。由于残余的 PMMA 使石墨烯表面模糊不清。箭头指示的黑点是少量氧化铜纳米颗粒

样也可以通过高分辨率 TEM 结合拉曼对接近同一位置的表面清洁度进行检测。为了通过拉曼散射对表面清洁度进行量化,对 PMMA 转移的石墨烯薄膜在不同气氛环境下进行退火以得到不同表面清洁度的石墨烯。

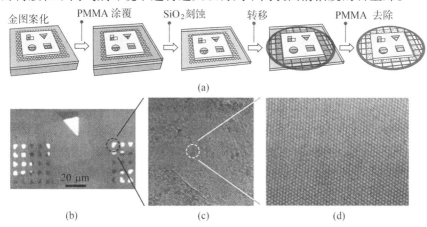

图 11.3 (a)在带有可识别孔的金属支架上制备独立石墨烯的部分过程。(b)石墨烯/300 目 TEM 多孔栅格支架的光学形貌。(c)由(a)中描述的工艺制备的石墨烯 TEM 形貌。(d)干净区域石墨烯的高分辨率 TEM 图像

图 11.4 是相同条件下三种自立石墨烯的 TEM 图像和相应的拉曼光谱。以 PMMA 作为转移支撑或电子束光刻抗蚀剂会在石墨烯上留下残余

薄层,如图 11.4(a)所示。200~500 ℃的轻微退火可以去除部分污染物,得到部分区域清洁的表面(图 11.4(b)(c))。拉曼光谱可以说明表面清洁度等级,当石墨烯远离任何支撑基板呈悬浮态时,G 带性征表现尤为明显。这种情况下,在污染物和悬浮石墨烯之间会发生光的多个反射和干涉。

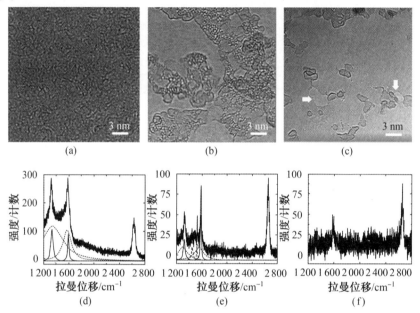

图 11.4 石墨烯表面不同清洁度的拉曼光谱

(a)~(c)是不同厚度 PMMA 残余物覆盖的石墨烯的 TEM 图像。(d)~(f)分别对应(a)~(c)中石墨烯的拉曼光谱。两个尖锐实线峰对应石墨烯的拉曼 G 和 D 模式,而两个点宽峰对应无定形碳的拉曼 G 和 D 模式。集中在 1 450 cm^{-1} 和 1 530 cm^{-1}(短划线)的峰,为激发波长 633 nm 下的 PMMA 两个特征峰。退火使 PMMA 致密化,会出现特征峰。(c)中箭头所指是采用强电子束照射产生的用于识别层数的洞穴[24]

图 11.4(f)显示了独立洁净的石墨烯的三种独特拉曼特征:①相对高的噪声;②相对低的拉曼强度;③低频(1 100~1 600 cm^{-1})内没有宽背底。

对于干净表面,入射光的多重反射受到抑制,信噪比和光谱中的峰值强度会减弱[37-40]。事实上,运用这一特性可快速探测表面清洁度。我们通过比较两个不清洁的石墨烯样品来证实这一点,如图 11.4(d)和图 11.4(e)所示。首先,从宽背底上极高强度的 D 带和 G 带叠加可以看到明显的差异。这个宽泛的信号范围为 1 100~1 600 cm^{-1},证实是无定形碳(sp^2 和

sp³ 键合碳的混合物)[41-42]。低频光谱被拟合成四部分:两个尖锐峰(点状的)是石墨烯的 G 带和 D 带,而相近位置其他两个宽峰(黑实线)源于无定形碳的 G 带和 D 带。无定形碳的拉曼信号随着 PMMA 覆盖面积增加而增强,当 PMMA 厚度达到几个纳米时饱和。然而只有消除光的衬底反射,才可以识别极薄的无定形碳的非共振拉曼信号,表明石墨烯是独立的。

  干净表面的另一个显著特点是拉曼光谱中的低信噪比。G 带位于 1 584 cm^{-1} 附近,说明费米级接近电荷中性点位置[43]。悬浮干净石墨烯的极高噪声和低峰强度可以通过干净的石墨烯片层的大量光反射来解释。图 11.5(a)和(b)所示分别是通过单层石墨烯和 PMMA/石墨烯叠层的光路。通过每个界面边界条件下多次反射和传输的重复转移矩阵形式,计算各层电场。通过 N 层堆积,得到电场向前(+)和向后(-)的振幅表达式为

$$\begin{bmatrix} E_0^- \\ E_0^+ \end{bmatrix} = H_{12}L_2 \cdots L_{N-1}H_{N-1,N} \begin{bmatrix} E_N^- \\ E_N^+ \end{bmatrix} \tag{11.1}$$

其中,H_{ij} 和 L_j 分别代表界面转移矩阵和层传播矩阵;H_{ij} 决定于传输系数 t_{ij} 和界面的反射系数 r_{ij},而 L_j 是传播层 j 的波相转移数。两个矩阵可写成

$$H_{ij} = \frac{1}{t_{ij}} \begin{bmatrix} 1 & r_{ij} \\ r_{ij} & 1 \end{bmatrix}, \quad L_j = \begin{bmatrix} e^{-i\alpha_j} & 0 \\ 0 & e^{i\alpha_j} \end{bmatrix} \tag{11.2}$$

其中,$t_{ij} = 2n_i/(n_i + n_j)$ 和 $r_{ij} = (n_i - n_j)/(n_i + n_j)$ 通过第 i 层和 j 层的折射率相关计算得到。对于正常入射光来说,当光穿过厚度为 d 的层时,产生的相转移为 $e^{-i2\pi n_j d/\lambda_0}$。通过简单的代数反射率可表示为

$$R = |r|^2 = |M_{12}/M_{22}|^2 \tag{11.3}$$

考虑正常入射光从空气($n_0 = 1$)到厚度 $d_2 = 0.335$ nm 的悬浮单层石墨烯[44]的情况,复数折射率 $n_2 \approx 2.0 \sim 1.1i$。虚数部分代表光吸收。通过转移矩阵 $M = H_{02}L_2H_{20}$,得到反射率 $R = (2r_{02}\sin \alpha/1 - r_{02}^2 e^{-i2\alpha})^2$。对于确定的 d_2,反射率与 d_2 呈正比例关系,如图 11.5 所示,也因此为区分石墨烯层数提供了一个与堆叠次序无关的简单方法。同样,两面都覆盖有 PMMA 的单层石墨烯,反射系数可以以 PMMA 折射率 $n_1 \approx 1.49$ 得到,跟光波长关系不大。图 11.5(c) 中插图是由于 PMMA/石墨烯/PMMA 三明治结构中光的干扰,随着 PMMA 厚度增加产生振荡的全反射。我们看到第一个最大强度出现在 $d_1 = 59$ nm。当 $d_1 < 59$ nm 时,可以利用拉曼 G 峰的强度来定性评价悬浮石墨烯的表面清洁度。

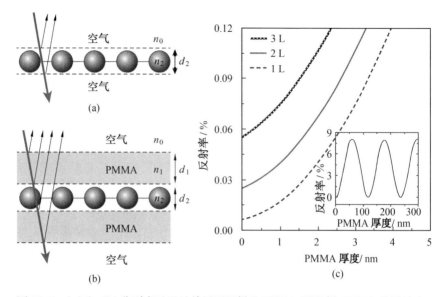

图 11.5 （a）和（b）分别表示通过单层石墨烯和 PMMA/石墨烯/PMMA 叠层的光的多反射。（c）悬浮石墨烯理论计算的反射率与 PMMA 厚度的函数关系。图中是 1~3 层石墨烯。插图是更广尺度内单层石墨烯中得到的相同结论[24]

11.4 吸附对石墨烯场效应晶体管的影响

前面章节论述了石墨烯转移过程会在表面留下一层薄薄的聚合物。本节的内容表明聚合物污染物只有通过热退火才可以除去。然而进行退火，污染也不会彻底清除干净，在石墨烯表面仍然会留有接近渗透阈值的聚合物残留。残余污染物的另一个来源是催化衬底（典型的铜或镍）上的金属纳米颗粒。虽然石墨烯可以防止水和氧的渗透，但是将石墨烯/金属暴露在空气中仍会引起金属表面部分氧化。金属氧化物通常会沿着样本边缘和石墨烯晶界缓慢生长，这些边界往往由各种晶格缺陷（如五边形-七边形对）组成（详见第 5 章）。用来去除金属性物质的刻蚀剂不太容易去除金属氧化物。经过湿转移过程后，上述两种残留物均会强烈吸附到石墨烯上，尤其是缺陷位置附近。只有在 C-C 断裂过程中聚合物和石墨烯之间形成了共价键，聚合物残留物（如 PMMA）才会对电荷传输产生很大影响。作为鲜明对比，金属残留物的长程库仑散射会明显降低电子和空穴迁移率，但当金属颗粒被氧化时，这种散射现象就会减弱。本章讨论了 PM-

MA 吸附物和金属颗粒对 CVD 石墨烯载流子迁移率的影响情况。

图 11.6 是 250 ℃ 退火前后,CVD 石墨烯场效应晶体管(FETs)的典型电阻率和栅极电压关系曲线图。用表达式 $\rho = 1/ne\mu + \rho_s$ 计算迁移率 μ,其中,n 和 ρ_s 分别表示电荷密度和额外电阻率。表达式中迁移率与载流子密度无关。分析得到退火前 CVD 石墨烯的电子迁移率 μ = 500 $cm^2 \cdot V^{-1} \cdot s^{-1}$。250 ℃ 下在氢气和氩气混合物中退火 2 h 后,电子迁移率 μ 降到 200 $cm^2 \cdot V^{-1} \cdot s^{-1}$。大部分 CVD 石墨烯退火后,迁移率都会降低,尤其是初始迁移率低的石墨烯。气体环境对迁移率影响不大,但是当退火温度增加时这种影响就会增加。通常情况下,电子和空穴的迁移率降低是对称的。这种结果可能来自衬底诱导的与石墨烯间的相互作用。然而 $T>400$ ℃ 时衬底效应开始明显[16]。这就排除了石墨烯-衬底相互作用增加是主要原因这一说法。另一个观点认为迁移率的下降是石墨烯与用来进行转移的聚合物之间的相互作用的结果。

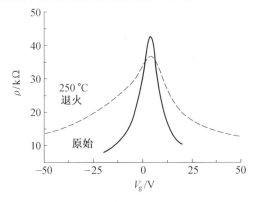

图 11.6 CVD 石墨烯退火前后的电阻和栅极电压关系

为了验证这一观点,利用密度泛函理论计算了聚合物(这里是 PMMA)覆盖理想的和有缺陷的石墨烯的吸附能及能带结构。采用了 SIESTA 码的完全自洽计算求解自旋极化 Kohn-Sham 方程[45-48]。局部态密度近似用于交换和相关项,能量网格截止为 200Ry。在 Monkhorst-Pack 体系中,布里渊区是一个 6×6×1K 点的网络[49]。因为聚甲基丙烯酸甲酯(PMMA 单体)是 PMMA 热降解的主要挥发性产物,所以我们只考虑 10×10 的石墨烯超级晶胞上的单体吸附。图 11.7 是三种优化的吸附构型,图 11.8 所示是它们相应的能带结构。对于理想的石墨烯,由于范德瓦耳斯作用使单体在表面被吸收。平衡距离为 0.283 nm 的吸附能是 -1.72 eV。吸附过程中石墨烯没有明显的构象变化。吸附方向不同会使平衡距离处的吸附能略微增

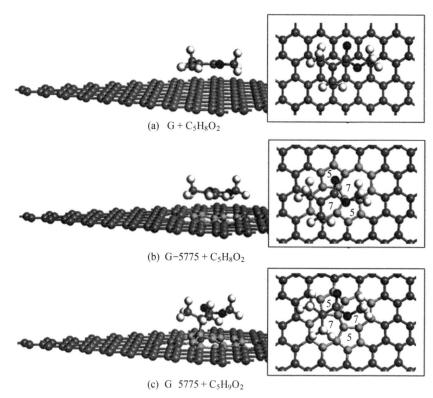

图 11.7 MMA($C_5H_8O_2$)最稳定结构的主视图和俯视图

(a)理想石墨烯。(b)石墨烯超级晶胞中的 Stone-Wales 缺陷。(c)$C_5H_9O_2$ 基团，在 Stone-Wales 缺陷顶部与石墨烯之间形成共价键[50]

加。图 11.8(a)所示清楚表明 MMA 的吸附对石墨烯的低能带结构没有影响，这与大部分的实验结论一致，MMA 的吸附也不会引起掺杂。即使存在 Stone-Wales 缺陷，也会引起与吸附物之间更加强烈的相互作用，MMA-石墨烯复合物的带结构与天然石墨烯的带结构也基本一致，如图 11.8(b)所示。而当石墨烯与 MMA 之间形成共价键时，碳从 sp^2 到 sp^3 的再次杂化，形成能隙，如图 11.8(c)所示，导致 K 点附近的费米速度明显降低。不同退火条件下悬浮石墨烯拉曼 2D 模式的系统蓝移现象表明费米速度降低，原则上可以推断出 MMA-石墨烯复合物中的载流子迁移率受到缺陷引发的短程散射极大影响。事实上经过 $T>250$ ℃退火后，在低初始迁移率的石墨烯器件中的迁移率会降低。

即使在理想石墨烯中，大分子的出现(如常用的 PMMA 和聚碳酸酯)

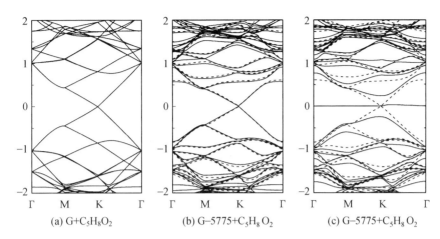

图 11.8　图 11.7 中各种 MMA-石墨烯复合物的能带结构。作为比较，用短划线表示天然石墨烯的能带结构

(a) 低能量下，吸附 MMA($C_5H_8O_2$) 的理想石墨烯的能带结构与天然石墨烯的能带结构非常吻合，说明可忽略 MMA 和石墨烯之间的相互作用。(b) 在石墨烯超级晶胞中引入 Stone-Wales 缺陷没有引起能带结构极大变化。(c) 由于碳从 sp^2 到 sp^3 的重杂化引起的能隙，导致 K 点附近费米速度的急剧下降[50]

也会对电荷迁移率产生不可忽视的影响，退火后仍是如此。而金属吸附物则使传输性能产生更多的剧烈变化。图 11.9 是天然石墨烯和不同浓度钾吸附的石墨烯的 σ-V_g 曲线[6]。超高真空下钾的吸附，对传输性能有以下影响：①使迁移率下降；②高浓度载流子下 σ-V_g 线性关系的增长；③最小电导率 $V_{g,min}$ 的栅极电压的偏移；④提高与 $V_{g,min}$ 相关的电子-空穴对称性。

后面两个特征是电子掺杂的直接结果。随着钾浓度的增加，$V_{g,min}$ 更多转向负栅极电压，说明电子从钾转移到石墨烯。纯净石墨烯中远离最小点的电导率，随着 V_g 增加出现次线性增长，但是在钾吸附时变得更加线性。此外，电荷转移引起电子电导和空穴电导的不对称。这归因于石墨烯电极里电子态密度非对称的金属诱导宽化[28]。导致电子掺杂抑制空穴电导，而空穴掺杂对电子也有同样的作用。

上述①和②的影响结果可以由带电杂质引起的长程库仑散射进行解释。基于玻耳兹曼传输理论的计算说明长程散射会导致 σ 对载流子密度的线性依赖，而短程散射则导致次线性依赖[13,51]。钾的吸附使电子从钾转移到石墨烯，导致带正电的静电中心通过长程库仑作用与移动电荷载流

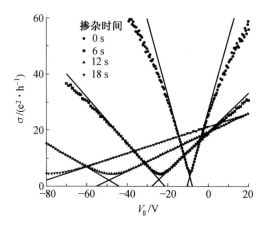

图 11.9 石墨烯和三个不同掺杂浓度的钾掺杂石墨烯的电导(σ)和栅极电压(V_g)关系[6]

子相互作用。结果如图 11.9 所示,随着越来越多的钾原子引入表面,由带电杂质引起的库仑相互作用成为体系中主要散射机制。载流子迁移率下降,$\sigma(n)$ 从次线性向线性发生变化。杂质密度可由石墨烯的长程库仑散射理论得到,即迁移率与散射中心的密度 $\mu n_{imp} = 5 \times 10^{15} V^{-1} \cdot s^{-1}$ 成反比例。

实际上,由于暴露在空气中,吸附的金属颗粒常以氧化物的形式存在。因为绝缘子里的电子被紧束缚,氧化物吸附物对石墨烯没有任何掺杂影响。

图 11.10 是另一个吸附实验,天然石墨烯、吸附 Ti 和 TiO_2 的石墨烯的 σ-V_g 比较情况。通过氧化吸附 Ti 的石墨烯样本可以获得吸附 TiO_2 的石墨烯样本,二者具有相同的浓度。Ti 颗粒的沉积会引起 σ-V_g 曲线的强烈

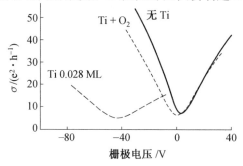

图 11.10 天然石墨烯、钛吸附石墨烯和一个单层(ML)是 1.908×10^{15} 个原子/cm^2 的二氧化钛吸附石墨烯的电导率(σ)和阈值电压关系的比较[15]

变化,类似于图 11.9 中钾掺杂的情况。暴露在氧气中对传输性能会有直接影响。$V_{g,min}$ 转变回到初始值,说明从金属 Ti 到绝缘 TiO_2 的转变降低了石墨烯的电子掺杂水平。$V_{g,min}$ 恢复到初始值与迁移率的增加相对应,就像在氧气下 σ-V_g 曲线变窄的情况一样。Ti 被氧化后,电荷杂质引起的散射逐渐减弱。这符合短程散射是石墨烯上绝缘吸附物的主要散射机制[26]。

11.5 石墨烯中聚合物残留物的去除

前面讨论了分子和金属吸附如何影响石墨烯载流子迁移率。本节解决吸附物去除的重要问题,尤其是物理吸附的大分子。最常遇到的大分子是电子束光刻技术、转移石墨烯、钝化处理或器件制造所用的 PMMA。与其他典型极性分子相比,如聚碳酸酯或水,PMMA 是与石墨烯具有强偶极子相互作用的长链分子。11.3 节论述了 PMMA 残余物的薄层紧贴在石墨烯表面,用任何熟知的有机溶剂都很难去除。使石墨烯晶格完好无损又能去除 PMMA 的方法是在合适的气氛和持续时间下进行热处理。这就需要利用 11.3 节中的 TEM 和拉曼合并方法,在不同退火条件下对 PMMA 分解程度及石墨烯晶格完整情况进行系统评价。

高分子的热分解是一个复杂的自由基链反应。PMMA 的分解反应一直是众多研究的课题,该过程通常包括三步重量损失[52]。因为 H—H 键解离能低于 C—C 支链,所以 H—H 链断开引发第一步(约 165 ℃)。第二步由不饱和端的断裂引发,包括乙烯基团 β 断裂。最后一步需要最大活化能,由聚合物链内随机断裂引发。C—C 的随机断裂是分解 PMMA 的关键过程,甲基丙烯酸甲酯(单体)是主要产物[53-54]。单体的生成伴随大量低分子量的稳定物的形成(H_2、CO、CO_2、CH_4、C_2H_4、C_2H_6 和 $HCOOCH_3$)[55]。在典型随机断裂中,大自由基经 β 断裂释放单体,聚合物分子链解离。然而两个键同时断裂释放一个单体却不太可能[56]。在这种情况下,单体和其他稳定物产生之前,随机断裂先产生大量自由基。自由基可能连接到与周围聚合物分子链相互作用的其他长的大分子碎片上。碎片的相互作用和惰性能完全阻止键的断裂,尤其当碎片在分子链内部或与衬底(如石墨烯)相连接[57]。

X 射线光电子能谱是定性分析退火后悬浮石墨烯上的 PMMA 残余物信息的实用灵敏的方法。图 11.11 是不同温度下退火的石墨烯的 C1s 芯能级谱,其中采用 Shirley 算法扣除 C1s 背底以便更好地拟合。每个图中最强

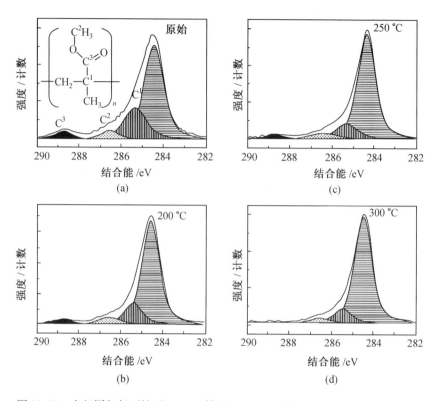

图 11.11 在相同气氛环境下,PMMA 转移 CVD 石墨烯的 C 1s 芯能级 XPS 谱
(a)退火前和退火后;(b)200 ℃;(c)250 ℃;(d)300℃。sp^2 C—C 键(横条纹)结合能是 284.4eV,PMMA 中 C^1、C^2 和 C^3 的化学位移为+1.4 eV (+2.4±0.1)eV 和(+4.3±0.1)eV,如(a)中插入的示意图所示[50]

的峰(横条纹)基本是由石墨烯中 sp^2 和 sp^3 碳组成,可进一步分峰成两个相关的次峰。sp^3 杂化结合能相对于 sp^2 时发生约 0.7 eV 的偏移,这与先前研究报道的石墨和金刚石之间的结合能差接近[58-59]。能谱中其他尾峰与 PMMA 或石墨烯上大量官能团中的不同碳原子的结合能相对应,相对于 284.4 eV 的主峰发生能量偏移为+1.4 eV、(+2.4±0.1)eV 和(+4.3±0.1)eV[60-62]。这些尾峰的强度变化说明在高于 200 ℃下退火 PMMA 残余物会减少,也证实了引发反应可能包括甲基羰基(MC═COOCH$_3$)支链的断裂。

影响表面清洁度最重要的退火参数是温度、气氛和时间。在特定温度和气氛下退火 30~60 min,表面清洁度通常不会得到改善。因此在下面论述中,将退火时间调整为 60 min,并改变温度和气氛。对于气氛来说,真空退火提供了最温和的路线,高温下进行也不会对石墨烯有太大损害。氧气下退火是最粗暴的方法,需要控制好退火时间和温度。这里介绍两个悬浮态石墨烯退火的例子。类型Ⅰ,在 H_2 和 Ar 混合气流中退火 2 h。类型Ⅱ,石墨烯片先在空气中退火 1 h,接着在 H_2(200 cm^3/min)和 Ar(400 cm^3/min)混合气流下再退火 1 h。

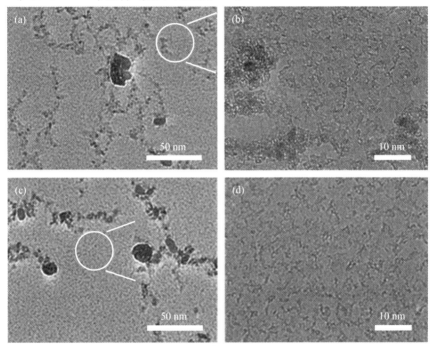

图 11.12 200 ℃ 退火后有 PMMA 残留物的 CVD 石墨烯的 TEM 图
(a)类型Ⅰ石墨烯。(c)类型Ⅱ石墨烯。(b)和(d)分别表示(a)和(c)中圆圈部分的放大图像

图 11.12 是类型Ⅰ和类型Ⅱ样本在 200 ℃ 下退火的石墨烯的 TEM 图像。200 ℃,悬浮石墨烯层能承受长时间退火且没有微小损坏[50]。正如前面论述的,延长退火时间(大于 2 h)对表面清洁度没有任何改善。为了更多了解石墨烯表面的 PMMA 的分解情况,我们按照厚度来标识 PMMA 残余物。如 11.3 节中介绍的,纯净 PMMA 吸附物厚度为 1~2 nm。PMMA 按照两步进行分解:外部 PMMA(对着空气)在低于 200 ℃ 下开始降解,退火后出现许多蠕虫状的条样;PMMA 内部大部分(与石墨烯接触)在该温度下保持完好无损,说明 PMMA 残余物内部的断裂反应只在高温下才会发生。

对于电子束光刻或 PMMA 辅助转移,外部 PMMA 通常是 3~5 层厚,而内部 PMMA 直接与石墨烯相连,只有单层覆盖。经过退火后,外部 PMMA 用包裹的颗粒形成带状。证实这些颗粒是非结晶的 CuO_x,带电杂质散射被抑制[15],使迁移率减缓降低。这解释了退火后迁移率会增加的原因。从 PMMA 薄膜的分解中可知[52],初始断裂发生在 160 ℃,但实验显示 200 ℃才是确保石墨烯外部 PMMA 残余物分解的关键。然而即使在更严格的退火条件下(如 350 ℃,5 h),外部 PMMA 残余物也不会彻底去除。图 11.12(b)是接近渗流阈值的极薄的内部 PMMA 残余物的网络结构的 TEM 图像,说明内部 PMMA 的初始断裂需要更高温度(大于 200 ℃)。

退火温度增加到 250 ℃ 能够提高表面的清洁度,如图 11.13(a)和图 11.13(b)所示。大部分内部的 PMMA 被烧掉,留下干净表面,但却留下条状的外部 PMMA 残余物。与外部 PMMA 一样,内部 PMMA 即使在更高的

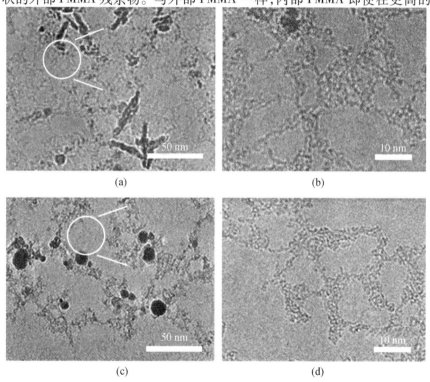

图 11.13 250 ℃ 退火后具有 PMMA 残留物的 CVD 石墨烯的 TEM 图像
(a)类型 I 石墨烯。(c)类型 II 石墨烯。(b)和(d)分别是(a)和(c)中圆圈部分的放大图像

温度下(大于 250 ℃)也不会彻底分解。这可能因为 C—C 键的随机断裂产生大量 PMMA 自由基与周围聚合物长链或石墨烯缺陷相连,阻止了 PMMA 膜的分解。300 ℃ 和 350 ℃ 下的退火表明更高的温度不会得到更加清洁的表面,反而可能在 TEM 网格或支撑石墨烯的金属支架弯曲时损坏石墨烯。

图 11.13(b)是类型 II 样本经过退火后的石墨烯表面。这些干净的表面经过空气中 250 ℃ 退火,又在 H_2 和 Ar 混合气体中进行第二轮退火得到。这种方法使干净的面积从 100 nm^2 扩大到约 $8×10^3$ nm^2。有氧气时退火 30 min 会引起各种缺陷。虽然洁净面积很小,但为了不降低石墨烯的质量,可以在这个温度下在形成气体(稀释的氢气)中进行单轮退火。实际上,除去石墨烯上的聚合物吸附物比看起来更困难、更复杂。尽管如此,还是可以依照以下对石墨烯进行退火:

(1) 退火温度高于 250 ℃,可能不会损坏衬底上的石墨烯,但高于 350 ℃ 时,石墨烯和衬底之间的相互作用会使电子迁移率降低[16]。

(2) 对于独立的石墨烯,温和气氛下可以使用更高温度。例如,真空下 700 ℃ 退火几个小时后,石墨烯没有明显的损坏。

虽然退火温度超过 250 ℃ 能够稍微提高 PMMA 的降解,但需要在表面清洁度和迁移率间权衡,如 11.4 节中提到的。图 11.14 是 PMMA 最终随机断裂的结构示意图。高温退火可能促使石墨烯缺陷和 PMMA 之间的共价键的形成。图 11.15(a)(c)中拉曼 2D 蓝移也说明了这一点。独立和悬浮石墨烯片退火后,2D 频带的蓝移清楚可见。蓝移范围 Δ_{2D} = 3 ~ 23 cm^{-1},主要由退火温度决定。对于独立石墨烯,Δ_{2D} 总体平均是 6 cm^{-1}、11 cm^{-1} 和 12 cm^{-1},分别对应 200 ℃、250 ℃ 和 300 ℃ 的退火,如图 11.15(c)所示。

2D 蓝移可能受到各种机制的制约。第一种与引发 2D 模式散射的压缩应变相关联。由于石墨烯片的独立结构,我们先不考虑其起源。另一种可以解释 2D 偏移产生的原因是掺杂。而独立的石墨烯呈电荷中性。两种掺杂类型(电子或空穴)都会引起明确的 G 峰上移和降低的 I_{2D}/I_G[63-67]。超过 80% 的测试样品没有 G 峰偏移,其余样品表现出 1 ~ 3 cm^{-1} 的偏移,这一结果可能是实验误差或由气体分子吸附引起的无意掺杂造成的。因此我们可以通过掺杂排除任何导致产生偏移的因素。一种可能产生 2D 蓝移的情况,如图 11.15(d)所示。C—C 键的随机断裂会产生大量的 PMMA 自由基。11.4 节中的 DET 算法表明 PMMA 自由基可能与石墨烯缺陷形

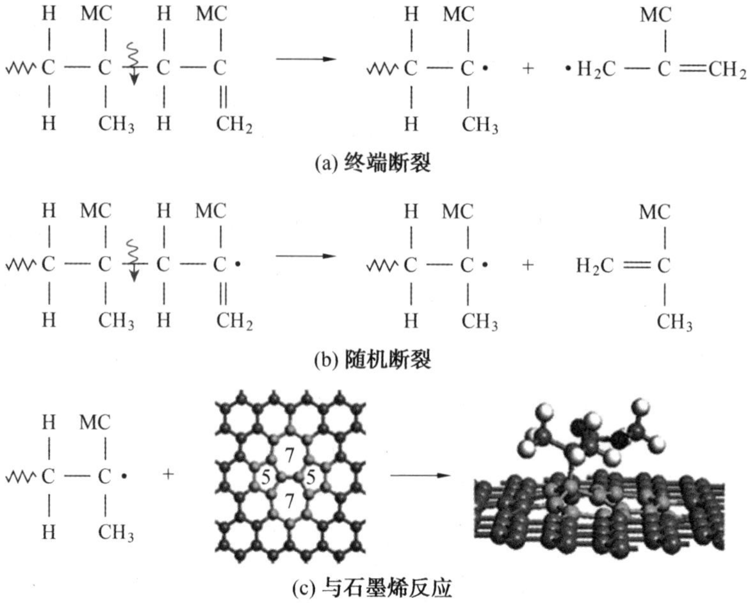

图 11.14　(a) PMMA 的终止断裂。MC 表示甲氧甲酰(MC-COOCH$_3$)。(b) PMMA 的随机断裂,主要产生 MMA。(c) PMMA 重复单元和具有 5-7-7-5 成对缺陷的石墨烯之间的共价键的形成。先形成吸附物[50]

成共价键,通常发生在晶界处。碳原子发生 sp^2 到 sp^3 的转变,进而会改变石墨烯在费米能级附近的谱带结构。局部能带最重要的变化是费米速率的降低[68-69],导致 2D 蓝移。然而,理想的石墨烯结构并不利于与 PMMA 自由基形成共价键。此外,PMMA 在石墨烯上的物理吸附残留物既不会引起能带结构变化,也不会引起可觉察的电荷传输。

11.6　结　论

聚合物和金属颗粒残留物是 CVD 石墨烯上两种主要的无意吸附物,如果吸附物和石墨烯之间形成共价键,那么通过带电杂质的长程库仑相互作用和缺陷附近的费米速率的重正化,两种吸附物会降低载流子迁移率。虽然退火很难彻底除去污染物,但它是去除石墨烯聚合物污染的简单方法。适当条件下的退火可以有效地降低具有小晶粒尺寸的石墨烯中 sp^2 到 sp^3 的杂化转变。通过拉曼光谱可以判断退火样本的表面清洁度,并给出

第 11 章 吸附物对石墨烯电子传输的影响

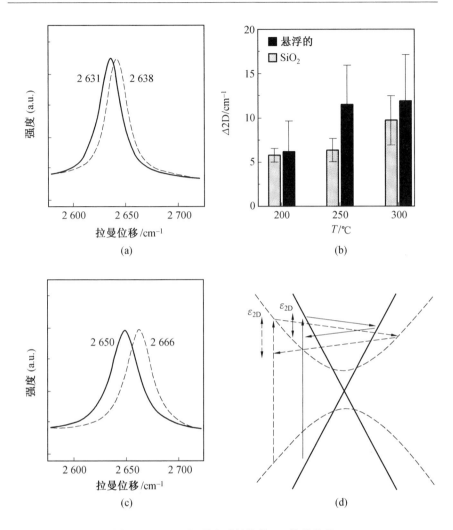

图 11.15 250 ℃ 退火后的拉曼 2D 带的蓝移

(a) SiO_2/Si 衬底上的石墨烯退火前(实线)和退火后(虚线)的拉曼 2D 带。(b) 独立石墨烯退火前(实线)和退火后(虚线)的拉曼 2D 带。(c) 不同温度下 2D 带蓝移柱状图。(d) 天然(直线)和退火(双曲线)的低能带结构示意图。箭头指示的是两种石墨烯中的双共振过程[50]

大面积范围内丰富的表面信息。为了避免石墨烯的表面吸附,采用无催化剂生长或无聚合物转移的技术,对于石墨烯的发展是非常必要的。

11.7 参考文献

[1] BOLOTIN K I, SIKES K J, JIANG Z, et al. Ultrahigh electron mobility in suspended graphene[J]. Solid State Communications, 2008, 146:351.

[2] MOROZOV S V, NOVOSELOV K S, KATSNELSON M I, et al. Giant intrinsic carrier mobilities in graphene and its bilayer[J]. Physics Review Letters, 2008, 100:016602.

[3] NI Z H, PONOMARENKO L A, NAIR R R, et al. On resonant scatterers as a factor limiting carrier mobility in graphene[J]. Nano Letters, 2010, 10:3868.

[4] LAFKIOTI M, KRAUSS B, LOHMANN T, et al. Graphene on a hydrophobic substrate: doping reduction and hysteresis suppression under ambient conditions[J]. Nano Letters, 2010, 10:1149.

[5] PONOMARENKO L A, YANG R, MOHIUDDIN T M, et al. Effect of a high-κ environment on charge carrier mobility in graphene[J]. Physics Review Letters, 2009, 102:206603.

[6] CHEN J H, JANG C, ADAM S, et al. Charged-impurity scattering in graphene[J]. Nature Physics, 2008, 4:377.

[7] CHO S, FUHRER M S. Charge transport and inhomogeneity near the minimum conductivity point in graphene[J]. Physics Review B, 2008, 77:081402.

[8] ZHANG Y, TAN Y W, STORMER H L, et al. Experimental observation of the quantum Hall effect and Berry's phase in graphene[J]. Nature, 2005, 438:201.

[9] FRADKIN E. Critical behavior of disordered degenerate semiconductors. II. Spectrum and transport properties in mean-field theory[J]. Physics Review B, 1986, 33:3263.

[10] LEE P A. Localized states in a d-wave superconductor[J]. Physics Review Letters, 1993, 71:1887.

[11] LUDWIG A W W, FISHER M P A, SHANKAR R, et al. Integer quantum hall transition: an alternative approach and exact results[J]. Physics Review B, 1994, 50:7526.

[12] SHON N H, ANDO T. Quantum transport in two-dimensional graphite

system[J]. Journal of the Physical Society of Japan,1998,67:2421.
[13] HWANG E H,ADAM S,SARMA S D. Carrier transport in two-dimensional graphene layers[J]. Physics Review Letters,2007,98:186806.
[14] MARTIN J,AKERMAN N,ULBRICHT G,et al. Observation of electron-hole puddles in graphene using a scanning single-electron transistor[J]. Nature Physics,2008,4:144.
[15] MCCREARY K M,PI K,KAWAKAMI R K. Metallic and insulating adsorbates on graphene[J]. Applied Physics Letters,2011,98:192101.
[16] CHENG Z,ZHOU Q,WANG C,et al. Toward intrinsic graphene surfaces: a systematic study on thermal annealing and wet-chemical treatment of SiO_2-supported graphene devices[J]. Nano Letters,2011,11:767.
[17] FERRALIS N,MABOUDIAN R,CARRARO C. Evidence of structural strain in epitaxial graphene layers on 6H-SiC(0001)[J]. Physics Review Letters,2008,101:156801.
[18] HE R,ZHAO L,PETRONE N,et al. Large physisorption strain in chemical vapor deposition of graphene on copper substrate[J]. Nano Letters,2012,12:2408.
[19] FRANK O,TSOUKLERI G,PARTHENIOS J,et al. Compression behavior of single-layer graphenes[J]. ACS Nano,2010,4:3131.
[20] NI Z,WANG Y,YU T,et al. Reduction of Fermi velocity in folded graphene observed by resonance Raman spectroscopy[J]. Physics Review B,2008,77:235403.
[21] CHAN K T,NEATON J B,COHEN M L. First-principles study of metal adatom adsorption on graphene [J]. Physics Review B, 2008, 77:235430.
[22] WEHLING T O,KATSNELSON M I,LICHTENSTEIN A I. Impurities on graphene: midgap states and migration barriers[J]. Physics Review B,2009,80:085428.
[23] LI X,ZHU Y,CAI W,et al. Transfer of large-area graphene films for high-performance transparent conductive electrodes[J]. Nano Letters,2009,9:4359.
[24] LIN Y C,JIN C,LEE J C,et al. Clean transfer of graphene for isolation and suspension[J]. ACS Nano,2011,5:2362.

[25] REGAN W, ALEM N, ALEMáN B, et al. A direct transfer of layer-area graphene[J]. Applied Physics Letters,2010,96:113102.
[26] OHTA T, BOSTWICK A, SEYLLER T, et al. Controlling the electronic structure of bilayer graphene[J]. Science,2006,313:951.
[27] PONOMARENKO L A, SCHEDIN F, KATSNELSON M I, et al. Chaotic dirac billiard in graphene quantum dots[J]. Science,2008,320:356.
[28] FARMER D B, GOLIZADEH-MOJARAD R, PEREBEINOS V, et al. Chemical doping and electron-hole conduction asymmetry in graphene devices[J]. Nano Letters,2009,9:388.
[29] LU C C, JIN C, LIN Y C, et al. Characterization of graphene grown on bulk and thin film nickel[J]. Langmuir,2011,27:13748.
[30] REINA A, JIA X, HO J, et al. Large area, few-layer graphene films on arbitrary substrates by chemical vapor deposition[J]. Nano Letters,2009,9:30.
[31] SUTTER P, FLEGE J I, SUTTER E A. Epitaxial graphene on ruthenium[J]. Nature Materials,2008,7:406.
[32] SUTTER P, SADOWSKI J T, SUTTER E. Graphene on Pt(111): growth and substrate interaction[J]. Physics Review B,2009,80:245411.
[33] KIM K S, ZHAO Y, JIANG H, et al. Large-scale pattern growth of graphene films for stretchable transparent electrodes[J]. Nature,2009,457:706.
[34] LI X, CAI W, AN J, et al. Large-area synthesis of highquality and uniform graphene films on copper foils[J]. Science,2009,324:1312.
[35] JIAO L, FAN B, XIAN X, et al. Creation of nanostructures with poly(methyl methacrylate)-mediated nanotransfer printing[J]. The Journal of American Chemical Society,2008,130:12612.
[36] GERINGER V, SUBRAMANIAM D, MICHEL A K, et al. Electrical transport and low-temperature scanning tunneling microscopy of microsoldered graphene[J]. Applied Physics Letters,2010,96:082114.
[37] BLAKE P, HILL E W, CASTRO NETO A H, et al. Making graphene visible[J]. Applied Physics Letters,2007,91:063124.
[38] CASIRAGHI C, HARTSCHUH A, LIDORIKIS E, et al. Rayleigh imaging of graphene and graphene layers[J]. Nano Letters,2007,7:2711.
[39] RODDARO S, PINGUE P, PIAZZA V, et al. The optical visibility of gra-

phene: interference colors of ultrathin graphite on SiO_2[J]. Nano Letters,2007,7:2707.

[40] WANG Y Y,NI Z H,SHEN Z X,et al. Interference enhancement of raman signal of graphene[J]. Applied Physics Letters,2008,92:043121.

[41] CAPANO M A,MCDEVITT N T,SINGH R K,et al. Characterization of amorphous carbon thin films[J]. Journal of Vaccum Science & Technology A,1996,14:431.

[42] FERRARI A C,ROBERTSON J. Raman spectroscopy of amorphous, nanostructured,diamond-like carbon, and nanodiamond[J]. Philosophical Transactions of the Royal Society of London A,2004,362,2477.

[43] YAN J,ZHANG Y,KIM P,et al. Electric field effect tuning of electron-phonon coupling in graphene[J]. Physics Review Letters,2007,98:166802.

[44] NI Z H,WANG H M,KASIM J,et al. Graphene thickness determination using reflection and contrast spectroscopy[J]. Nano Letters,2007,7:2758.

[45] BRANDBYGE M,MOZOS J L,ORDEJóN P,et al. Densityfunctional method for nonequilibrium electron transport[J]. Physics Review B,2002,65:165401.

[46] QUANTUM WISE A/S,Atomistix Tool Kit,2010,Ch. 10. 8. 2.

[47] SOLER J M,ARTACHO E,GALE J D,et al. The SIESTA method for ab initio order-N materials simulation[J]. Journal of Physics: Condensed Matter,2002,14:2745.

[48] TAYLOR J,GUO H,WANG J. Ab initio modeling of quantum transport properties of molecular electronic devices[J]. Physics Review B,2001,63:245407.

[49] MONKHORST H J,PACK J D. Special points for Brillouin-zone integrations[J]. Physics Review B,1976,13:5188.

[50] LIN Y C,LU C C,YEH C H,et al. Graphene annealing: how clean can it be? [J]. Nano Letters,2012,12:414.

[51] ADAM S,HWANG E H,GALITSKI V M,et al. A self-consistent theory for graphene transport[J]. Proceedings of the National Academy of Science USA,2007,104:18392.

[52] KASHIWAGI T,INABA A,BROWN J E,et al. Effects of weak linkages

on the thermal and oxidative degradation of poly(methyl methacrylates)[J]. Macromolecules,1986,19:2160.
[53] ARISAWA H,BRILL T B. Kinetics and mechanisms of flash pyrolysis of poly(methyl methacrylate) (PMMA)[J]. Combust Flame,1997,109:415.
[54] COSTACHE M C,WANG D,HEIDECKER M J,et al. The thermal degradation of poly(methyl methacrylate) nanocomposites with montmorillonite,layered double hydroxides and carbon nanotubes[J]. Polymers for Advanced Technologies,2006,17:272.
[55] MADORSKY S L. Thermal degradation of organic polymers[J]. New York:Interscience Publisher,1964.
[56] STOLIAROV S I,WESTMORELAND P R,NYDEN M R,et al. Reactive molecular dynamics model of thermal decomposition in polymers:1. Poly(methyl methacrylate)[J]. Polymer,2003,44:883.
[57] MORGAN A B,ANTONUCCI J M,VANLANDINGHAM M R,et al. Thermal and flammability properties of a silica-PMMA nanocomposites[J]. Polymeric Materials:Science and Engineering,2000,83:57.
[58] CHEN C T,SETTE F. High resolution soft x-ray spectroscopies with the dragon beamline[J]. Physica Scripta,1990,T31:119.
[59] MORAR J F,HIMPSE F J,HOLLINGER G,et al. C 1s excitation studies of diamond (111). I. Surface core levels[J]. Physics Review B,1986,33:1340.
[60] CHIANG T C,SEITZ F. Photoemission spectroscopy in solids[J]. Annals of Physics,2001,10:61.
[61] GROSS T,LIPPITZ A,UNGER W E S. Some remarks on fitting standard-and high resolution C 1s and O 1s x-ray photoelectron spectra of PMMA[J]. Applied Surface Science,1993,68:291.
[62] YUMITORI S. Correlation of C 1s chemical state intensities with the O 1s intensity in the XPS analysis of anodically oxidized glass-like carbon samples[J]. Journal of Materials Science,2000,35:139.
[63] CASIRAGHI C. Doping dependence of the Raman peaks intensity of graphene close to the Dirac point[J]. Physics Review B,2009,80:233407.
[64] DAS A,PISANA S,CHAKRABORTY B,et al. Monitoring dopants by Raman scattering in an electrochemically top-gated graphene transistor

[J]. Nature Nanotechnology,2008,3:210.
[65] LIN Y C,LIN C Y,CHIU P W. Controllable graphene N-doping with ammonia plasma[J]. Applied Physics Letters,2010,96:133110.
[66] MEDINA H,LIN Y C,OBERGFELL D,et al. Tuning of charge densities in graphene by molecule doping[J]. Advanced Functional Materials, 2011,21:2687.
[67] MOHIUDDIN T M G,LOMBARDO A,NAIR R R,et al. Uniaxial strain in graphene by Raman spectroscopy: G peak splitting,Grüneisen parameters,and sample orientation[J]. Physics Review B,2009,79:205443.
[68] NI Z H,YU T,LU Y H,et al. Uniaxial strain on graphene: raman spectroscopy study and band-gap opening[J]. ACS Nano,2008,2:2301.
[69] PONCHARAL P,AYARI A,MICHEL T,et al. Raman spectra of misoriented bilayer graphene[J]. Physics Review B,2008,78:113407.

第12章 石墨烯中的单电荷传输

本章论述了石墨烯器件中的单电荷隧穿行为,并将石墨烯压缩物、图案化的单电子晶体管和石墨烯纳米带中的库仑阻塞现象与具有金属岛的传统单电子晶体管中的进行比较。讨论了在遵循量子霍尔理论下,石墨烯中形成压缩量子点时发生的现象,包括由单电荷遂穿进出量子点引起的微观电导波动。

12.1 引 言

通常,电子器件的宏观传输测试中看不到电子离散现象。平均来看,独立电子行为都是隐藏的,只能获得统计测量值。要弄清独立电荷传输行为,就要研究粒子束缚波函数的量子态,可先从以下研究入手:①金属岛中的库仑斥力,即单电子晶体管(SETs);②由尺寸效应引起的量子约束,即量子点(QDs);③高磁场下不可压缩的凝聚中的电荷填充,即量子霍尔效应(QHE)。

前两项研究要求是纳米尺寸的小器件,而量子霍尔效应[1]要求是几百微米级的宏观器件。正如近期研究报道的,量子霍尔局域化与 SETs 或 QDs[2]的物理学本质都基于同一原则。单电荷行为的研究主要针对巨大粒子、传统金属和半导体的体系。对结晶石墨进行机械剥离获得石墨烯,该无质量粒子的二维电子体系(2DES)带动了对单个无质量电荷传输行为的研究。

本章组成:12.2 节介绍单电荷隧穿行为;12.3 节讨论石墨烯的电学性能;12.4 节回顾有关石墨烯中无质量粒子的单电荷传输行为的研究;12.5 节论述石墨烯中量子霍尔局域化行为,尤其是控制电荷隧穿进入可压缩 QDs 的库仑阻塞物理研究。

12.2 单电荷隧穿

12.2.1 单电荷隧穿和库仑阻塞

当导电隧道中的电子路径被电容 C 的绝缘能隙阻塞,如果绝缘载流子的隧道电阻 R_T 比量子电阻($h/e^2 \approx 25.813 \text{ k}\Omega$)大很多,由于量子隧穿,则会发生穿过能隙的电导,这里 h 是普朗克常量,e 是电子电荷。充电能级($E_c \equiv e^2/C$)对穿过势垒的电导起着重要作用,其中热能 k_BT 远小于 E_c。这里 k_B 是玻耳兹曼常数,T 是温度。因此低温下会发生单电荷遂穿。单电子晶体管(SET)是研究单电荷隧穿行为的常用器件,由金属岛和三个电极(源极、漏极和栅极)组成,如图 12.1(a) 所示。这些组成部分与传统晶体管相同。岛除了与三个电极的弱耦合,与外部环境完全分离。

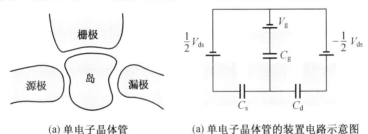

(a) 单电子晶体管　　(a) 单电子晶体管的装置电路示意图

图 12.1　单电子晶体管及其装置电路示意图

为了将附加电子连接到岛上,例如,从 N_e 到 (N_e+1),其中,N_e 是岛中满态的电子数目,由于库仑斥力需要有附加能量,使 (N_e+1) 和 N_e 电子的电化学势产生差异,$\Delta E_{add} = \mu_{N+1} - \mu_N$。其中电化学势[3,4]为

$$\mu_N = E_C\left(N_e - \frac{1}{2}\right) + e\alpha_g V_g \tag{12.1}$$

栅极影响因子是 $\alpha_g \equiv C_g/C_\Sigma$,其中总电容 $C_\Sigma = C_s + C_d + C_g$,且 C_s、C_d 和 C_g 分别是源极、漏极和栅极的电容。当岛尺寸足够小,能级间距超过热致宽时,会出现离散能量态。岛被认为是人造原子或 QD,按照泡利不相容原理,不能忽略 QD 中 N 电子的单粒子能量 E_N。电荷注入 QD 所需要的能量等于相互作用引发能量(充电能量)和单粒子能量间距(常被称为量子约束)的总和;$\Delta_N = \varepsilon_{N+1} - \varepsilon_N$,其中 ε_N 是单粒子能量,$\Delta E_{add} = E_c + \Delta_N$。当岛的电化学势通过具有较小源-漏偏压 V_{ds} 的栅极电压 V_g 的调整而进行调

谐,岛中的离散能级就会逐渐升高或降低。当源极和漏极($\mu_s \approx \mu_d$)的化学势处于岛的离散能级之间的能隙区域时,由于强烈的库仑斥力,电子无法通过从漏极到岛的隧道。如图12.2(a)所示。只有当岛的电位(a level of the island)接近μ_s和μ_d,电流才能流动,如图12.2(b)所示。这就是库仑阻塞,它会引起与栅极电压有函数关系的电导的振荡,如图12.2(c)所示。 由于泡利不相容原理,($N_e + 1$)态可被一个电子占据,所以一次只能发生一个电子隧穿。这就是为什么库仑阻塞表现为单电荷隧穿的原因;单电荷隧穿会出现电导。热能$k_B T$使库仑峰变宽。振荡周期ΔV_g和单电荷隧穿所需能量ΔE_{add}的关系为$e\alpha_g \Delta V_g = \Delta E_{add} - \Delta_N$。如果岛的尺寸足够大,可以忽略量子约束,等式简化为$e\alpha_g \Delta V_g = \Delta E_{add}$,从而可以确定栅极影响因子,即体系总能量与栅极单电子充电能量$e\Delta V_g$的比值。

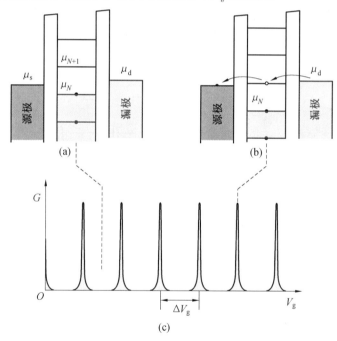

图 12.2 单电子晶体管中单电荷隧穿行为示意图
(a)当源极和漏极的化学势位于N_e电子能级和($N_e + 1$)电子能级之间的能隙中时,电导被阻断。(b)当源极和漏极的费米能级相当时,会发生每次一个电子的隧穿。(c)G带的库仑振荡与V_g函数关系

源-漏偏压变化也会影响库仑振荡。电子从源极到岛(ΔE_s)和从岛到漏极(ΔE_d)的隧穿后,能量随V_{ds}的变化关系为

$$\Delta E_{\mathrm{s}} = \frac{e}{C_{\Sigma}}\left[e\left(N+\frac{1}{2}\right) - V_{\mathrm{ds}}\left(C_{\mathrm{s}}+\frac{C_{\mathrm{g}}}{2}\right) + C_{\mathrm{g}}V_{\mathrm{g}}\right] \quad (12.2)$$

$$\Delta E_{\mathrm{d}} = \frac{e}{C_{\Sigma}}\left[e\left(N+\frac{1}{2}\right) - V_{\mathrm{ds}}\left(C_{\mathrm{s}}+\frac{C_{\mathrm{g}}}{2}\right) + C_{\mathrm{g}}V_{\mathrm{g}}\right] \quad (12.3)$$

Grabert 和 Devoret（1992）及 Beenakker（1991）给出了详细计算。假定 0 ℃ 下，当两种能量都是负值时，$\Delta E_{\mathrm{s}} < 0$ 和 $\Delta E_{\mathrm{d}} < 0$，电子从源极到漏极隧穿，否则电导为 0。这为电导阻断提供了条件：

$$\begin{cases} e\left(N-\frac{1}{2}\right) < C_{\mathrm{g}}V_{\mathrm{ds}} + \left(C_{\mathrm{s}}+\frac{1}{2}C_{\mathrm{g}}\right)V_{\mathrm{g}} < e\left(N+\frac{1}{2}\right) \\ e\left(N-\frac{1}{2}\right) < C_{\mathrm{g}}V_{\mathrm{ds}} + \left(C_{\mathrm{s}}+\frac{1}{2}C_{\mathrm{g}}\right)V_{\mathrm{g}} < e\left(N+\frac{1}{2}\right) \end{cases} \quad (12.4)$$

当 $V_{\mathrm{g}} - V_{\mathrm{ds}}$ 平面进行投影，明确了电导受阻机理，该机理过程具有菱形的结构，就是所说的库仑菱形（图 12.3）。在阴影区域中，在 V_{g} 和 V_{ds} 函数关系下测得的 SET 电导为 0，在菱形边缘存在库仑峰。按照边缘的正负斜率，C_{d} 和 C_{s} 值可以按照式（12.4）进行估算。

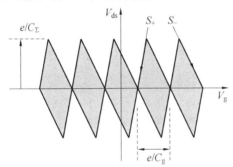

图 12.3　库仑菱形结构
（阴影菱形是电导被抑制的区域）

12.3　石墨烯的电性能

12.3.1　石墨烯的电子结构

石墨烯是石墨和碳纳米材料（如碳纳米管）的基本构建模块。石墨烯由具有蜂窝晶格的单层碳原子组成，所以是纯二维体系。它有以下独特性能：高热导率[5]，室温下电子和空穴电导有着极高近乎相同的迁移率，可达 200 000 $cm^2 \cdot V^{-1} \cdot s^{-1}$[6-8] 和微米尺寸的弹道传输[6-8]。由于线性能量色散关系,石墨烯的这些特性源于电荷载流子无质量性[9-18],使石墨烯对于纳米电子器件来讲是一种大有前途的可选材料。

石墨烯的晶格结构以及它的能带结构如图12.4(a)所示。晶格有两个子晶格A和B,导致出现除常见的自旋简并之外的能谷简并(常称为赝自旋)。在狄拉克点附近,赝自旋使手性与波函数的两部分有关联[9]。不考虑二阶跃迁,能量散射关系为

$$E(k_x, k_y) = \pm \gamma_1 \sqrt{1 + 4\cos\frac{\sqrt{3}ak_y}{2}\cos\frac{ak_y}{2} + 4\cos^2\frac{ak_y}{2}} \quad (12.5)$$

其中,k_x和k_y是二维动量要素;γ_1是最近邻跳跃能量;a是晶格常数($a = \sqrt{3}a_0$,a_0是石墨烯蜂窝晶格中C—C距离[9,11]。图12.4(b)和(c)是能带结构图。零能量附近的散射关系的简化形式是$E(q) \approx \pm v_F q$,其中,v_F是费米速率(约为$1 \times 10^8 \mathrm{cm \cdot s^{-1}}$),$q$是狄拉克点($K$或$K'$)的相对力矩。图12.4(c)是线性色散。由于零能量附近的线性色散关系,使电荷载流子遵循狄拉克方程而不是薛定谔方程。狄拉克方程为$-iv_F \boldsymbol{\sigma} \cdot \nabla \Psi(r) = E\Psi(r)$,其中$\boldsymbol{\sigma}$是泡利矩阵,$\Psi(r)$是粒子的波函数。

12.3.2 石墨烯的传输性能

传输测试通常处理较低能量。在石墨烯的传输性能测试中,线性色散关系成立。在传统的半导体中,粒子遵循抛物线带,态密度(DOS)保持恒定不变且与能量无关。然而,在具有线性能带的石墨烯中,DOS随着能量绝对值(图12.5)的增加而呈线性增加。导带和价带只有在狄拉克点才会彼此接触,导致费米面在零能量下消失。尽管是零DOS,零能量处也没有能带间隙。

通过石墨烯晶体管的传输测试可以观察到DOS的行为。图12.6上面插图是悬浮石墨烯器件的扫描隧道显微镜图像。下面两个插图是从高定向热解石墨(HOPG)剥离并转移到硅衬底(左侧)器件的光学图像和在金属衬底上生长并转移到硅上(右侧)的器件的光学图像。硅片由厚度约为300 nm的热生长氧化层所覆盖。氧化层使单层石墨烯可见[13]并作为栅极介质。高掺杂硅用作背栅以调整石墨烯中电荷载流子密度(n)。对于厚度为300 nm的SiO_2,载流子密度可以按照下列关系来估计:$n = \alpha V_{bg}$,其中栅极效率因数$\alpha \approx 7.2 \times 10^{10} \mathrm{cm^{-2} \cdot V^{-1}}$[14]。图12.6为背栅极电压$V_{bg}$与电阻$R$的函数关系,即传输特性。由于电荷中性点附近的DOS消失,出现了电阻峰。大致对称的传输曲线说明电子-空穴的对称性。

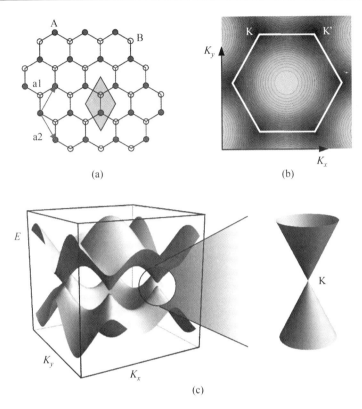

图 12.4 （a）石墨烯的原子结构。阴影菱形是晶格的单胞。A 和 B 是子晶格。(b) 和 (c) 是石墨烯能带结构。白色六边形表示(b)中布里渊区。三维图像是导带和价带在 K 和 K' 点接触

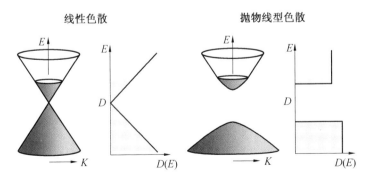

图 12.5 线性和抛物线型色散的粒子的态密度与能量 $D(E)$ 函数关系

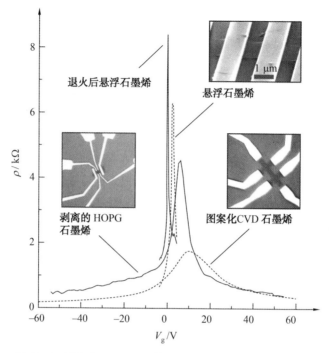

图 12.6 图案化 CVD 石墨烯、HOPG 剥离的石墨烯及电流退火前后悬浮石墨烯器件的传输特性。插图是器件图像(上面是 SEM 图像,下面是光学显微镜图像)。由每个峰的半峰宽获得的密度不均匀性为 $1.8\times10^{12}\ cm^{-2}$、$5\times10^{11}\ cm^{-2}$、$1.7\times10^{11}\ cm^{-2}$ 和 $6\times10^{10}\ cm^{-2}$,分别对应图案化 CVD 石墨烯、剥离 HOPG 石墨烯、悬浮石墨烯和退火后悬浮石墨烯

值得注意的是,电荷中性点不一定与能带的狄拉克点相对应。自然界中体系均存在无序。石墨烯中出现无序可能由于结构缺陷[15-18]、带电杂质[19-21]、褶皱[22-23]、波纹[24]、衬底粗糙度[25]等原因引起,或是这些缺陷共同作用的结果。不管无序怎么产生,都会引起样本的电位变化(图 12.7)。根据无序电位分布,局部频带结构会发生变化。当费米能量(E_F)在电荷中性点附近时,会出现以电子和空穴为主的两个载流子区域,这是无序电位和石墨烯无能带间隙性作用的结果。石墨烯独有的现象是所谓的电子-空穴坑的形成[26],而在传统半导体 2DESs 中只有带正电荷或负电荷密度坑。由于存在电子-空穴坑,电荷中性点附近应该有极高的电阻。而由于 Klein 隧穿理论[27-28],电荷没有完全约束在单个坑里,因此坑边缘没有清楚

的边界。结果,出现有限的电阻峰。$R-n$ 曲线中峰的半峰宽(FWHM)是对密度不均性(Δn_d)的粗略估计。由图 12.7 可知,化学气相沉积法(CVD)生长的石墨烯($\Delta n_d \approx 1.8 \times 10^{12}\ cm^{-2}$)比高定向热解石墨(HOPG)并剥离的石墨烯($\Delta n_d \approx 5 \times 10^{11}\ cm^{-2}$)具有更大的无序密度。因为衬底粗糙度或 SiO_2 衬底嵌入的电荷使石墨烯晶体管产生无序[26],所以当去掉衬底时,石墨烯器件就会悬浮[29,30],电阻峰变窄,表现出低无序密度 $\Delta n_d \approx 6 \times 10^{10}\ cm^{-2}$。电阻峰值变得更高,如图 12.7 所示。

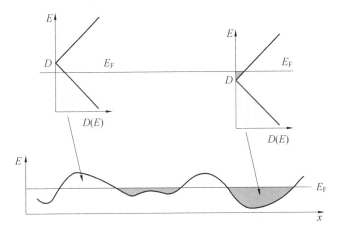

图 12.7　无序电位示意图
(浅灰和深灰面积分别对应过多空洞和电子区域)

当有磁场时,石墨烯中出现类似于传统 2DESs 的量子霍尔效应(QHE),表现却不同于 2DESs。传统的 2DESs 具有大量遵循抛物线能带结构的载流子,于是整数填充因子(整数 QHE)出现量子霍尔平台。而具有线性色散关系和零能带间隙的石墨烯,在填充因子序列 $\nu = \pm 2$,± 6,± 10,…,霍尔电导率 $\sigma_{xy} = \nu e^2/h$ 出现量子霍尔平台,不考虑相互作用,这就是所谓的"半整数 QHE"[10,12]。值得注意的是,还要考虑自旋和谷对称的四重简并。石墨烯的典型量子霍尔行为如图 12.8 所示。独特的半整数 QHE 源自零能量朗道能级(LL)简并。石墨烯的 LL 能量由 $E_N = \pm \hbar \omega_c (N + 1/2)$ 得到,其中 $\hbar$ 是 2π 时的普朗克常量,$\omega_c = eB/m^*$,是回旋频率,N 是量子数,B 是磁场,m^* 是有效质量。零磁场下的 DOS 在确定磁场下转变成 LLs。LL 简并是 $4eB/h$。因子 4 指四重简并。然而,固定在零能量的最低 LL,本身就比较特殊,电子和空穴共享该能态且两者都是半简并的。当无序减弱时,出现由相互作用引起的其他更弱的 QH 状态。例如,在 $\nu = 0$,

±4，±8，…，自发对称性破缺产生双倍对称(自旋或赝自旋)，以及 $\nu = \pm 1$，±3，…，产生的完全破缺对称态[31-36]。此外，复合准粒子形成的结果是出现分数量子霍尔态[30,37,29,38,39]。

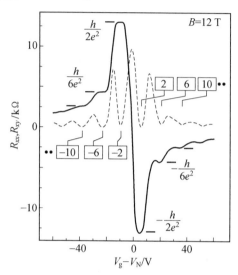

图12.8　石墨烯的典型量子霍尔行为

在12 T 处测得的纵向电阻(R_{xx})和霍尔电阻(R_{xy})与 V_g-V_N 的函数关系图，其中 V_N 是电荷中性的电压

12.4　石墨烯中的单电荷遂穿

当石墨烯中的电子被约束在一维(石墨烯纳米带)或二维(石墨烯量子点)，无质量狄拉克粒子会出现单电荷隧穿行为。被限定的狄拉克粒子背后有着丰富的物理学。石墨烯优于其他纳米材料，如碳纳米管(CNTs)或半导体纳米线，是由于它的二维性和它可由蚀刻技术的光刻过程实现图案化。因为基于传统的自上而下的方法，所以这一过程很容易实现。本章节论述了石墨烯纳米带和石墨烯单电子晶体管中量子约束效应的实验测试与理论的研究。

12.4.1　石墨烯能带间隙的打开

石墨烯最吸引人的电子应用之一是场效应晶体管，因为石墨烯的优良电学性能如具有电子-空穴对称性的极高的载流子迁移率[10,14,40-41]和微米尺寸的弹道传输[42]。此外,石墨烯还表现出超过金刚石和石墨的高热导率[43],可用于制造优于传统硅基器件的散热性更低的集成器件。然而,石墨烯有个重要问题就是它只有一个零能带间隙。在电荷中性点总是存在载流子,导致断态电流在电荷中性点附近没有完全受到抑制。Martin 等[26]很好地揭示了石墨烯中电子-空穴坑的这种定域现象。

打开石墨烯的能带间隙是探索高性能石墨烯电子器件的主要问题之一。已经提出几种打开石墨烯能带间隙的方法:石墨烯超晶格、偏置双层石墨烯和石墨烯纳米带。石墨烯超晶格方法是除了蜂窝晶格外又引入另一个周期性结构来打破能谷对称性。已经出现了一些获得石墨烯超晶格的方法,如在金属衬底上生长石墨烯中常出现的周期波纹[44]、反点阵[45-47],周期性电位[48]、图案化氢吸附[49]、BN 衬底引发的超晶格[50]、光刻图案法[51-52]等。偏双层石墨烯已经实现能带间隙大于 0.1 eV 的场效应晶体管。为了使上下层原位能量不对等[53-55],研究人员对 SiC 热分解法生长的石墨烯进行分子掺杂[56],在剥离的双层石墨烯片中形成双栅极结构[57-60]或双分子掺杂到双层石墨烯的上下两层[61]。

12.4.2 石墨烯纳米带

能带间隙打开的另一种方法是对石墨烯中电子波函数的空间限制。例如,可将石墨烯场效应器件的宽度减小至纳米尺寸,成为石墨烯纳米带,纳米带宽度和边缘原子结构决定了纳米带的金属性或半导体行为。这类似于碳纳米管,其边界条件由石墨烯的卷曲方向(手性矢量)决定,而在石墨烯纳米带中,两个边缘决定了边界条件。石墨烯纳米带获得深入研究的另一种应用途径是场效晶体管[62-71],因为它们表现出可调能带间隙。石墨烯纳米带不仅可用于高性能场效应晶体管,在量子力学方面也可作为单电荷隧穿行为的基础研究体系,特别是无质量狄拉克粒子的研究。当纳米带或纳米结构与两个金属体系(金属电极或大片石墨烯)相互连接时,所形成的能隙(或沿着传输方向分布的各种能隙)成为单电荷隧穿的势垒。

石墨烯纳米带电子结构[72-75]的理论研究(1996)比单原子厚的剥离石墨烯早(2004)[14]。石墨烯纳米带的基本电子性能甚至可以用紧束模型来计算。Fujita 等 (1996)报道了锯齿边缘和扶手椅边缘的石墨烯纳米带的能带结构。发现在锯齿边缘石墨烯的两个相反边缘之间出现具有反平行自旋取向的独特的边缘铁磁性,对于扶手椅边缘石墨烯,出现能带间隙。Nakada 等[74]和 Ezawa[72]证实扶手椅边缘的石墨烯纳米带具有通过改变宽度 w 而可调的能隙 Δ。w 的大小通常以纳米带中垂直于平移方向的线的数量为单位来确定,N_a 和 N_z 分别对应扶手椅边缘和锯齿边缘的石墨烯纳米带(图 12.9)。受边缘条件影响,如果一维模式经过狄拉克点,石墨烯带会表现出金属行为,否则就是具有一定能隙的半导体。当 $N_a=3p$ 或 $3p+1$,扶手椅边缘的石墨烯具有能带间隙;当 $N_a=3p+2$,则表现为金属性,其中 p 是正整数。而锯齿边缘的纳米带在零能量下,总表现出平整带,与纳米带宽度无关,如图 12.10 所示。

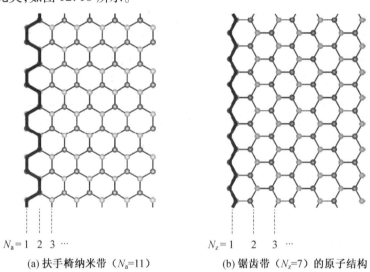

(a) 扶手椅纳米带($N_a=11$)　　(b) 锯齿带($N_z=7$)的原子结构

图 12.9　扶手椅纳米带($N_a=11$)和锯齿带($N_z=7$)的原子结构
(深灰点和浅灰点表示子晶格,边缘位置由粗黑线表示)

按照最近由 Son 等[76]报道的从头计算结果,即使是锯齿边缘石墨烯纳米带,由于六方晶格的边缘磁化引起的交错子晶格电位,也会出现能隙(图 12.11)。

制造只有扶手椅形锯齿形清晰边缘的石墨烯纳米带还没有实现。无论是图案化纳米带还是剥离片层,往往都具有较粗糙的锯齿形和扶手椅形混合的边缘,正如 Krauss[77]所报道的。有几种可以制造石墨烯纳米带的方法。通常采用的是通过电子束光刻法和等离子刻蚀技术减少石墨烯尺寸[78-80]。这种方法用的石墨烯通常用胶带机械剥离天然石墨晶体或 HOPG 得到[14],再沉积在 300 nm 厚热氧化层的硅衬底上,由于增加了光学路径和石墨烯显著的不透明度,可以在光学显微镜下看见单层石墨烯[13]。转移片层可通过光学显微镜确认。电子束抗蚀剂被喷涂到带有转移石墨烯的 SiO_2 基板上。将蚀刻图案通过电子束光刻技术刻画在石墨烯

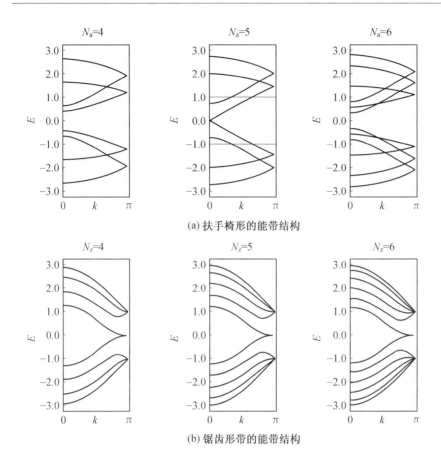

图 12.10 由紧束缚模型计算的不同宽度下扶手椅形和锯齿形带的能带结构[74]

片顶部的抗蚀剂上,暴露的抗蚀图案通过化学试剂完成。部分石墨烯是暴露的,整个器件放置在反应离子蚀刻(RIE)室中进行氧气或氩气部分等离子刻蚀。最后去除抗蚀剂,图案化的石墨烯纳米带或纳米结构制备完成,通过进一步的电子束光刻和金属化过程实现制造器件的电触点。石墨烯纳米带中的能隙与纳米带宽呈反比例关系(图 12.12)[79],发现 $\Delta \approx 0.2/(w-16)$。比如,$w \approx 18$ nm 的石墨烯纳米带可以有 0.1 eV 的能隙。虽然制备程序很常见、可靠,但是由于等离子光刻过程的各向异性以及离子束对石墨烯和抗蚀剂的刻蚀选择性差,该工艺并不适合获得原子级清晰的边缘结构。

虽然人们希望石墨烯纳米带具有相关传输性能[81-82],但还无法制备边缘清晰的石墨烯纳米带。而低温下栅极相关电导(G)通常出现振荡或波

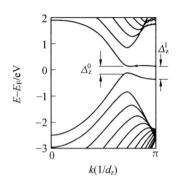

图 12.11 $N_a = 12$ 锯齿形石墨烯纳米带的能带结构
(Δ_z^0 和 Δ_z^1 分别是直接能带间隙[76])

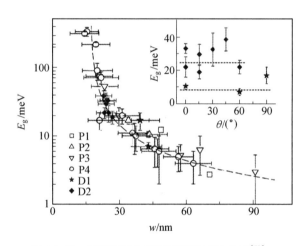

图 12.12 能带间隙与器件宽度的函数关系[79]

动,图 12.13 可能是由石墨烯 QDs 网络结构和(或)粗糙边缘结构引起的。[83-85]石墨烯纳米带中的库仑菱形结构也很常见[83-84,86]。由于石墨烯的弹道电荷和能谷简并的提升,无序边缘的纳米带(锯齿形和扶手椅形边缘的混合)表现出 $2e^2/h$ 整数倍的电导量子化[85,87]。

控制边缘更有效的方法是基于 SiO_2 在氩环境约 700 ℃下经过碳热还原成 SiO[88]。石墨烯边缘的碳原子参与上述过程,片边缘则以晶体取向排列。拉曼光谱证实了石墨烯边缘的取向[77]。由氢等离子刻蚀[89]、气相化学刻蚀过程[90]或由用纳米球作为掩罩进行的等离子体刻蚀[91],也能得到类似的结论。用这些方法可以制造边缘可控的长石墨烯纳米带晶体管(图 12.14)[89-93]。而对于边缘可控纳米带或器件的单电荷隧穿行为的研究还

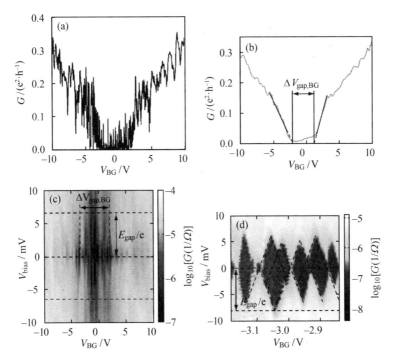

图 12.13 （a）电导率与 V_g 的函数关系。（b）为（a）中曲线的光滑处理。（c）石墨烯纳米带器件的稳定图。（d）为（c）中曲线的放大图[83]

没有相关报道。

12.4.3 石墨烯单电子晶体管

通过将部分石墨烯片分离并使之与相反侧的两个压缩物（或纳米带）相连，得到石墨烯 SETs，可以作为源极和漏极电容器。侧栅极可以图案化到石墨烯上或 Si 背栅也可以用来调整岛的静电能。为了把 SET 结构在石墨烯上图案化，往往涉及等离子体刻蚀法。石墨烯 SETs 的基本结构与基于金属或半导体的传统 SETs 一样。然而它们之间有着明显的不同。由于体系中电荷的无质量性，量子限制不同于传统的 SETs，传统的 SETs 中的电荷遵循质量抛物线型色散关系。

当岛尺寸足够小时，也就是可以把它当作 QD，量子限制能将具有重要作用。对于质量载流子，例如 2DES，岛量子限制能由二维盒子问题得到 $\Delta_N \approx h^2/(8m_e d^2)$，其中，$m_e$ 是有效质量，d 是岛直径。说明 Δ_N 与 $1/d^2$ 成比

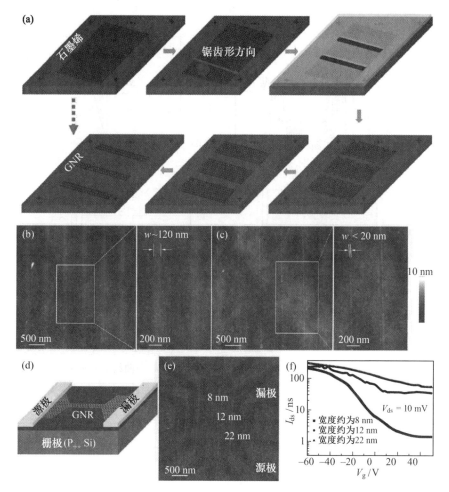

图 12.14 由各向异性刻蚀制造的石墨烯纳米带(GNR)[93]

例。然而,对于具有无质量电荷载流子的石墨烯 SETs 具有不同光谱,$\Delta_N \approx v_F h/(2d)$。$\Delta_N$ 与 d 呈反比例关系[94],证实 $1/d$ 取决于石墨烯 SETs 的量子限制能。研究还发现振荡周期 ΔV_g 随机变化(由于质量载流子自旋简并,没有出现偶数-奇数序列),并且混沌变异 $\delta(\Delta V_g)$ 比非石墨烯 QDs 中的典型变化高出几个数量级(图 12.15)。

按照 $\delta(\Delta V_g) \propto 1/d^2$,随着 d 降低而明显变宽。该狄拉克弹球行为使石墨烯 QD 形状对体系单电荷隧穿的影响极其重要。按照理论计算[95],规则形状(圆盘状)石墨烯 QDs 具有尖锐共振性,减少了有效栅极区域;但不规则形状的 QDs 由于混沌动力学而不支撑束缚态。具有图案化岛的 SET

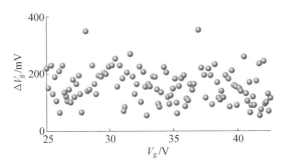

图 12.15 石墨烯量子点中测得零偏压(ΔV_g)下最近邻峰的分离。峰间距有很大变化(因子是 5 或更大)[94]

结构中发现了不规则的库仑振荡[96],纳米压缩物中存在库仑振荡行为[84,97]。规则的 ΔV_g 存在于正常 SETs 中(图 12.16),或存在于在量子霍尔机制下运用扫描隧道显微镜(STM)直接测试可压缩的 QDs 时(发现了由于自旋和能谷简并引起的 4 倍周期)[98]。除了基态库仑菱形,双 QD 体系中的激发态和现象也在实验中得到证实[96,99-100]。尽管做了深入研究,受限无质量粒子背后有趣的物理学还尚待研究。例如,通过结晶学蚀刻法

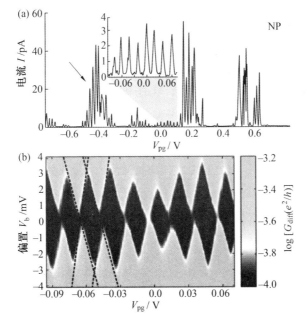

图 12.16 (a)源-漏电流和栅极电压函数关系。(b)库仑菱形结构[103]

用边缘可控的压缩物或用六角形石墨烯岛制造 SET，应该是一个有趣的实验。此外，用边缘控制其自旋态可能提供有关自旋量子比特的信息，边缘可能存在长程耦合[101]。

12.5 石墨烯的电荷局域化

12.5.1 量子霍尔机制下的局域化

存在于库仑阻塞物理边界中的量子霍尔机制下的电荷局域化，会形成量子霍尔平台[2,102]。对于电荷限制，不可压缩条的绝缘势垒起了重要作用。正如 12.3.2 节中所论述的，石墨烯中无序电位分布情况会引起整个样品中的密度分布。载流子的重新分布使得无序电位趋于均衡（图 12.17）。而当施加垂直磁场时，这种线性筛选是无效的。在一些机制中，当 n 增加时，能级简并被完全填满，载流子不能筛选出机制中这些无序电位。由于 LLs 之间的能带间隙，载流子不会被注入该区，也就是区域变得完全不可压缩。这些不可压缩区作为绝缘势垒包围了可压缩点。这些进入可压缩 QDs 的电荷被库仑阻塞物理控制。由于电子（空洞）发生一次进入 QDs 的离散级别都会对应一次充电。这导致局域压缩性（$dn/d\mu$）的突然跳跃，μ 是化学势。因为 LL 简并与 $B(eB/h)$ 成比例，所以在高场强 B 下，高 n 情况下会出现相同的密度分布。可压缩性的峰值会产生平行于具体填充因数的线，如图 12.17 所示。QH 机制中可压缩 QDs 的存在，可通过 STM 仪器的悬臂测量点的直传输证实[98]。

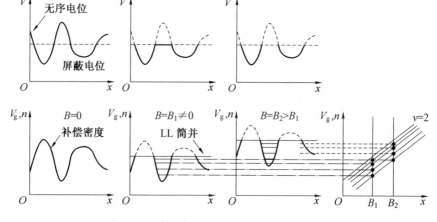

图 12.17　高磁场强下可压缩的 QD 形成机理图

12.5.2 量子霍尔机制中的可压缩量子点

可压缩 QDs 带来的对石墨烯器件宏观电导的影响可通过电导波动的磁传输测试进行研究[104]。在高磁场强下,电导曲线与 V_{bg} 呈函数关系,常有波动,干扰了主要电导量子化行为。而这些波动被认为是电荷局域化的表现。虽然由于平均化的作用,局部过程没有在磁传输中显示出来,但波动性却反映了可压缩的 QDs 中的单电荷隧穿发生情况。在 15 T 时,电导与 n 的函数关系如图 12.18(a)所示。除了量化步骤和量化特征,波动对于多次测试具有可重复性。如果测试了互导 g_m,这些波动更明显。在这里,将交流调制 δV_{bg} 施加于背栅,只测试源-漏电流的交流部分 δI_{ds},则 $g_m = \delta I_{ds}/\delta V_{bg}$。两种测量采用直流偏压 V_{ds} = 500 μV 进行。δV_{bg} 的频率是 433 Hz,它的均方振幅是 10 mV。图 12.18 中的插图是测试电路图。

对于不同磁场重复相同的背栅扫描。(n,B) 平面的绘制结果如图 12.19(a)所示。图中有许多直线特征,将其划分成组,组中直线彼此平行。每组中的直线与特定的填充因子直线平行。由于存在 $N=0$ 和 $N=1$ LLs 之间的最大能带间隙,因此当填充因数为 2 时最有意义,与 $\nu = \pm 2$ 平行的线最强、最清晰。此外还可以分辨出 $\nu = 6$ 的线,它是另一个半整数填充因子。此外,可以在波动图中看到更脆弱的对称性破缺态和部分量子霍尔态,有些状态应该可见但却看不到,或只能看到不清晰的特征[104]。

图 12.19 所示的线性特征类似于在石墨烯[102]及早期 GaAs 的 2DESs[2]上用扫描 SET 测得的局部可压缩图。因此,把可压缩尖锐线和电导波动联系起来是自然的。远离能级简并的完全填充,线性筛选保持不变。通过调整背栅电压来增加 n,如 12.5.1 节描述的,会出现可压缩量子点,不可压缩带对隔离点起了一定作用。如图 12.20 所示,由于边缘电位的原因,电导通道往往出现在样本边缘附近,电流可以流经这些边缘态。在调整 n 的过程中,由于库仑阻塞允许对点充电,导致一些点会出现可压缩的尖峰。因此,通过点组成的网络的附加传输隧道可能出现或消失,或发生路径改变。在每个可压缩尖峰,充电会增加源-漏电流的波动。即使在不同场强下,相对于完全填充发生相同的偏离 n,会有相同的波动行为。结果是 (n,B) 平面中电导波动峰或谷与填充因子线平行。值得注意的是,在 GaAs 的 2DESs[105]中已经发现电导波动的线性特征,最近在石墨烯中[106-107]也发现该特征,过程中测量了电阻,结果发现平行于单粒子态的线条,石墨烯的半整数 QH 态和 GsAs 整数 QH 态。对于传输电导测量,运用锁定技术揭示了更多弱相互作用引发态(对称破缺态和部分量子霍尔态)

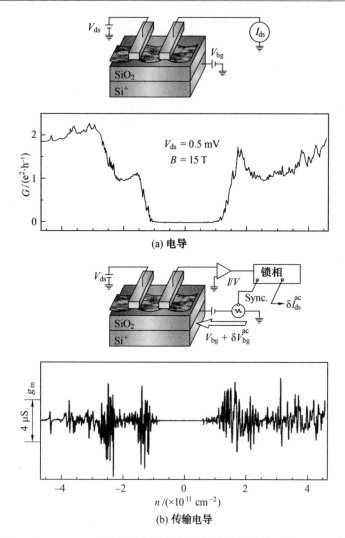

图 12.18　$B=15$ T 时测量的电导和传输电导的曲线及测试原理示意图[105]

存在[104]。例如,对称破缺态的 $\nu=0$ 和 $\nu=1$ 线条和分数填充因子 ($\nu=1/3$) 线条如图 12.19 所示。

为了将波动线条的强度可见,以及确定线的填充因子,将窗口中的数据组转变为与填充因子 $C(\nu)$ 相关的光谱,表明平面 (n,B) 中数据与一定斜率的符合程度,以及如何区分得到的具有斜率的线。图 12.21 所示是 $C(\nu)$ 数据分析示意图。把 n 和 B 的函数 g_m 数据组看成一个矩阵 $D(n,$

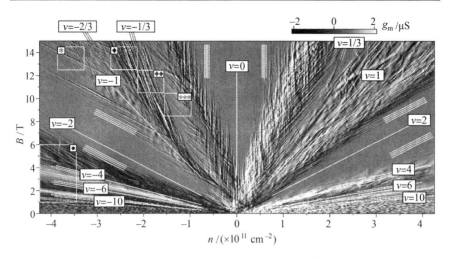

图 12.19 (n,B) 平面中 g_m 的灰度显现[104]

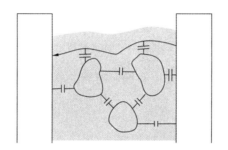

图 12.20 石墨烯器件中可压缩点和边缘隧道的原理显示图

$B)$。数据窗口在 $n_{min} \leq n \leq n_{max}$ 和 $B_{min} \leq B \leq B_{max}$ 是关闭的。矩阵元素是 $D_{ij} \equiv D(n_j, B_i)$。$n_j(B_i)$ 是 $n(B)$ 从 $n_{min}(B_{min})$ 开始的第 j(第 i) 个元素。相关函数与填充因子的函数关系可表示为

$$C(v) = \frac{\sum_{k,l,p,q,k \neq p} D_{kl} D_{pq} \delta_v}{\sum_{k,l,p,q,k \neq p} \delta_v} \quad (12.6)$$

其中,δ_v 是补偿函数,如果 $r(\theta_m) < r_a$,$\delta_v = 1$,否则 $\delta_v = 0$。$r(\theta_m)$ 是 D_{pq} 点和过 D_{kl} 点具有一定斜率的直线之间的数据像素(Δ_n 和 Δ_B)的单位距离,$s = \tan(\pi/2 - \theta_m)$,也就是填充因子线,$v(\theta_m) = (\Phi_0/s)(\Delta_n/\Delta_B)$。等式 (12.6) 的分母是归一化因子。补偿因子 $r_a = 1$。对于平行于线性特征 g_m 峰或谷的 $v(\theta_m)$ 值来说,D_{kl} 和 D_{pq} 的总和更大,而当 $v(\theta_m)$ 线不平行于图中线性特征时,总和就会被取消。相关函数分析如图 12.22 所示。

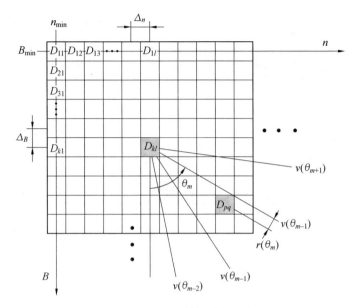

图 12.21　$C(\nu)$ 数据分析示意图

在图 12.21 左下角矩形中数据所得到的 $C(\nu)$ 光谱中,填充因子 ν = -2、-4、-6 和 -10。ν = -4 态尤其有趣,因为它是由自旋或能谷的对称破缺引起的,而且它没有出现在电导数据中[104]。分数量子霍尔态也会出现。从上面矩形窗口中的数据得到的 $C(\nu)$ 清楚表明填充因子 ν = -1/3 或 -2/3,这是石墨烯中最明显的分数量子霍尔态。这些在电导测试中是看不到的。

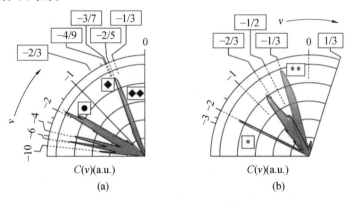

图 12.22　相关函数分析[104]

只有当无序电位的非线性筛选产生了由绝缘势垒隔离的可压缩 QDs，以上描述的可压缩 QD 才是有效的。尽管用 STM 针尖在可压缩 QD 观察到了单电荷隧穿行为[98]，但为了更好地理解 QDs 网络结构对电导波动的影响，通过对 QDs 的网络结构进行直接的传输测试也是非常必要的。然而，平行于 QDs 网络结构的边缘隧道的出现阻碍了直接传输测试。在高磁场下，量子霍尔绝缘行为使 QDs 的网络结构的直接传输测试成为可能。由于零能量态的能谷对称破缺，零边缘态会产生真正的绝缘行为[35]。电导微分 dI_{ds}/dV_{ds} 作为 V_{bg} 和 V_{ds} 的函数，如图 12.23 所示。电导微分的稳定性示意图显示了库仑菱形的结构，模糊地证实了使 QDs 隔离的隧道势垒的存在。由菱形结构估测的 QDs 长度尺寸范围为 160~400 nm，如图 12.3 所示，因为 QDs 形成是由无序电位的非线性屏蔽引起的，所以这个尺寸可被当作是无序的长度。从菱形结构中估算 $d^2 = 4C_g/(\pi e\alpha)$，其中 $C_g = C_\Sigma/(S_p^{-1} + S_n^{-1})$，$S_p$ 和 S_n 是菱形结构的正负斜率。无序长度尺寸也与体系的平均自由程(l_{mfp})接近。通过没有磁场作用的悬浮样品在电荷中性点附近的电导 $3.5e^2/h$，可以得出 $l_{mfp} = 350$ nm[108]，这与库仑菱形中估计的长度尺寸相一致。菱形结构的不规则性或是 QDs 网络结构的结果，或是 QDs 会出现混沌共振的非圆盘形状的结果[95]。

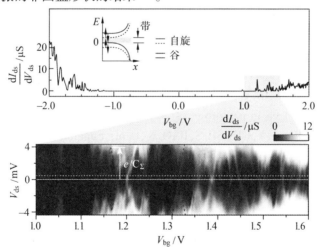

图 12.23　电导微分 $dI_{ds} = dV_{ds}$ 与栅极电压的关系以及 $dI_{ds} = dV_{ds}$ 与 V_{ds} 和 V_{bg} 的关系[104]

12.6 结 论

最近关于控制石墨烯电子特性的方法发展显著,不仅满足了石墨烯在电子器件中的应用(如晶体管),而且也可以应用于量子计算机设备。然而石墨烯中单电荷行为背后的物理学还没有被完全理解。比如,还无法获得较大能带间隙的石墨烯。12.4 节论述了通过控制尺寸实现石墨烯单电子晶体管或量子点的方法。此外,12.5 节讨论了单电荷隧穿如何影响介观传输,如何在宏观器件中实现独立电荷的局域化。无质量粒子遵循线性色散关系是理解石墨烯中单电荷传输的关键,这一认识给纳米电子未来的潜力发展提供了视角。

12.7 参考文献

[1] VON KLITZING K, DORDA G, PEPPER M. New method for high-accuracy determination of the fine-structure constant based on quantized Hall resistance[J]. Physical Review Letters, 1980, 45: 494.

[2] ILANI S, MARTIN J, TEITELBAUM E, et al. The microscopic nature of localization in the quantum Hall effect[J]. Nature, 2004, 427: 328.

[3] BEENAKKER C W J. Theory of coulomb-blockade oscillations in the conductance of a quantum dot[J]. Physical Review B, 1991, 44: 1646.

[4] GRABERT H, DEVORET M H. Single charge tunneling: coulomb blockade phenomena in nanostructres[M]. New York: Plenum Press, 1992.

[5] BALANDIN A A, GHOSH S, BAO W, et al. Superior thermal conductivity of single-layer graphene[J]. Nano Letters, 2008, 8: 902.

[6] BOLOTIN K I, SIKES K J, JIANG Z, et al. Ultrahigh electron mobility in suspended graphene[J]. Solid State Communications, 2008, 146: 351.

[7] DEAN C R, YOUNG A F, MERIC I, et al. Boron nitride substrates for high-quality graphene electronics[J]. Nature Nanotechnology, 2010, 5: 722.

[8] MOROZOV S V, NOVOSELOV K S, KATSNELSON M I, et al. Giant intrinsic carrier mobilities in graphene and its bilayer[J]. Physical Review Letters, 2008, 100: 016602.

[9] CASTRO NETO A H, GUINEA F, PERES N M R, et al. The electronic

properties of graphene[J]. Reviews of Modern Physics,2009,81:109.

[10] NOVOSELOV K S,GEIM A K,MOROZOV S V,et al. Two-dimensional gas of massless Dirac fermions in graphene[J]. Nature,2005,438:197.

[11] WALLACE P R. The band theory of graphite[J]. Physical Review, 1947,71:622.

[12] ZHANG Y B,TAN Y W,STORMER H L,et al. Experimental observation of the quantum Hall effect and Berry's phase in graphene[J]. Nature, 2005,438:201.

[13] BLAKE P,HILL E W,NETO A H C,et al. Making graphene visible[J]. Applied Physics Letters,2007,91:063124.

[14] NOVOSELOV K S,GEIM A K,MOROZOV S V,et al. Electric field effect in atomically thin carbon films[J]. Science,2004,306:666.

[15] GASS M H,BANGERT U,BLELOCH A L,et al. Free-standing graphene at atomic resolution[J]. Nature Nanotechnology,2008,3:676.

[16] HASHIMOTO A,SUENAGA K,GLOTER A,et al. Direct evidence for atomic defects in graphene layers[J]. Nature,2004,430:870.

[17] MEYER J C,KISIELOWSKI C,ERNI R,et al. Direct imaging of lattice atoms and topological defects in graphene membranes[J]. Nano Letters, 2008,8:3582.

[18] WARNER J H,RUMMELI M H,GE L,et al. Structural transformations in graphene studied with high spatial and temporal resolution[J]. Nature Nanotechnology,2009,4:500.

[19] ANDO T. Screening effect and impurity scattering in monolayer graphene [J]. Journal of the Physical Society of Japan,2006,75:4716.

[20] CHEN J H,JANG C,ADAM S,et al. Charged-impurity scattering in graphene[J]. Nature Physics,2008,4:377.

[21] NI Z H,PONOMARENKO L A,NAIR R R,et al. On resonant scatterers as a factor limiting carrier mobility in graphene[J]. Nano Letters,2010, 10:3868.

[22] GERINGER V,LIEBMANN M,ECHTERMEYER T,et al. Intrinsic and extrinsic corrugation of monolayer graphene deposited on SiO_2[J]. Physical Review Letters,2009,102:076102.

[23] LOCATELLI A,KNOX K R,CVETKO D,et al. Corrugation in exfoliated graphene:an electron microscopy and diffraction study[J]. ACS Nano,

2010,4:4879.

[24] BAO W,MIAO F,CHEN Z,et al. Controlled ripple texturing of suspended graphene and ultrathin graphite membranes[J]. Nature Nanotechnology,2009,4:562.

[25] ISHIGAMI M,CHEN J H,CULLEN W G,et al. Atomic structure of graphene on SiO_2[J]. Nano Letters,2007,7:1643.

[26] MARTIN J,AKERMAN N,ULBRICHT G,et al. Observation of electron-hole puddles in graphene using a scanning single-electron transistor[J]. Nature Physics,2008,4:144.

[27] CHEIANOV V V,FAL'KO V I. Selective transmission of Dirac electrons and ballistic magnetoresistance of n-p junctions in graphene[J]. Physical Review B,2006,74:041403.

[28] KATSNELSON M I,NOVOSELOV K S,GEIM A K. Chiral tunnelling and the Klein paradox in graphene[J]. Nature Physics,2006,2:620.

[29] DU X,SKACHKO I,DUERR F,et al. Fractional quantum Hall effect and insulating phase of Dirac electrons in graphene[J]. Nature,2009,462:192.

[30] BOLOTIN K I,GHAHARI F,SHULMAN M D,et al. Observation of the fractional quantum Hall effect in graphene[J]. Nature,2009,462:196.

[31] CHECKELSKY J G,LI L,ONG N P. Zero-energy state in graphene in a high magnetic field[J]. Physical Review Letters,2008,100:206801.

[32] GOERBIG M O,MOESSNER R,DOUÇOT B. Electron interactions in graphene in a strong magnetic field. Physical Review B,2006,74:161407.

[33] GUSYNIN V P,MIRANSKY V A,SHARAPOV S G,et al. Edge states, mass and spin gaps,and quantum Hall effect in graphene[J]. Physical Review B,2008,77:205409.

[34] JUNG J,MACDONALD A H. Theory of the magnetic-field-induced insulator in neutral graphene sheets[J]. Physical Review B,2009,80:235417.

[35] YANG K. Spontaneous symmetry breaking and quantum hall effect in graphene[J]. Solid State Communications,2007,143:27.

[36] ZHANG Y,JIANG Z,SMALL J P,et al. Landau-level splitting in graphene in high magnetic fields[J]. Physical Review Letters,2006,96:

136806.

[37] DEAN C R, YOUNG A F, CADDEN-ZIMANSKY P, et al. Multicomponent fractional quantum Hall effect in graphene[J]. Natrue Physics, 2011, 7:693.

[38] PAPIC Z, GOERBIG M O, REGNAULT N. Theoretical expectations for a fractional quantum Hall effect in graphene[J]. Solid State Communications, 2009, 149:1056.

[39] TÖKE C, JAIN J K. Erratum: SU(4) composite fermions in graphene: fractional quantum Hall states without analog in GaAs[J]. Physical Review B, 2007, 75:245440.

[40] GEIM A K, NOVOSELOV K S. The rise of graphene[J]. Nature Materials, 2007, 6:183.

[41] SEMENOFF G W. Condensed-matter simulation of a three-dimensional anomaly[J]. Physical Review Letters, 1984, 53:2449.

[42] MIAO F, WIJERATNE S, ZHANG Y, et al. Phasecoherent transport in graphene quantum billiards[J]. Science, 2007, 317:1530.

[43] SEOL J H, JO I, MOORE A L, et al. Two-dimensional phonon transport in supported graphene[J]. Science, 2010, 328:213.

[44] GUINEA F, LOW T. Band structure and gaps of triangular graphene superlattices[J]. Philosophical Transactions of the Royal Society A-Mathematical Physical and Engineering Sciences, 2010, 368:5391.

[45] FURST J A, PEDERSEN J G, FLINDT C, et al. 2009. Electronic properties of graphene antidot lattices[J]. New Journal of Physics, 2009, 11:095020.

[46] LIU W, WANG Z F, SHI Q W, et al. Band-gap scaling of graphene nanohole superlattices[J]. Physical Review B, 2009, 80:233405.

[47] PEDERSEN T G, FLINDT C, PEDERSEN J, et al. Graphene antidot lattices: designed defects and spin qubits[J]. Physical Review Letters, 2008, 100:136804.

[48] TIWARI R P, STROUD D. Tunable band gap in graphene with a noncentrosymmetric superlattice potential[J]. Physical Review B, 2009, 79:205435.

[49] BALOG R, JORGENSEN B, NILSSON L, et al. Bandgap opening in graphene induced by patterned hydrogen adsorption[J]. Nature Materials,

2010,9:315.
[50] GIOVANNETTI G,KHOMYAKOV P A,BROCKS G,et al. Substrate-induced band gap in graphene on hexagonal boron nitride: ab initio density functional calculations[J]. Physical Review B,2007,76:073103.
[51] MARTINAZZO R,CASOLO S,TANTARDINI G F. Symmetry-induced band-gap opening in graphene superlattices[J]. Physical Review B,2010,81:245420.
[52] MEYER J C,GIRIT C O,CROMMIE M F,et al. Hydrocarbon lithography on graphene membranes[J]. Applied Physics Letters,2008,92:123110.
[53] CASTRO E V,NOVOSELOV K S,MOROZOV S V,et al. Biased bilayer graphene: semiconductor with a gap tunable by the electric field effect[J]. Physical Review Letters,2007,99:216802.
[54] MCCANN E. Asymmetry gap in the electronic band structure of bilayer graphene[J]. Physical Review B,2006,74:161403.
[55] MIN H,SAHU B,BANERJEE S K,et al. Ab initio theory of gate induced gaps in graphene bilayers[J]. Physical Review B,2007,75:155115.
[56] OHTA T,BOSTWICK A,SEYLLER T,et al. Controlling the electronic structure of bilayer graphene[J]. Science,2006,313:951.
[57] OOSTINGA J B,HEERSCHE H B,LIU X L,et al. Gate-induced insulating state in bilayer graphene devices[J]. Nature Materials,2008,7:151.
[58] TAYCHATANAPAT T,JARILLO-HERRERO P. Electronic transport in dual-gated Bilayer graphene at large displacement fields[J]. Physical Review Letters,2010,105:166601.
[59] XIA F,FARMER D B,LIN Y-M,et al. Graphene field-effect transistors with high on/off current ratio and large transport band gap at room temperature[J]. Nano Letters,2010,10:715.
[60] ZHANG Y,TANG T-T,GIRIT C,et al. Direct observation of a widely tunable bandgap in bilayer graphene[J]. Nature,2009,459:820.
[61] PARK J,JO S B,YU Y-J,et al. Single-gate bandgap opening of bilayer graphene by dual molecular doping[J]. Advanced Materials,2012,24:407.
[62] KEDZIERSKI J,HSU P L,HEALEY P,et al. 2008. Epitaxial graphene

transistors on SIC substrates[J]. IEEE Transactions on Electron Devices,2008,55:2078.

[63] KEDZIERSKI J,HSU P L,REINA A,et al. Graphene-on-insulator transistors made using C on Ni chemical-vapor deposition[J]. IEEE Electron Device Letters,2009,30:745.

[64] LEMME M C,ECHTERMEYER T J,BAUS M,et al. A graphene fieldeffect device[J]. IEEE Electron Device Letters,2007,28:282.

[65] LI X,CAI W,AN J,et al. Large-area synthesis of high-quality and uniform graphene films on copper foils[J]. Science,2009,324:1312.

[66] MERIC I,HAN M Y,YOUNG A F,et al. Current saturation in zero-bandgap,top-gated graphene field-effect transistors[J]. Nature nanotechnology,2008,3:654.

[67] LIAO L,BAI J W,QU Y Q,et al. High-kappa oxide nanoribbons as gate dielectrics for high mobility top-gated graphene transistors[J]. Proceedings of the National Academy of Sciences of the United States of America,2010,107:6711.

[68] LIN Y-M,JENKINS K A,VALDES-GARCIA A,et al. Operation of graphene transistors at gigahertz frequencies[J]. Nano Letters,2008,9:422.

[69] LIN Y-M,DIMITRAKOPOULOS C,JENKINS K A,et al. 100-GHz transistors from wafer-scale epitaxial graphene[J]. Science,2010,327:662.

[70] MOON J S,CURTIS D,HU M,et al. Epitaxial-graphene RF field-effect transistors on Si-face 6H-SiC substrates[J]. IEEE Electron Device Letters,2009,30:650.

[71] YANG H,HEO J,PARK S,et al. Graphene barristor,a triode device with a gate-controlled schottky barrier[J]. Science,2012,336:1140.

[72] EZAWA M. Peculiar width dependence of the electronic properties of carbon nanoribbons[J]. Physical Review B,2006,73:045432.

[73] FUJITA M,WAKABAYASHI K,NAKADA K,et al. Peculiar localized state at zigzag graphite edge[J]. Journal of the Physical Society of Japan,1996,65:1920.

[74] NAKADA K,FUJITA M,DRESSELHAUS G,et al. Edge state in graphene ribbons: nanometer size effect and edge shape dependence[J]. Physical Review B,1996,54:17954.

[75] WAKABAYASHI K, FUJITA M, AJIKI H, et al. Electronic and magnetic properties of nanographite ribbons[J]. Physical Review B, 1999, 59: 8271.

[76] SON Y-W, COHEN M L, LOUIE S G. Energy gaps in graphene nanoribbons[J]. Physical Review Letters, 2006, 97: 216803.

[77] KRAUSS B, NEMES-INCZE P T, SKAKALOVA V, et al. Raman scattering at pure graphene zigzag edges[J]. Nano Letters, 2010, 10: 4544.

[78] CHEN Z H, LIN Y M, ROOKS M J, et al. Graphene nano-ribbon electronics[J]. Physica E Low Dimensional Systems and Nanostructures, 2007, 40: 228.

[79] HAN M Y, ÖZYILMAZ B, ZHANG Y, et al. Energy band-gap engineering of graphene nanoribbons[J]. Physical Review Letters, 2007, 98: 206805.

[80] TODD K, CHOU H-T, AMASHA S, et al. Quantum dot behavior in graphene nanoconstrictions[J]. Nano Letters, 2008, 9: 416.

[81] DARANCET P, OLEVANO V, MAYOU D. Coherent electronic transport through graphene constrictions: subwavelength regime and optical analogy[J]. Physical Review Letters, 2009, 102: 136803.

[82] MUÑOZ-ROJAS F, JACOB D, FERN NDEZ-ROSSIER J, et al. Coherent transport in graphene nanoconstrictions[J]. Physical Review B, 2006, 74: 195417.

[83] MOLITOR F, JACOBSEN A, STAMPFER C G, et al. Transport gap in side-gated graphene constrictions[J]. Physical Review B, 2009, 79: 075426.

[84] TERRES B, DAUBER J, VOLK C, et al. Disorder induced Coulomb gaps in graphene constrictions with different aspect ratios[J]. Applied Physics Letters, 2011, 98: 228.

[85] TOMBROS N, VELIGURA A, JUNESCH J, et al. Quantized conductance of a suspended graphene nanoconstriction[J]. Natrue Physics, 2011, 7: 697.

[86] SAFRON N S, BREWER A S, ARNOLD M S. Semiconducting two-dimensional graphene nanoconstriction arrays[J]. Small, 2011, 7: 492.

[87] BREY L, FERTIG H A. Electronic states of graphene nanoribbons studied with the Dirac equation[J]. Physical Review B, 2006, 73: 235411.

[88] NEMES-INCZE P, MAGDA G, KAMARAS K, et al. Crystallographically selective nanopatterning of graphene on SiO_2[J]. Nano Research, 2010, 3:110.

[89] XIE L, JIAO L, DAI H. Selective etching of graphene edges by hydrogen plasma[J]. Journal of the American Chemical Society, 2010, 132: 14751.

[90] WANG X, DAI H. Etching and narrowing of graphene from the edges [J]. Nature Chemistry, 2010, 2:661.

[91] LIU L, ZHANG Y L, WANG W L, et al. Nanosphere lithography for the fabrication of ultranarrow graphene nanoribbons and on-chip bandgap tuning of graphene[J]. Advanced Materials, 2011, 23:1246.

[92] LU Y, GOLDSMITH B, STRACHAN D R, et al. High-on/off-ratio graphene nanoconstriction field-effect transistor[J]. Small, 2010, 6:2748.

[93] YANG R, ZHANG L C, WANG Y, et al. An anisotropic etching effect in the graphene basal plane[J]. Advanced Materials, 2010, 22:4014.

[94] PONOMARENKO L A, SCHEDIN F, KATSNELSON M I, et al. Chaotic dirac billiard in graphene quantum dots[J]. Science, 2008, 320:356.

[95] BARDARSON J H, TITOV M, BROUWER P W. Electrostatic confinement of electrons in an integrable graphene quantum dot[J]. Physical Review Letters, 2009, 102:226803.

[96] SCHNEZ S, MOLITOR F, STAMPFER C, et al. Observation of excited states in a graphene quantum dot[J]. Applied Physics Letters, 2009, 94: 012107.

[97] DROSCHER S, KNOWLES H, MEIR Y, et al. Coulomb gap in graphene nanoribbons[J]. Physical Review B, 2011, 84:073405.

[98] JUNG S, RUTTER G M, KLIMOV N N, et al. Evolution of microscopic localization ingraphene in a magnetic field from scattering resonances to quantum dots[J]. Nature Physics, 2011, 7:245.

[99] LIU X L, HUG D, VANDERSYPEN L M K. Gate-defined graphene double quantum dot and excited state spectroscopy[J]. Nano Letters, 2010, 10:1623.

[100] MOLITOR F, KNOWLES H, DROESCHER S, et al. Observation of excited states in a graphene double quantum dot[J]. Europhysics Letters, 2010, 89:67005.

[101] TRAUZETTEL B,BULAEV D V,LOSS D,et al. Spin qubits in graphene quantum dots[J]. Nature Physics,2007,3:192.
[102] MARTIN J,AKERMAN N,ULBRICHT G,et al. The nature of localization in graphene under quantum Hall conditions[J]. Nature Physics,2009,5:669.
[103] STAMPFER C,SCHURTENBERGER E,MOLITOR F,et al. Tunable graphene single electron transistor[J]. Nano Letters,2008,8:2378.
[104] LEE D S,SKAKALOVA V,WEITZ R T,et al. Transconductance fluctuations as a probe for interaction-induced quantum Hall states in graphene[J]. Physical Review Letters,2012,109:056602.
[105] COBDEN D H,BARNES C H W,FORD C J B. Fluctuations and evidence for charging in the quantum Hall effect[J]. Physical Review Letters,1999,82:4695.
[106] BRANCHAUD S,KAM A,ZAWADZKI P,et al. Transport detection of quantum Hall fluctuations in graphene[J]. Physical Review B,2010,81:121406.
[107] VELASCO J,JR,LIU G,et al. Probing charging and localization in the quantum Hall regime by graphene p-n-p junctions[J]. Physical Review B,2010,81:121407.
[108] MUCCIOLO E R,LEWENKOPF C H. Disorder and electronic transport in graphene [J]. Journal of Physics-Condensed Matter, 2010, 22:273201.

第13章 石墨烯自旋电子学

本章介绍了石墨烯内自旋电子学的物理概念,包括如何在石墨烯内实现长自旋相干(a long spin coherence);综述了石墨烯自旋电子学相关的重要研究以及最新的研究进展;总结了自旋-轨道相互作用的相关理论,并描述了采用电产生纯自旋电流,实现了石墨烯的自旋注入的相关应用。

13.1 引 言

电子具有两个自由度:电荷和自旋。这两个自由度分别在半导体物理和磁性领域得到应用,已形成大量的实际应用,也为基础科学研究提供了基础。自旋电子学是半导体物理和磁学的结合领域,这是由于自旋电子学涉及了两个自由度。1986年,Gruenberg及其合作者报道了Fe/Cr/Fe层上Fe的反铁磁耦合,其磁阻比率为百分之几[1]。1988年,Fert等采用Fe/Cr多层膜得到第一巨磁阻(GMR)约为40%,也因此开辟了自旋电子学的研究领域[2]。在此研究之后,类似的GMR现象也在Co/Cr多层膜等体系内被发现[3-4]。

GMR是一个奇特的物理特性,这是由于其MR值非常高,因此可以实现一些特殊的应用,例如放音磁头。对于GMR,Cr层起自旋通道的作用,如导电自旋在Cr层的注入和转移。相比之下,绝缘层而非导电层插入到两层铁磁性之间,从而可形成自旋相关的隧穿传输,即所谓的隧道磁阻效应(tunneling magnetoresistance (TMR) effect)。Miyazaki、Tezuka[5]及Moodera等[6]于1995年分别发现了TMR效应,其TMR值在常温下高达20%。他们将Al-O绝缘层引入到两层铁磁层间形成三明治结构。颗粒状态TMR现象也被报道[7-8],其隧道电阻存在特殊的温度依赖性,这是由于电荷效应(a charging effect)而表现出来的[9-10]。2004年,MgO单晶被引入到TMR器件中作为隧穿势垒,从而使Fe/MgO/Fe体系具有相干自旋隧穿的可能[11-12],且室温TMR值迄今为止达到了600%[13]。

除了金属自旋电子学的发展,如上面所讨论的现象,无机半导体(GaAs、Si等)的自旋电子学研究也得到了广泛关注。在这些无机半导体中,GaAs 可用于实现自旋晶体管,如 Datta 和 Das 等人所报道[14]。在该过程中,注入 GaAs 的自旋被栅极电压通过强的自旋-轨道相互作用(spin-orbit interaction)所旋转。另外,Si 是一种相对较轻的元素,具有晶格反演对称性,因此可获得自旋金属半导体场效应晶体管(spin metal-on-semiconductor field effect transistors,MOSFETs)。有关自旋注入和自旋传输的研究广受关注[15-16],该领域被称为自旋电子学的第二支柱——半导体自旋电子学。自 1999 年,分子自旋电子学作为自旋电子学的第三大组成部分受到了自旋电子学和分子电子学研究者的广泛关注。一个分子具有一个相当小的自旋-轨道相互作用。自旋-轨道相互作用被定义为所能引起自旋相干损失的相互作用,因此,为了实现量子计算系统和所谓的 Sugahara-Tanaka 型自旋,MOSFETs 需要一个具有较小自旋-轨道相互作用的材料[17]。为了促进该领域的发展,以纳米碳分子(石墨烯、碳纳米管和富勒烯)和有机分子为研究对象,进行了广泛的研究[18]。本章以石墨烯纯自旋电流为对象研究自旋电子学。

尽管有关固体内自旋注入的报道很多(金属、无机半导体和分子),但是仍然需要谨慎地确定注入是否确定发生。这是由于各种噪声信号,例如通常的 MR 测试中各向异性的 MR 信号,可能导致误判。这些因素对于分子自旋电子学的精确测量影响很大(如文献[18]所描述)。为了避免这些阻碍精确测量的因素,自旋电子学研究中引入了一个非局域化的四端测量(a non-local four-terminal measurement,NL-4T)。该 NL-4T 方法是由 Johnson 和 Silsbee 首先提出的[19],他们认为电产生的为纯自旋电流,而非自旋极化电流(spin-polarized current)。自 2002 年,Jedema 及其合作者采用实验方法提出了非局域化 MR 以来[20],该现象成为普遍现象,从而基于纯自旋电流发展了一种新型的电子器件,被称为自旋电流电子学(spin currentronics),成为研究的热点。

本章的目的在于:①解释纯自旋电流的物理本质;②介绍与石墨烯的自旋注入有关的重要成果。由于石墨烯线性能带结构的物理特性在其他文献中已有大量描述[21],本章就不做重点讨论。

13.2 理论基础及重要概念

本节将介绍自旋-轨道相互作用的本质。自旋-轨道相互作用是一个相对的效应,它来源于狄拉克方程:

$$i\frac{\partial}{\partial t}\psi(x,t) = [-i(\sum_{i=1}^{3}\boldsymbol{\alpha}_i\frac{\partial}{\partial x_i}) + \boldsymbol{\beta}m]\psi(x,t) \qquad (13.1)$$

其中

$$\boldsymbol{\alpha}_i = \begin{pmatrix} 0 & \sigma_i \\ \sigma_i & 0 \end{pmatrix}, \boldsymbol{\sigma}_1 = \begin{pmatrix} 0 & 1 \\ 1 & 0 \end{pmatrix}, \boldsymbol{\sigma}_2 = \begin{pmatrix} 0 & -i \\ i & 0 \end{pmatrix}, \boldsymbol{\sigma}_3 = \begin{pmatrix} 1 & 0 \\ 0 & -1 \end{pmatrix}$$

$$\boldsymbol{\beta} = \begin{pmatrix} 1 & 0 \\ 0 & -1 \end{pmatrix}$$

t 为时间;$\psi(x,t)$ 为波函数;x 为三维空间中的坐标;m 为质量;$\boldsymbol{\alpha}$ 和 $\boldsymbol{\beta}$ 均为 4×4 矩阵;$\boldsymbol{\alpha}_i$ 为 2×2 泡利自旋矩阵(Pauli's spin matrix)。空间坐标和时间在相对论的量子力学中是等效的,波函数具有四个组成部分(3 + 1),其中两个组成部分对应正能量解,而另外两个组成部分对应负能量解。换句话说,正能量解和负能量解分别对应电子和正电子的解。应该指出的是,我们可以用狄拉克方程为旋量描述一个自旋的自由度。在这里,电子和正电子的波函数在狄拉克方程中被混合,这是由于 σ_1 在矩阵 α_1 中的存在。为了研究狄拉克电子在非相对论极限的相互作用,需要将狄拉克哈密顿量进行适当的对角线化。从公式(13.1)中,狄拉克哈密顿量可以描述为

$$H = \begin{pmatrix} m\boldsymbol{I} & \boldsymbol{\sigma p} \\ \boldsymbol{\sigma p} & -m\boldsymbol{I} \end{pmatrix} = \boldsymbol{\alpha p} + \boldsymbol{\beta}m \qquad (13.2)$$

其中,$\boldsymbol{I}$ 为 2×2 单元矩阵;$\boldsymbol{p}$ 为动量。将 U_F 以 S 的形式表示为 $U_F = e^{+iS}$,其中,S 为厄米特(hermitian)且无明确的时间依赖性,那么正变换(the unitary transformation)可表示为

$$\psi' = e^{+iS}\psi \qquad (13.3)$$

$$i\frac{\partial}{\partial t}\psi = e^{+iS}H\psi = e^{+iS}He^{-iS}\psi' = H'\psi' \qquad (13.4)$$

其中,H' 为包含非对角成分的转换。这种转变被称为 Foldy-Wouthuysen 变换。$U_F = e^{+iS}$ 表示为

$$e^{iS} = \exp[\boldsymbol{\beta\alpha p}\theta(\boldsymbol{p})] = \cos|\boldsymbol{p}|\theta + \frac{\boldsymbol{\beta\alpha p}}{|\boldsymbol{p}|}\sin|\boldsymbol{p}|\theta \qquad (13.5)$$

狄拉克哈密顿量经过么正变换变为

$$H' = (\cos|\boldsymbol{p}|\theta + \frac{\boldsymbol{\beta\alpha p}}{|\boldsymbol{p}|}\sin|\boldsymbol{p}|\theta)(\boldsymbol{\alpha p} + \boldsymbol{\beta}m)(\cos|\boldsymbol{p}|\theta - \frac{\boldsymbol{\beta\alpha p}}{|\boldsymbol{p}|}\sin|\boldsymbol{p}|\theta) \qquad (13.6)$$

因为 a 和 b 具有 $\{a,b\} = 0$ 的关系,则

$$H' = (\cos|\boldsymbol{p}|\theta + \frac{\boldsymbol{\beta\alpha p}}{|\boldsymbol{p}|}\sin|\boldsymbol{p}|\theta)(\boldsymbol{\alpha p} + \boldsymbol{\beta}m)(\cos|\boldsymbol{p}|\theta - \frac{\boldsymbol{\beta\alpha p}}{|\boldsymbol{p}|}\sin|\boldsymbol{p}|\theta) =$$

$$(\alpha p + \beta m)\left(\cos|p|\theta - \frac{\beta\alpha p}{|p|}\sin|p|\theta\right)^2 =$$

$$(\alpha p + \beta m)\exp(-2\beta\alpha p\theta) =$$

$$\alpha p\left(\cos 2|p|\theta - \frac{m}{|p|}\sin 2|p|\theta\right) + \beta(m\cos 2|p|\theta + |p|\sin 2|p|\theta) \tag{13.7}$$

当我们选择 a 为

$$\tan 2|p|\theta = \frac{|p|}{m} \tag{13.8}$$

则非对角线元素 a 可以被消除,可得

$$H' = \alpha p\left(1 - \frac{m}{|p|}\tan 2|p|\theta\right)\cos 2|p|\theta + \beta(m + |p|\tan 2|p|\theta)\cos 2|p|\theta =$$

$$\beta\left(m + \frac{|p|^2}{m}\right)\frac{m}{\sqrt{p^2 + m^2}} =$$

$$\beta\sqrt{p^2 + m^2} \tag{13.9}$$

为了使其更具普遍性,我们引入一个电磁场:

$$\begin{cases} H = \sigma(p - eA) + \beta m + e\varphi = \beta m + \theta + \varepsilon \\ \theta = \sigma(p - eA) \\ \varepsilon = e\varphi \end{cases} \tag{13.10}$$

式中,e 为一个电荷;A 为矢量势;φ 为标量势。这里,θ 具有非对角线元素,β 和 θ 二者不可交换,而 β 和 ε 是可交换的。在这里再次考虑么正变换:

$$i\frac{\partial}{\partial t}e^{-iS}\psi = H\psi = He^{-iS}\psi' = e^{-iS}\left(i\frac{\partial}{\partial t}\psi'\right) + \left(i\frac{\partial}{\partial t}e^{-iS}\right)\psi'$$

则

$$i\frac{\partial}{\partial t}\psi' = \left[e^{iS}\left(H - i\frac{\partial}{\partial t}\right)e^{-iS}\right]\psi' = H'\psi' \tag{13.11}$$

采用 Baker-Hausdorff 公式进行三次 Foldy-Wouthuysen 变换,然后哈密顿函数可以写为

$$H' = \beta\left\{m + \frac{(p - eA)^2}{2m} - \frac{p^4}{8m^3}\right\} + e\varphi - e\frac{1}{2m}\beta(\sigma B) -$$

$$\frac{ie}{8m^2}\sigma\mathrm{rot}(E) - \frac{e}{4m^2}\sigma(Ep) - \frac{e}{8m^2}\mathrm{div}(E) \tag{13.12}$$

这就是在非相对论限制磁场时一个电子狄拉克方程的哈密顿量,且

$$-\frac{ie}{8m^2}\sigma\mathrm{rot}(E) - \frac{e}{4m^2}\sigma(Ep) \tag{13.13}$$

式(13.13)为自旋-轨道哈密顿量。当我们采用一个球对称势场$V(r)$进行简化考虑,公式(13.13)中的第一项为0。最终获得了自旋-轨道哈密顿量,表示为

$$H_{\text{spin-orbit}} = \frac{e^2}{4m} \cdot \frac{1}{r} \cdot \frac{\partial V}{\partial r} \boldsymbol{\sigma} L \tag{13.14}$$

其中,$L = rp$。

上述讨论即使在原子序数$Z \neq 1$时也是有效的。值得注意的是,这时哈密顿量与Z^4成比例。由于势场$V(r)$为原子核与核外电子之间的库仑相互作用力,相对影响下的自旋-轨道哈密顿量描述为

$$H_{\text{spin-orbit}} = \frac{Ze^2}{4mr^3} \boldsymbol{\sigma} L \tag{13.15}$$

当计算一个经典的自旋-轨道哈密顿量时,其值为公式(13.15)所得数值的两倍,这是由 Thomas 运动(Thomas precession)引起的。在这里,r与原子质量Z成反比,这是由于原子序数越大,其外层电子半径越大。因此,自旋-轨道相互作用与Z^4成比例,而且较轻元素具有较小的自旋-轨道相互作用[14]。由于石墨烯内仅有碳原子,且99%的碳原子没核自旋,碳的质子数为6,因此石墨烯内具有极好的自旋相干。

现在来解释纯自旋电流的本质。图13.1给出了纯自旋电流的概念。电流为电荷的流动,在该过程中,向上和向下自旋电流的数量相同。自旋极化电流为向上和向下自旋数量不等形成。仅存在一种自旋且这些自旋同向时才能形成完美的极化电流。纯自旋电流与上述几种情况完全不同。相同数量的向上和向下自旋电流,从而使得纯自旋电流时没有电荷流动。值得注意的是,向下的自旋所产生的向左电流与向上的自旋产生的向右电流是相等的,这归因于时间反演对称性(时间的反转会引起运动和自旋方向同时反转)。因此,只存在一个方向的自旋角动量流动,而没有电荷流动。此外,理想的纯自旋流具有时间反演对称性,产生一个无耗散流(在实验过程中,经常会获得一个扩散的纯自旋流,即为耗散流)。一个无耗散运动的例子为无摩擦谐波振子,可以用如下公式描述该运动:

$$m \frac{\mathrm{d}^2 x}{\mathrm{d}t^2} = -kx \tag{13.16}$$

$$m \frac{\mathrm{d}^2 x}{\mathrm{d}t^2} = -kx - \kappa \frac{\mathrm{d}x}{\mathrm{d}t} \tag{13.17}$$

其中,m为质量;x为位置;t为时间;k为弹簧常数;κ为摩擦系数。公式(13.16)描述了无摩擦谐振子的运动,而公式(13.17)描述了摩擦存在时

的谐振子运动。因此,公式(13.16)在 $t \rightarrow -t$ 变化下具有时间反演对称性,而公式(13.17)不具有时间反演对称性。因此,可以说具有时间反演对称性的运动为无耗散运动。

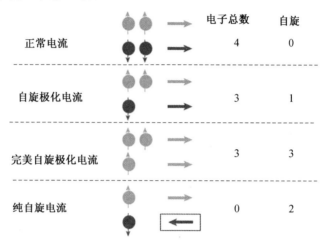

图13.1 纯自旋电流的概念[18]

根据已有文献报道,迄今为止已有几种方法实现纯自旋电流,如电学方法、动力学方法和热学方法。电学方法获得纯自旋电流是按照如下实现的:如图13.2(a)所示,一种金属丝(如Al)和两个铁磁钴电极构成装置,通过一个电路(左侧部分)获得电流并施加到装置上。Co1下方的Al线内产生自旋累积,但是该过程在非磁性连接点不会发生。左侧电路累积的自旋电流是由于电场产生的,其由于自旋的各向同性扩散,扩散至右侧电路;假设大多数自旋为向上的,那么出现向上的自旋向右扩散,如图13.2(c)所示。在这里,Al与Co2间不存在磁性和电流。因此,向下的自旋与向上的自旋数量相等,从而保持电荷中性和自旋平衡。由于向上自旋电流向右和向下自旋流向左,从而实现由电场力产生纯自旋流。当Co2自旋方向通过外磁场发生变化时,Co2处可检测到Al线内向上自旋向右传播或者向下自旋向左传播的电化学势,导致所输出电压的反向,如图13.2(d)所示。石墨烯内自旋流的产生和检测的例子如图13.3所示。采用NL-4T方法可以消除假象的影响(如各向同性MR和局部霍尔效应),这是由于检测电流中没有使用电流。这是在纯固体中采用该方法获得自旋注入的显著优势。采用电学方法获得纯自旋电流广泛应用于金属或半导体体系,例如 $Al^{[15]}$、$Cu^{[22-23]}$、$Si^{[24]}$、$GaAs^{[25]}$、室温多层或单层石墨烯[26,27]及4 K温度下单壁碳纳米管[28]。所获得的可重复性结果被用于讨论和研究这些现象所具有的物理本质。

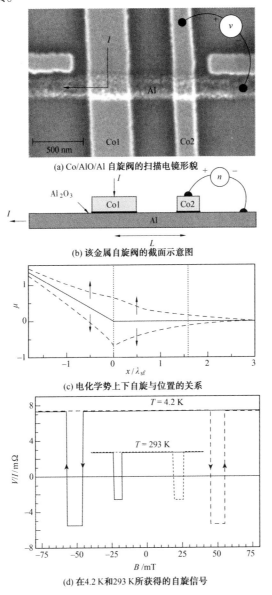

图 13.2　Co/AlO/Al 自旋阀的扫描电镜形貌、该金属自旋阀的截面示意图、电化学势上下自旋与位置的关系及在 4.2 K 和 293 K 所获得的自旋信号

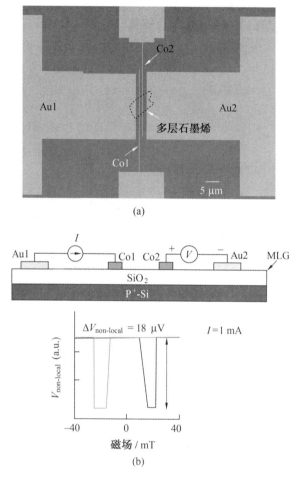

图13.3 石墨烯中自旋流产生和检测实例

13.3 纯自旋电流的试验获得及其性能

2007年对于石墨烯自旋电子学而言是重要的一年,因为这一年里发表了一些开拓性的研究成果。Ohishi等报道了在室温下采用自旋注入在多层石墨烯内获得纯自旋电流,这是室温下该体系内获得纯自旋电流的首次报道[26]。他们发现在NL-4T方法中,除了局部MR外,采用局部的方法仅能获得各向异性的MR(所谓各向异性MR为因铁磁体的磁翻转引起的伪信号,如文献[18]所列举实例)。图13.4给出了所检测到的信号;在消除杂散伪信号方面,NL-4T方法与局部方法相比具有明显的优势。随后,

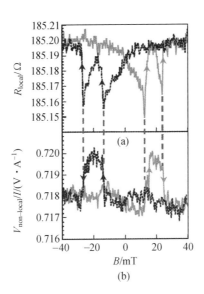

图 13.4 (a) 石墨烯自旋阀内局部磁致电阻为各向异性磁致电阻(一个假效应)。(b) 由于自旋注入、自旋传输以及石墨烯内纯自旋电流产生的非局部磁致电阻[27]

Tombros 等报道了在室温下单层石墨烯内获得了纯自旋电流[27],且报道了 Hanle 型自旋进动(图 13.5)。他们估算室温下自旋相干长度和时间分别为 1.6 μm 和 200 ps,且栅极电压和自旋信号存在依赖关系。该依赖关系在其他报道中也被发现[29],但是结果存在不一致性。从理论上而言,该依赖性可由铁磁和石墨烯间界面状态进行解释,即如果为欧姆接触,则自旋信号在狄拉克点具有最小值;如果为隧道接触,则为最大值[30]。虽然 Al-O 隧穿势垒层存在于 Co 和石墨烯之间,自旋信号在狄拉克点仍然为最小,这归因于 Al-O 存在孔隙(图 13.5)。Cho 等[29]没有真实反映栅极电压和自旋信号间的依赖性(图 13.6)。Han 等报道了自旋信号的栅极电压依赖性[31],这与传统理论相一致,得到了自旋信号在透明、带有孔隙和隧穿机制时的栅极电压依赖性,如图 13.7 所示,类似的结果也被其他研究者报道[32]。

自石墨烯的自旋注入技术建立之后,石墨烯自旋注入后的各种物理性能被研究。由于石墨烯为零带隙半导体,自旋的漂移效应(the spin drift effect)是不可被忽略的。Jozsa 等报道石墨烯内自旋漂移的定量研究[33-34]。有趣的是,石墨烯内自旋松弛不是各向同性的[35],其中,相对于平行于石墨烯片层的自旋,垂直于石墨烯片层自旋的自旋弛豫时间发生了 20% 的

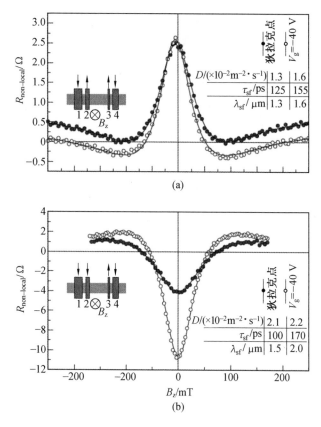

图 13.5 室温下单层石墨烯的典型 Hanle 型自旋运动信号[28]

下降(图 13.8)。Tombros 等提出 Elliot-Yafet 型自旋松弛是该现象的主要机制[35],而且这是很多有关单层和双层石墨烯的自旋弛豫机理的源头。Han 等[36]和 Yang 等[37]发现双层石墨烯内自旋弛豫时间明显大于单层石墨烯,且双层石墨烯内弛豫机制为 D'yakonov-Perel 型,其内在物理本质仍在研究当中。他们的实验结果表明,双层石墨烯内自旋寿命与载流子迁移能力(扩散系数)成反比,而单层石墨烯则表现出相反的趋势。一个可能的原因是双层石墨烯内自旋传导层被衬底保护,从而导致自旋相干变长。另一项研究也表明了双层石墨烯内自旋相干变长[38-39],因此,随着石墨烯厚度增加,自旋相干有望增强。需要指出的是,在研究自旋散射机制的过程中可实现有关石墨烯化学掺杂的研究[40-41]。研究发现,自旋相干与理论预测相比相差很远,然而为了实现石墨烯自旋器件,则需要更强的自旋相干。单层石墨烯由于具有非常良好的柔性,表现出良好的自旋相干,从而

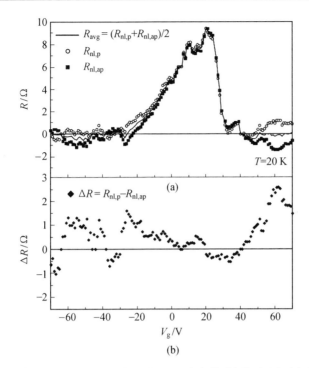

图 13.6 自旋信号对栅极电压的关系图,可以看出其对栅极电压不存在明显的依赖性[29]

在电子器件中的应用受到关注。Józsa 等仔细研究了单层石墨烯内动量和自旋散射间的关系[42],结果表明无论电子还是空穴均表现出线性关系。其研究结果指出,通过抑制载流子散射,可以在单层石墨烯内获得更好的自旋相干,而上述过程可以采用悬浮石墨烯而实现。

石墨烯纯自旋电流的另一个具有吸引力的特性在于,其自旋信号相对于偏置电流(电压)具有稳定性(或称为鲁棒性,robustness)。在常见的由非磁性金属和无机半导体制备的自旋器件中,存在一个被广泛认可的现象,即自旋信号随着偏置电流(电压)增加而出现单调下降,该现象是这些器件所要解决的首要问题。然而,石墨烯自旋信号在偏置电压变化时表现出较小的依赖性[43-45],从而使其在自旋电子学领域表现出极大的优势(图13.9)。该独特性质归因于所注入自旋具有稳定性(或称为鲁棒性)[43]。

最后,石墨烯自旋电子学另一个重要的突破在于石墨烯内自旋和电荷电流间非线性相互作用[46-47]。如图 13.2 和图 13.3 所示,自旋流的检测是采用一个铁磁电极置于检测电流的方法。然而,如果电荷和自旋电流之间存在相互作用,那么即使使用非磁性电极也可获得非局域自旋信号。Vera-Marun等成功地证实了该非线性效应的存在[46-47],也因此打开了自旋电子学新的研究领域。

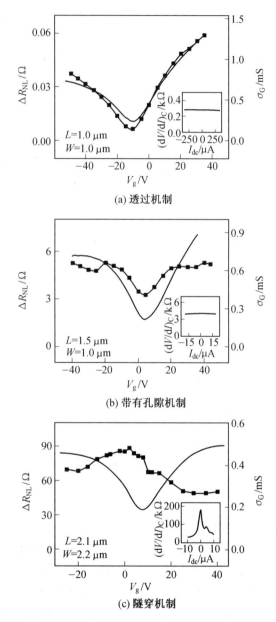

图 13.7　单层石墨烯自旋信号对栅极电压的依赖性[31]

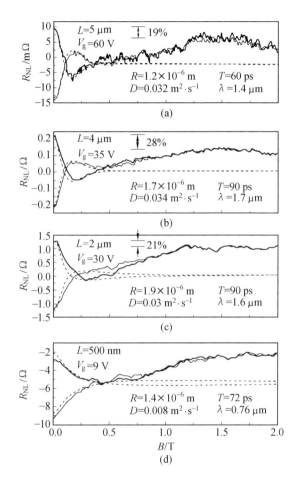

图 13.8　单层石墨烯的各向异性自旋弛豫
[垂直于石墨烯平面的自旋表现出更强的自旋弛豫(约 20%)[35]]

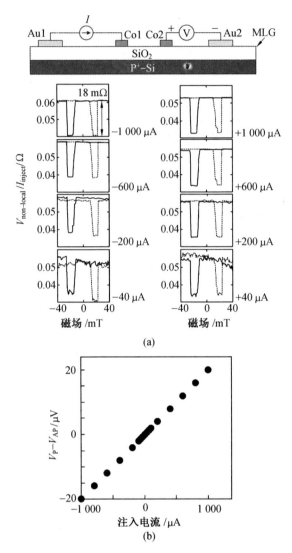

图 13.9 石墨烯自旋信号的鲁棒性

自旋信号表现出对偏置电流的线形依赖,这归因于自旋极化的鲁棒性[43]

13.4 结论及展望

自 2007 年以来,石墨烯自旋电子学作为后起之秀,无论在分子自旋电子学还是在自旋电子学方面,都取得了巨大的进步,而且在自旋传输方面展现出了一些有趣的物理特征。石墨烯自旋电子学的发展将取决于以下几方面的发展:①较长自旋相干的实现;②自旋功能新型器件的制备,如自旋 FETS 等;③化学气相沉积(CVD)生长石墨烯大规模应用于器件生产。

对于自旋相干,有关自旋相干时间和长度的数值仍然很有限,远远低于理论预测数值,且该数值仍然没有定论。因此,亟须新的方式实现石墨烯内自旋传输,进而推动石墨烯自旋电子学的发展。其中一种可能的方法为动力学自旋注入法(dynamical spin pumping method)[48]。通过电学和动力学相结合,石墨烯内自旋弛豫机理将有望进一步明确和深刻理解。为了实现与石墨烯有关的自旋器件的制备和实际应用,大面积石墨烯是不可缺少的,而 CVD 方法被认为是最有望实现大面积石墨烯应用的方法。一些研究小组已经成功实现芯片级大小石墨烯自旋器件的制备[49,50],从而为石墨烯自旋电子学开辟了一个新的研究前沿。作为一种新型的石墨烯器件,自旋功能石墨烯器件无论在理论方面(石墨烯自旋逻辑)还是未来发展方面,都为广大研究者提供了无限的可能。

13.5 参考文献

[1] BINASCH G, GRÜNBERG P, SAURENBACH F, et al. Enhanced magnetoresistance in layered magnetic structures with antiferromagnetic interlayer exchange[J]. Physical Review B 1989,39:4828.

[2] BAIBICH M N, BROTO J M, FERT A, et al. Giant magnetoresistance of (001)Fe/(001)Cr magnetic superlattices[J]. Physics Review Letters, 1988,61:2472.

[3] MOSCA D H, PETROFF F, FERT A, et al. Oscillatory interlayer coupling and giant magnetoresistance in Co/Cu multilayers[J]. Journal of Magnetism & Magnetic Materials,1991,94:L1.

[4] SCHAD R, POTTER C D, BELIEN P, et al. Bruynseraede giant magnetoresistance in Fe/Cr superlatticces with very thin Fe layers[J]. Applied Physics Letters,1994,64:3500.

[5] MIYZAKI T, TEZUKA N. Giant magnetic tunneling effect in Fe/Al_2O_3/Fe junction[J]. Journal of Magnetism & Magnetic Materials, 1995, 139: L231.

[6] MOODERA J S, KINDER L R, WONG M. Large magnetoresistance at room temperature in ferromagnetic thin film tunnel junctions[J]. Physics

Review Letters,1995,74:3273.
[7] FUJIMORI H,MITANI S,OHNUMA S. Tunnel-type GMR in metal-nonmetal granular alloy thin films[J]. Materials Science and Engineering B,1995,31:219.
[8] MITANI S,TAKAHASHI S,TAKANASHI K,et al. Enhanced magnetoresistance in insulating granular systems: evidence for higher-order tunneling[J]. Physics Review Letters,1998,81:2799.
[9] SHENG P,ABELES B,ARIE Y. Hopping conductivity in granular metals[J]. Physics Review Letter,1973,31:44.
[10] HERMAN J S,ABELES B. Tunneling of spin-polarized electrons and magnetoresistance in granular Ni films[J]. Physics Review Letters,1976,37:1429.
[11] YUASA S,NAGAHAMA T,FUKUSHIMA A,et al. Giant room-temperature magnetoresistance in single-crystal Fe/MgO/Fe magnetic tunnel junctions[J]. Nature Materials,2004,3:868.
[12] PARKIN S S P,KAISER C,PANCHULA A,et al. Giant tunnelling magnetoresistance at room temperature with MgO(100) tunnel barriers[J]. Nature Materials,2004,3:862.
[13] IKEDA S,HAYAKAWA J,ASHIZAWA Y. Tunnel magnetoresistance of 604% at 300K by suppression of Ta diffusion in CoFeB/MgO/CoFeB pseudo-spin-valves annealed at high temperature[J]. Applied Physics Letters,2008(93):082508.
[14] DATTA S,DAS B. Electronic analog of the electro-optic modulator[J]. Applied Physics Letters. 1990(56):665.
[15] OHNO Y,YOUNG D K,BESCHOTEN B,et al. Electrical spin injection in a ferromagnetic semiconductor heterostructure[J]. Nature,1999,402:790.
[16] APPELBAUM I,HUANG B,MONSMA D J. Electronic measurement and control of spin transport in silicon[J]. Nature,2007,447:295.
[17] MATSUNO T,SUGAHARA S,TANAKA M. Novel reconfigurable logic gates using spin metal-oxide-semiconductor field-effect transistors[J]. Japanese Journal of Applied Physics,2003,43:6032.
[18] SHIRAISHI M,IKOMA T. Molecular spintronics[J]. Physica E,2011,43:1295.

[19] JOHNSON M, SILSBEE R H. Interfacial charge-spin coupling: injection and detection of spin magnetization in metals[J]. Physics Review Letters, 1985, 55: 1790.

[20] JEDEMA F J, HEERSCHE H B, FILIP A T, et al. Electrical detection of spin precession in a metallic mesoscopic spin valve[J]. Nature, 2002, 416: 713.

[21] GEIM A K, NOVOSELOV K S. The rise of graphene[J]. Nature Materials, 2007, 6: 183.

[22] KIMURA T, HAMRLE J, OTANI Y. Estimation of spin-diffusion length from the magnitude of spin-current absorption: multiterminal ferromagnetic/nonferromagnetic hybrid structures[J]. Physical Review B, 2005, 71: 014461.

[23] KIMURA T, HAMRLE J, OTANI Y. Switching magnetization of a nanoscale ferromagnetic particle using nonlocal spin injection[J]. Physics Review Letters, 2006, 96: 037201.

[24] SUZUKI T, SASAKI T, OIKAWA T, et al. Room-temperature electron spin transport in a highly doped Si channel[J]. Applied Physics Express, 2011, 4: 023003.

[25] UEMURA T, AKIHO T, HARADA M, et al. Non-local detection of spin-polarized electrons at room temperature in $Co_{50}Fe_{50}$/GaAs Schottky tunnel junctions[J]. Applied Physics Letters, 2011, 99: 082108.

[26] OHISHI M, SHIRAISHI M, NOUCHI R, et al. Spin injection into a graphene thin film at room temperature[J]. Japanese Journal of Applied Physics, 2007, 46: L605.

[27] TOMBROS N, JOZSA C, POPINCIUC M, et al. Electronic spin transport and spin precession in single graphene layers at room temperature[J]. Nature, 2007, 448: 571.

[28] TOMBROS N, VAN DER MOLEN S J, VAN WEES B J. Separating spin and charge transport in single-wall carbon nanotubes[J]. Physical Review B, 2006, 73: 233403.

[29] CHO S, CHEN Y-H, FUHRER M S. Gate-tunable graphene spin valve[J]. Applied Physics Letters, 2007, 91: 123105.

[30] TAKAHASHI S, MAEKAWA S. Spin injection and detection in magnetic nanostructures[J]. Physical Review B, 2003, 67: 052409.

[31] HAN W, PI K, MCCREARY K M, et al. Tunneling spin injection into single layer graphene[J]. Physics Review Letters,2010,105:167202.

[32] POPINCIUC M, JóZSA C, ZOMER P J, et al. Electronic spin transport in graphene field-effect transistors [J]. Physical Review B, 2009, 80: 214427.

[33] JOZSA C, POPINCIUC M, TOMBROS N, et al. Electronic spin drift in graphene field-effect transistors[J]. Physics Review Letters,2008,100: 236603.

[34] JOZSA C, POPINCIUC M, TOMBROS N, et al. Controlling the efficiency of spin injection into graphene by carrier drift[J]. Physical Review B, 2009,79:081402.

[35] TOMBROS N, TANABE S, VELIGURA A, et al. Anisotropic spin relaxation in graphene[J]. Physics Review Letters,2008,101:046601.

[36] HAN W, KAWAKAMI R K. Spin relaxation in single layer and bilayer graphene[J]. Physics Review Letters,2011,107:047207.

[37] YANG T-Y, BALAKRISHNAN J, VOLMER F, et al. Observation of long spin-relaxation times in bilayer graphene at room temperature[J]. Physics Review Letters,2011,107:47206.

[38] GOTO H, KANDA A, SATO T, et al. Gate control of spin transport in multilayer graphene[J]. Applied Physics Letters,2008,92:212110.

[39] MAASSEN T, DEJENE F K, GUIMARAES M H D, et al. Comparison between charge and spin transport in few-layer graphene[J]. Physical Review B,2011,83:115410.

[40] PI K, MCCREARY K M, BAO W, et al. manipulation of spin transport in graphene by surface chemical doping[J]. Physical Review B,2009,80: 075406.

[41] MCCREARY K M, PI K, SWARTZ A G, et al. Effect of cluster formation on graphene mobility[J]. Physical Review B,2010,81:115453.

[42] JÓZSA C, MAASSEN T, POPINCIUC M, et al. Linear scaling between momentum and spin scattering in graphene [J]. Physical Review B, 2009,80:241403 (R).

[43] SHIRAISHI M, OHISHI M, NOUCHI R, et al. Robustness of spin polarization in graphene-based spin valves[J]. Advanced Functional Materials,2009,19:3711.

[44] HAN W, WANG W H, PI K, et al. Electron-hole asymmetry of spin injection and transport in single-layer graphene[J]. Physics Review Letters, 2009, 102:137205.

[45] MURAMOTO K, SHIRAISHI M, NOZAKI T, et al. Analysis of degradation in graphene-based spin valves[J]. Applied Physics Express, 2009, 3:123004.

[46] VERA-MARUN I J, RANJAN V, VAN WEES B J. Nonlinear detection of spin currents in graphene with non-magnetic electrodes[J]. Physical Review B, 2011, 84:241408 (R).

[47] VERA-MARUN I J, RANJAN V, VAN WEES B J. Nonlinear detection of spin currents in graphene with non-magnetic electrodes[J]. Nature Physics, 2012, 8:313.

[48] TANG Z, SHIKOH E, AGO H, K. et al. Dynamically generated pure spin current in single-layer graphene[J]. Physical Review B, 2013, 87:140401.

[49] AVSAR A, YANG T-Y, BAE S, et al. Toward Wafer Scale Fabrication of Graphene Based Spin Valve Devices[J]. Nano Letters, 2011, 11:2363.

[50] MAASSEN T, VAN DEN BERG J J, Ijbema N, et al. Long spin relaxation times in wafer scale epitaxial graphene on SiC(0001)[J]. Nano Letters, 2012, 12:1498.

[51] DERY H, WU H, CIFTCIOGLU B, et al. Nano spintronics based on magneto-logic gates[J]. IEEE Transaction on Electron Devices, 2012, 59:259.

[52] DERY H, DALAL P, CYWINSKI L, et al. Spin-based logicin semiconductors for reconfigurable large-scalecircuits[J]. Nature, 2007, 447:573.

第 14 章 石墨烯的纳米电机学

本章综述了利用石墨烯开发纳米尺度机械结构;介绍了石墨烯谐振器的发展以及石墨烯谐振器的制备与表征所涉及的技术;重点关注了传感器技术的系列应用。

14.1 引 言

尽管石墨烯仅有一个原子层厚度,它被认为是最强的材料[1],在较大的载荷下可以表现出极高的失效强度。此外,石墨烯具有极好的柔韧性和阻隔气体的能力[2],这使它在诸多领域具有应用,也包括了生命科学领域。因此石墨烯以其优良的机械性能在纳米微米电机学系统(nanomicroelectro-mechanical systems (NEMS/MEMS))具有优势。这些应用方面包括响应、感知等[3-4]。将石墨烯的电学性能和机械性能进行结合,我们可获得具有强大功能的器件或设备。

石墨烯为具有蜂窝状结构的单原子层,石墨烯内碳-碳键长 0.142 nm 以 sp^2 杂化方式进行结合。单层石墨烯可以通过堆垛的方式形成多层石墨烯,层间距为 0.335 nm[5]。采用 SEM 研究石墨烯发现其具有起伏。这些起伏可能是由于杂质或衬底引起的。尽管如此,与其他材料相比较,石墨烯仍具有卓越的机械性能。在接下来的叙述中,我们将介绍如何利用这些性能制备纳米机械系统。

本章包含以下内容:14.1 节为引言,14.2 节对比了石墨烯和硅两种材料;14.3 节介绍石墨烯的机械性能,包括弹性模量和泊松比;14.4 节介绍制备方法及相关内容;14.5 节将讨论石墨烯纳米谐振结构;14.6 节介绍石墨烯所构成的纳米传感器的实例;14.7 节讨论石墨烯纳米器件的发展趋势。

14.2 石墨烯与硅

单晶硅是制造 MEMS 和 NEMS 过程中具有优异性能的材料,因此比较石墨烯和硅具有重要的意义。硅材料在过去的二十年里主导了 MEMS 技术,产生了数十亿美元的市场。基于石墨烯卓越的性能,石墨烯被认为在促进 MEMS 和 NEMS 技术发展甚至替代硅材料方面具有发展前景。表 14.1 中对比了这两种材料,包含了二者的系列性能参数。石墨烯的弹性模量高于硅,这使得石墨烯可以用于制备更高频率的谐振器。例如,根据表 14.1 中所列数值,由多层石墨烯所构成的纳米束所表现出的谐振频率是相同尺寸硅材料的 3 倍。

表 14.1 硅和石墨烯材料的参数

	弹性模量/TPa	电子迁移率/($cm^2 \cdot V^{-1} \cdot s^{-1}$)	热导率/($W \cdot m^{-1} \cdot K^{-1}$)	泊松比
石墨烯	1[1]	>15 000[6]	5 000[7]	0.165[1]
硅	0.13[8]	<1 400[9]	1 500[10]	0.28[10]

石墨烯具有非常高的热导率和高束流承载能力,这使得石墨烯有望替代铜,用作电子芯片的连接线。石墨烯的高迁移率(high values of graphene mobility)使其可作为金属氧化物半导体场效应晶体管(MOSFETs)的通道。薄体硅(Thin-body silicon)是制备 NEMS 的候选材料,而且在发展以硅为基础的新型高速器件方面以及提高集成密度方面,均有很好的前景。采用绝缘衬底可制备多种结构并集成到同一个衬底上的器件,如可移动结构(the moveable structure)和 MOSFETs 的结合[11-14]。原则上,上述混合集成的方法对于石墨烯而言可以进一步简化,因为石墨烯本身就是可移动结构,可以作为传感器存在,而不需要额外的侧晶体管。石墨烯一般采用如下三种方法制备:石墨的机械剥离;SiC 衬底外延生长;不同金属衬底化学气相沉积[15-21]。通过 4HSiC 衬底上碳终端面或硅终端面表面进行热分解的方法可以获得石墨烯层。碳化硅衬底上外延生长石墨烯的最大优势在于,采用标准的微电子工艺过程可获得较大面积的石墨烯层。例如,在晶片量级的范围内获得大尺寸的电子结构[22]。采用块状石墨为原料进行机械剥离被认为是获得高质量石墨烯片的唯一方法。因此,石墨烯可以借助已成熟的硅材料 IC 技术,制备石墨烯 MEMS/NEMS。此外采用自下而上(bottom-up)的方法也可以获得石墨烯的 MEMS/NEMS。

14.3 石墨烯的力学性能

自石墨烯发现以来,研究人员针对石墨烯的弹性模量、强度和泊松比进行了大量的实验研究。该部分将讨论研究这些力学参数所涉及的实验方法。

14.3.1 弹性模量

线性材料(服从虎克定律)的弹性模量 E 被定义为拉伸应力和拉伸应变的比值,对于各向同性的材料可由以下公式表示:

$$\sigma = E\varepsilon \tag{14.1}$$

其中,ε 为单轴应变。弹性模量可通过拉伸实验测量获得:向材料某一方向施加一定的载荷,则可测量获得材料相应的位移,弹性模量可以通过载荷-位移曲线的斜率获得。对于硅 MEMS 和 NEMS 而言,为了测量尺度从几十微米到几十纳米的材料,研究者开发了多种拉伸实验装置[14,23]。原子力显微镜(AFM)已被广泛用作纳米尺度试样的测试手段[24-25]。这些测试手段可广泛应用于单晶硅制备的 MEMS 元件、碳纳米管和纳米线[23,26,27]。

弹性参数是指与应变和应力张量有关的系列刚性张量。对于单层石墨烯,这些参数是针对二维材料而言的,因此应力所用单位为 $N \cdot m^{-1}$。在近年的发展中,Lee 等在单层石墨烯上孔洞处采用 AFM 纳米压痕法测定单层石墨烯的弹性参数。该研究中所用石墨烯为氧化硅衬底上所剥离的石墨烯。所测得的力-位移图可以描述为

$$F = \sigma_0(\pi r)\left(\frac{\delta}{r}\right) + E(q^3 r)\left(\frac{\delta}{r}\right)^3 \tag{14.2}$$

式中,δ 为位移;r 为薄膜的半径;σ_0 为预张力;E 为弹性模量;q 为与泊松比 ν 有关的参数($q = 1/(1.049 - 0.15\nu - 0.16\nu^2)$)。将实验数据与公式(14.2)进行拟合,可以获得弹性模量 $E \approx 342 \, N \cdot m^{-1}$。基于该数值(除以石墨层间距约 0.335 nm),可知有效弹性模量为 1 TPa。

在其他相关研究中,如 Bunch 等[28]测量了悬浮石墨烯(suspended graphene beams)和 SiO_2 沟槽上石墨烯的弹性模量。他们采用了静电以及光学的方式测试悬浮石墨烯的共振频率。这些结构的厚度不同,介于单层石墨烯的厚度到几个纳米之间。考虑到实验的不确定性,研究结果表明弹性模量的数值为 0.5~2 TPa。这些数值是通过共振频率所得理论数据和实验数据间的拟合获得的。所获得的弹性模量数值往往高于 SOI 中硅谐振器,但是与碳纳米管接近[29]。

14.3.2 泊松比

在大多数情况下,石墨烯的泊松比被认为是 0.16,这是基于石墨内底

层平面[30]。该数据也得到了第一原理计算、分子动力学以及连续介质理论[31-34]的支撑。Wei 等针对单层石墨烯采用密度泛函理论从头计算（ab initio），从而得出二阶弹性张量的非线性弹性常数。在图14.1中给出了采用从头计算所获得的应力-应变曲线，并将其与连续非线性理论（continuum nonlinear theory）进行了拟合。连续非线性理论和从头计算结果吻合很好。所得到的泊松比为0.17。Liu 等[31]在小应变下获得的泊松比数值为0.186，非常接近 Wei 和其合作者所获得的数值。在大应变时，泊松比表现为各向异性，这取决于应变沿锯齿形方向还是扶手椅形方向。

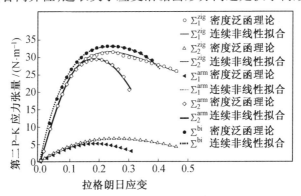

图14.1 采用从头计算（ab initio）方法在扶手椅型方向和锯齿型方向所获得的单轴应变和等双轴应变

不同曲线表示不同的应力张量。例如，Σ_1^{arm} 表示在扶手椅型方向施加单轴应变时第二 P-K 应力张量的第一部分。图中符号为采用从头计算（ab initio）所得的第二 P-K 应力和拉格朗日应变（Lagrangian strian）数据，图中曲线为采用最小二乘方法拟合所获得的结果[35]

14.3.3 机械强度

第一个实验测定单层石墨烯断裂强度是由 Lee 等[1]完成的。在其研究中，悬浮的单层石墨烯失效的最大应力采用 AFM 针尖获得。该强度数值对于无缺陷石墨烯而言为 42 N·m^{-1}。该实验过程所得强度仅仅取决于压痕，而与石墨烯膜的大小无关。该数值与第一原理所得数值很接近。Liu 等[31]采用从头计算的方式获得了理想石墨烯在两个不同方向的强度：锯齿形方向和扶手形型方向（单轴应变）。研究发现，强度具有各向异性，扶手椅形方向数值为 40.4 N·m^{-1}，而锯齿形方向为 36.7 N·m^{-1}。

14.4 石墨烯微机电系统(MEMS)制备技术

制备石墨烯纳米器件涉及一系列技术,这取决于器件所应用的领域。我们可以在已成熟的硅材料 MEMS 的自上而下的技术中得到很多启发和参考,这使得石墨烯可以与现有的制备 MEMS 和 NEMS 材料很好地结合。例如,石墨烯作为中间层形成三明治结构,进而通过 3D 芯片的设计,获得具有机械和电学性能的器件[36]。由于石墨烯为化学惰性,可以承受多种腐蚀剂。这使得石墨烯和硅进行集成具有极大的潜力[36],然而大规模生产石墨烯是该应用的前提。现阶段,最可行的石墨烯生产方式为化学气相沉积制备石墨烯并转移至其他衬底[37]。然而,转移石墨烯往往引起高的缺陷密度[38]。

14.4.1 湿法和干法刻蚀

为了获得石墨烯器件,需要涉及不同的工艺过程,即薄膜沉积、光刻和刻蚀。薄膜沉积是为了保护石墨烯或在石墨烯场效应器件中实现栅极氧化。该类材料的例子有氧化铪(HfO_2)、氧化铝(Al_2O_3)及氧化硅(SiO_2)。金属膜沉积是为了获得接触式导线和接触点(如 Au、Ti、Pd 和 Cr)。磁性材料的作用是在石墨烯自旋器件内作为源极或漏极。薄膜的制备一般通过热蒸镀、溅射、化学气相或原子层沉积。而石墨烯的图案化通常采用以下三种光刻技术:光刻印刷、电子束光刻和聚焦离子束。光刻印刷主要用于获得大于微米级的图案化,例如获得石墨烯 MEMS 谐振器。电子束光刻用于获得亚微米级结构,如石墨烯纳米带、量子点或 FET 晶体管的亚微米通道。多种光致抗蚀剂(如 S1813)在光刻过程中得到应用。双层聚 PMMA/MMA 抗蚀剂在电子束光刻技术中应用广泛,用于获得电气触点或亚微米级石墨烯图案。纳米压痕光刻技术也可用于获得石墨烯图案[39]。作为碳材料的一种,石墨烯可以很容易地被氧等离子体刻蚀。这是最广泛的一种干法刻蚀技术,该过程中需要借助具有一定硬度和抗蚀作用的模板,然而该技术在获得几十纳米宽度的石墨烯纳米带方面仍然存在困难。随后,抗蚀模板采用丙酮或异丙醇(IPA)去除。抗蚀模板去除后往往需要对石墨烯样品进行退火处理,这是由于抗蚀剂会在石墨烯表面有一定的残留。而这些残留对石墨烯器件的性能存在不利的影响,尤其是在讨论石墨烯的固有特性时。最近的研究[40]表明,采用透射电子显微镜(TEM)和拉曼光谱研究发现,物理吸附的 PMMA 在石墨烯退火过程中(约 300 ℃,H_2/Ar 气氛)不能完全去除。因此,为了恢复石墨烯加工前的原始状态,需要发展新的抗蚀剂以及新的清除方法。

虽然湿法刻蚀在制备石墨烯图案时并不常用,但该技术可用于刻蚀与石墨烯相邻近的材料。例如,BHF 可用于刻蚀石墨烯下层的氧化物,获得悬浮的石墨烯,而未影响石墨烯的结构和性能[41]。迄今为止,关于湿法刻蚀石墨烯方面还未形成系统的研究。这些研究对于将石墨烯与现有的 MEMS 和 NEMS 技术和材料相结合而言是必要的。

14.4.2 聚焦离子束

聚焦离子束(focused-ion beam,FIB)技术是一种分析手段,可以实现选择性的沉积和溅射[42]。在 FIB 技术中,一束聚焦的离子束作用于靶材,并导致靶材的溅射。商用系统中所用的离子束为镓离子。最近出现了氦离子的商用 FIB 系统,该系统中以聚焦的氦离子替代了镓离子[43]。

镓离子 FIB 可以用于膨胀石墨制备少层石墨烯[44],也可在 SiC 衬底上获得石墨烯纳米结构,形成周期性的局部缺陷[45]。聚焦的氦离子束可形成少层石墨烯的图案化,且精细程度优于电子束光刻技术,且氦离子束所引起的破坏程度小于镓离子束[43]。HIM 技术中氦离子束小于 0.7 nm,将氦离子束按照预先设计的图案路径扫描样品表面,从而选择性地去除了其溅射区域内的石墨烯材料[46]。氦离子束与电子束相比还表现出较低的邻近效应,从而可以实现高度局域化的写入。图 14.2 所示给出了纳米梁、酒杯盘状共振腔纳米结构和共振转矩纳米结构[47]。这些结构的分辨率为 1 nm/pixel(像素)。首先进行图案化处理,然后 HF 蒸汽刻蚀底层氧化物(300 nm 厚)获得上述结构。从图中可以看到,这些结构的边缘非常清晰,这是由于束流直径很小。而采用电子束光刻技术很难达到该精细程度。图 14.2(a)中纳米梁长度为 1 μm、宽度约为 200 nm,而腔体直径为 1 μm、间隙小于 30 nm。间隙可进一步缩小至 10 nm。这些结构是常见的 MEMS 和 NEMS 结构,被用作传感器或谐振器。该技术在制备纳米电子器件方面也有应用前景。该技术可以与电子束光刻技术相结合,电子束光刻技术构造几十纳米大小的形状,而 HIM 构造 10 nm 以下的结构[48]。

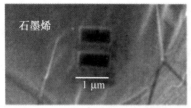
(a) 长度为1 μm，宽度为1 nm的纳米梁

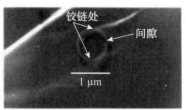
(b) 酒杯盘状共振腔纳米结构

(c) 共振转矩纳米结构

图14.2　采用氦离子束在双层石墨烯上构建的纳米结构

14.5　石墨烯纳米谐振器

目前大多数石墨烯上的NEMS器件为纳米谐振器。往往采用悬浮的石墨烯，或者采用将氧化硅层刻蚀除去的方法[49]，或者采用将石墨烯转移到预先图案化处理的衬底上[50]。这些悬梁或层采用静电或光进行驱动[28,51]，可以测量它们的共振频率和品质因子，研究并了解这些原子级别厚度的石墨烯谐振器是一个活跃的研究领域。尤其是精确获得原子级别厚度石墨烯结构的能量损失（阻尼）机理。石墨烯谐振器在射频、超灵敏质量检测以及量子信息处理领域具有潜在应用。下面将讨论这些器件是如何制备的，以及其机械性能测试（谐振频率和品质因子）所涉及的技术。

14.5.1　制造技术

一般而言，石墨烯悬浮结构制备或采用缓冲氢氟酸（buffered hydrofluoric acid，BHF）将氧化物衬底刻蚀的方法，或采用将石墨烯转移至预先带有沟槽的衬底上的方法[51]。在第一种方法中，石墨烯层采用石墨原料剥离制备，并转移到覆盖有90 nm或300 nm厚度二氧化硅的衬底上。为了达到片层悬浮，需要将其在两端夹住。在这种情况下，夹持垫采用导电触点。导电触点首先采用电子束光刻然后进行剥离工艺，通过蒸发沉积形成一薄层金属。经过导电触点沉积后,下层氧化物牺牲层采用 BHF 刻蚀掉。为了避免悬浮的结构黏附在衬底上,可以采用两种策略:临界干燥[52]和 HF 蒸汽刻蚀[47]。石墨烯悬浮之前也可以实现石墨烯结构器件构建,例如制备石墨烯纳米梁谐振器。在第二种方法中,在底层氧化物上预先构造沟槽,然后进行接触电极沉积,最后将剥离或 CVD 制备的石墨烯转移至沟槽结构之上[51]。

14.5.2 表征技术

测量石墨烯谐振器的谐振频率和品质因子时,会用到光学方法和电学方法。一个常用的电学技术为高频混合方法(high-frequency mixing approach[49]),该技术最初用来检测纳米管谐振运动[53]。在该方法中,一个直流栅极电压 V_g 和一个具有频率为 f 的高频栅压共同作用于器件,从而驱动器件,即 $V = V_g + a\cos(2\pi ft)$。另一个具有 $(f + \Delta f)$ 频率的 rf 电压施加到源极($\Delta f/f \ll 1$)。在谐振器的运动过程中,其电导率随着其与衬底间的距离而发生变化,不同频率 Δf 的混合电流即为所检测到的谐振器运动。该方法对于单层结构非常适用,但由于多层石墨烯内电导率随栅压调制响应较弱,不太适用。

图 14.3 给出了一种测量装置示意图。

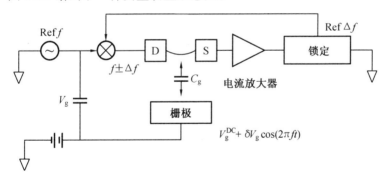

图 14.3　用于测量石墨烯纳米谐振器的谐振频率和品质因子的高频混合实验装置[49]

然而,该技术的缺点在于其测试带宽被限制在约 1 kHz[49],因此该技术不适合于较宽 rf 的测试。Xu 等[52]采用了另外一种技术,使偏离电容达到最小化,可以直接在网格分析仪上读出石墨烯谐振器的性能。所采用的驱动和检测石墨烯谐振器机械振荡的电路如图 14.4 所示,可以看出一个

直流电压 V_d 施加到漏极。一个振幅为 V_g、驱动频率为 f 的 rf 信号和直流电压 V_g 施加于栅极。采用偏置器,电流被分成 dc 和 rf 两部分,后者被输送到网格分析仪。该装置所测量的传输 $S_{12} = 50I_{rf}/V_g$,其中 I_{rf} 为电流的 rf 部分。

另一种检测石墨烯谐振器的方法为光学方法,如 Bunch 等[28]采用了该种方法。采用波长为 432 nm 的激光二极管作为光驱动。该信号通过网格分析仪进行频率调制后直接聚焦于可移动的石墨烯片层。石墨烯片层和衬底构成反射仪,谐振器的振动采用如下方式进行记录:采用一个波长为 632.8 nm 的第二束激光聚焦于谐振器并记录反射强度。在最近的研究中,拉曼光谱和 Fizeau 干涉仪也被用于检测石墨烯悬臂的机械谐振[55]。该方法的优势在于可实现亚微米(几百纳米)尺寸谐振器的检测。多层悬臂一端与金膜固定构成光学谐振腔的上层镜,下层氧化硅作为第二镜。腔的光学长度可以通过静电驱动悬臂进行调节,引起干涉条纹的变换,从而可以根据该变换检测到谐振器的振动。拉曼光谱可以用于检测谐振频率。由于光学声子的寿命(约为 1 ps) $\ll \tau_0 = \omega_0^{-1}$,其中 ω_0 为共振频率,散射光子可以提供振动悬臂内应力的信息[55]。

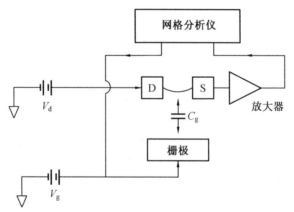

图 14.4　采用网格分析仪时的测试设置,可实现石墨烯谐振器直接读出[54]

14.5.3　振动光谱

根据形状、大小和层数的不同,石墨烯纳米谐振器的谐振频率可从几个 MHz 到几百 MHz 不等。振动梁的基本频率可根据 Stokey[64]公式表示为

$$f_0 = \frac{\lambda^2}{2\pi} \cdot \frac{1}{L^2} \cdot \sqrt{\frac{EI}{\rho A}} \tag{14.3}$$

其中,$\lambda = 4.73$；E、ρ、A、I 分别为弹性模量、密度、梁的截面积和横截面积惯性矩。如果梁承受一个小的张力 T，那么需要将上述公式增加一项，即[56]

$$f_0 = \frac{\lambda^2}{2\pi} \frac{1}{L^2} \sqrt{\frac{EI}{\rho A}} + \frac{0.28}{2\pi} T \sqrt{\frac{1}{\rho AEI}} \tag{14.4}$$

对于给定的截面积 A、长度 L、宽度 w 和厚度 t，则力矩 I 可表示为

$$I = \frac{wh^3}{12} \tag{14.5}$$

其中,w 和 h 为梁的宽度和高度。振动光谱由激发频率的变化情况决定。一般来说,对于线性响应,振动光谱与洛伦兹拟合结果吻合,因此根据该结果可以获得各种不同的参数。图 14.5(a) 给出了一个两侧夹住的单层石墨烯梁（长度 $L = 2~\mu m$，宽度 $w = 3~\mu m$）的光谱[51]。悬空的石墨烯梁采用 CVD 法在铜箔上图案化生长,然后该石墨烯被转移至预先进行沟槽处理的氧化硅衬底上。光谱的测量采用的是上述光学方法。单层石墨烯的共振频率是由梁的物理状态决定的,例如应变和石墨烯吸附物（这些吸附物可能是由于各种处理过程中引入的）。谐振频率可以根据以下公式建模[51]：

$$f_k = \frac{k}{2} \sqrt{\frac{E}{\rho_0 L^2}} \sqrt{\frac{s}{\alpha}} \tag{14.6}$$

其中,E 和 ρ_0 分别为单层石墨烯的弹性模量和密度；$k = 1,2,\cdots$；s 为面内应变；α 为吸附质量因子,即 $\alpha = \rho/\rho_0$；ρ 为包含了吸附质量的密度。公式 (14.6) 表明单层石墨烯的原始频率与 $1/L$ 存在直接关系,而该关系从图 14.5(b) 中得到了实验证明。

单层石墨烯谐振器的谐振频率还与栅极电压有关[51]。图 14.5(c) 所示给出了 $T = 30~K$ 时二者的依赖关系。在该测试过程中,采用了之前论述的混合电流的方法。当栅极电压从 0 V 变化到 8 V 时,谐振频率增加了两个指数。一个有趣的现象是,在低温时谐振频率与栅极电压间的依赖关系会变弱。虽然可以把这种依赖关系变弱归因于低温时石墨烯中张力的变化,但是具体机理我们知之甚少,而且关于降温过程中谐振器内降温与频率依赖性关系还没有定量的研究。Chen 等[49]在研究剥离石墨烯时发现了类似的趋势,他们还研究了附加质量对谐振器的影响。

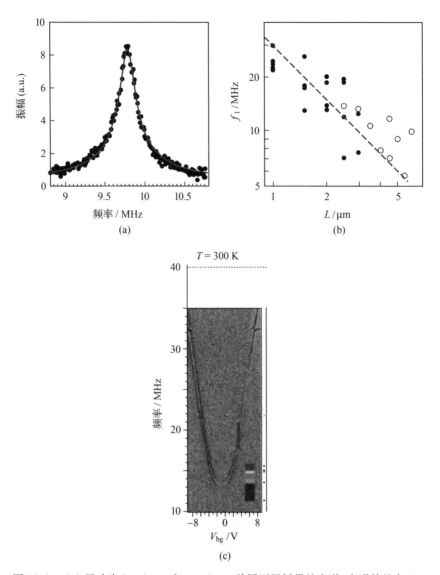

图 14.5　(a) 尺寸为 $L = 2\ \mu m$ 和 $w = 3\ \mu m$ 单层石墨烯带的光谱,表明其具有 $f_0 = 9.77\ MHz$ 的频率。(b) 第一谐振模式的频率受单层石墨烯谐振梁长度的影响。(c) 混合电流图,给出了谐振频率受栅极电压的影响[51]

14.5.4 品质因数

测试结果表明石墨烯谐振器的品质因数相对于其他半导体谐振器而言比较小。这是由于石墨烯谐振器内各种耗散机制引起的较大能量耗散,然而这一点的研究还不够深入。理论上存在几种可能的机制[57]。对于理想石墨烯片层,这些机制见表 14.2。表中还给出了对温度的依赖性(T)。可以看出,当存在温度依赖性时,它与品质因数的倒数 Q^{-1} 具有线性关系。Seoánez 等[57]认为在高温时,金属栅极和石墨烯层的欧姆损耗是主要的耗散机制。在这种情况下,品质因数的倒数与石墨烯层总电荷具有平方关系。后者可以很容易由栅极电压控制。在低温时,附加损耗(attachment losses)在耗散过程中起到很重要的作用。所谓附加损耗为能量辐射脱离谐振器的过程。另一个引起耗散的机制为"两能级系统"(two level system)谐振器耦合[57]。所谓两能级系统是指栅极处的原子或原子团具有两个简并能级结构。这些原子与振动发生相互作用引起两个简并能级间发生跃迁。在这种情况下,耗散是由于石墨烯振动与两能级系统耦合引起的。如果片层的振动很强烈,那么该种耗散就非常显著。

表 14.2　石墨烯谐振梁品质因子的倒数与耗散机制间的温度依赖性[57]

耗散机制	温度依赖性 $Q^{-1}(T)$
SiO_2 中的电荷	T
石墨烯中的电荷	T
薄膜及金属栅极	—
表面键的断裂和愈合	—
双能级系统	$A + BT$
附加损耗	—
热损耗	T

实验结果表明,品质因数可从低温时的几千变化到常温时的几百[28,49-51,58]。这些数值跟 Seoánez 等[57]的理论预测相比都相当小。这是由于实验过程中石墨烯均非理想晶体,而加工过程、起伏及掺杂这些因素均会影响到能量耗散。石墨烯的品质因数随着所施加到谐振器的静电力而降低。静电力的使用导致谐振器的弹簧指数的软化。这使得谐振器与衬底更加接近,从而导致耗散的增加[59]。品质因数随着栅极电压的增加而降低,在狄拉克点出现最小值(这归因于石墨烯在狄拉克点导电性最

低)。品质因数的最低点可解释为欧姆耗损,因为石墨烯谐振器在狄拉克点具有最大的电阻。

图 14.6 给出了品质因子与温度间关系的例子[49]。在该研究中,谐振器是由单层石墨烯两端夹住制备的。品质因子在 5 K 时达到了 14 000。这表明品质因子在低温时不具有可调性(调节栅极电压的方法)。品质因子的倒数 Q^{-1} 与温度具有指数 $\alpha = 0.36$ 的关系。然而其具体原因未进行解释。

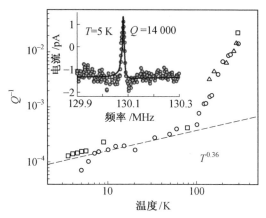

图 14.6　单层石墨烯谐振器品质因子的倒数对温度的依赖性。插图给出了温度为 5 K、品质因子为 14 000 时的振动光谱[49]

石墨烯纳米谐振器也具有非线性的共振现象[49,60],此时共振谱不再为洛伦兹谱。总体而言,阻尼过程为线性,随速率产生比例变化,而与振幅无关。Eichler 等[60]研究了石墨烯和碳纳米管谐振结构,发现了该类结构中具有强烈的非线性阻尼。他们还发现品质因子强烈依赖于振幅。他们考虑了三类谐振器:拉应力下的石墨烯、拉应力下的碳纳米管和松弛的碳纳米管。这些谐振器的动力学过程可以由以下微分方程描述:

$$m\frac{\mathrm{d}^2 x}{\mathrm{d}^2 t} = -kx - \gamma\frac{\mathrm{d}x}{\mathrm{d}t} - \alpha x^3 - \eta x^2\frac{\mathrm{d}x}{\mathrm{d}t} + A\cos(2\pi ft) \qquad (14.7)$$

其中,A 为激发振幅;k 为弹簧常数;γ 为线性阻尼系数;α 为 Duffing 非线性系数;η 为非线性阻尼系数。Eichler 等在 90 mK 时获得了品质因数 100 000,这是石墨烯谐振器迄今为止的最高值。

14.6 石墨烯纳米机械传感器

近年来,有关石墨烯生物传感器的研究受到了极大的关注,这是由于其极高的外部环境灵敏度和场效应调谐特性。石墨烯生物传感器是基于所吸附生物样品与石墨烯场效应通道间的电子交换。这种相互作用导致通道内电阻的变化,而该变化是可以被检测到的。例如,石墨烯传感器可以用于检测蛋白质、葡萄糖、气体分子[3-4,61]。这些传感器利用了单层石墨烯的高迁移率,这是由于检测的灵敏度是由石墨烯场效应通道的互通决定的。

14.6.1 生物传感膜

也许石墨烯薄膜最突出的应用就是作为生物传感膜检测DNA[62]。在该过程中,利用带有纳米孔的石墨烯将两部分含离子的溶液分隔开。当DNA分子通过该纳米孔时,测量两部分溶液间的离子电流。石墨烯膜被安装在带有孔隙的氮化硅膜上,被分割的两部分溶液采用Ag/AgCl电极进行连接。当DNA分子通过孔隙时,离子电流减小或消失。该电流的变化给出了分子大小的信息[62]。

采用石墨烯和Al_2O_3组成交替的含纳米孔的层状结构,也可用于检测DNA和DNA-蛋白复合物[63]。这种多层结构被称为纳米叠层膜(nanolaminate membrane),该结构中采用电子束形成纳米孔。该结构的优势在于低电噪声以及低pH电解质内的高灵敏度。该传感器为三端子纳米孔装置,因此对于折叠的DNA、非折叠的DNA以及RecA包覆的DNA均具有高的时间分辨率灵敏度。

14.6.2 基于石墨烯的质量传感器

当一个很小的质量δm施加到谐振梁时,其谐振频率会产生转变。为了获得该频率位移,需要涉及Rayleigh近似:梁被看作质量m和刚度k的谐波振荡器。等效质量和刚度可以由以下两个关系描述[64]:

$$m = \int_0^L \rho A \varphi(x)^2 \mathrm{d}x \quad (14.8)$$

$$k = \int_0^L EI[\varphi(x)'']^2 \mathrm{d}x \quad (14.9)$$

其中,$\phi(x)$为符合梁微分方程和响应边界条件的形状函数;ρ、E、I和A分别为梁的质量密度、弹性模量、惯性矩和梁横截面积。如果忽略静电弹簧软化(静电驱动过程会涉及),并假定附加的质量没有引起梁弹簧常数的变化,那么谐振频率可以近似地表示为

$$f_0 = \frac{\sqrt{k}}{2\pi}\frac{1}{\sqrt{m}} \tag{14.10}$$

因此,我们可以根据 δm 的微小变化计算出谐振频率的变化 δf:

$$\delta f = f_0 - f' = f_0\left(1 - \frac{1}{\sqrt{1+\frac{\delta m}{m}}}\right) \approx f_0\frac{\delta m}{2m} = R\delta m (\delta m \ll m) \tag{14.11}$$

其中,响应度 R 被定义为 $f_0/2m$。因此,为了增加梁的响应度,需要较小的质量和高谐振频率。这些要求对于碳纳米管而言很容易实现,这是由于碳纳米管质量很小(约为 10^{-21} kg)且其机械性能使得其具有非常高的谐振频率。事实上,Jensen 等[65]制备碳纳米管质量传感器达到了原子分辨率水平(0.104 MHz·zg^{-1},相当于几十个金原子)。同时,这些设备的灵敏度达到了每 Hz$^{1/2}$ 0.4 个金原子,考虑到检测是在室温下进行的,可见该数值已经非常低。在该研究中,梁采用单边固定,因此其行程比两端夹紧梁要高,即增加了动态范围。悬臂梁的另一个优点在于其相对于两端夹紧装置具有较低的耗散。

石墨烯纳米带和纳米片也可实现上述要求,所获得的器件在常温下具有较高的品质因数。到目前为止,高品质因数的石墨烯谐振器只能在低温下获得。

14.7 结论及发展趋势

考虑到石墨烯的发展时间较短,石墨烯纳米机电系统将具有广阔的发展前景。研究结果表明,石墨烯相关纳米尺度制造技术的发展使得其在开发和制备 NEMS 方面,具有其他 NEMS 材料所不具备的优势。石墨烯具有卓越的电学性能,如果与纳米尺度可移动器件相结合,有望获得可以与单原子发生相互作用的器件。碳纳米管同样具有巨大的发展潜力和非凡的机械性能。然而,由于碳纳米管的整合及其在芯片位置控制方面的问题,使其较难实现 NEMS 器件的批量制备。相反,有关石墨烯的制备方面,采用 CVD 法或 SiC 退火的方法可以获得芯片量级的石墨烯,且这方面的工

作已经进行较多。此外,石墨烯的工艺过程与其他由上而下的工艺技术具有很好的兼容性,这使得石墨烯与其他器件的集成存在优势。将石墨烯结构与具有机械功能的电子器件相结合,为我们提供了令人兴奋的结果。例如,可以采用振动模式开发和利用量子效应,研究悬浮石墨烯量子点上的单电子相互作用现象,构建可用于后 CMOS 时代的更强性能的传感器。将生物功能化的石墨烯纳米谐振器与原子级别厚度薄膜相结合,可构造出应用于不同生物技术领域的多种生物器件。

以石墨烯为基础的 NEMS 具有广阔的发展前景,而且全世界范围内正在进行的有关石墨烯的研究将为其提供广阔的应用可能性。

14.8 参考文献

[1] LEE C,WEI X,KYSAR J W,et al. Measurement of the elastic properties and intrinsic strength of monolayer graphene[J]. Science,2008,321: 385.

[2] BUNCH J S,VERBRIDGE S S,ALDEN J S,et al. Impermeable atomic membranes from graphene sheets[J]. Nano Letters,2008,8:2458.

[3] WU H,WANG J,KANG X,et al. Glucose biosensor based on immobilization of glucose oxidase in platinum nanoparticles/graphene/chitosan nanocomposite film[J]. Talanta,2009,80:403.

[4] ALWARAPPAN S,ERDEM A,LI C,et al. Probing the electrochemical properties of graphene nanosheets for biosensing applications[J]. The Journal of Chemical Physics C,2009,113:8853.

[5] DELHAES P. Graphite and precursors[M]. Florida:CRC Press,2001.

[6] NOVOSELOV K S,GEIM A K,MOROZOV S V,et al. Two-dimensional gas of massless Dirac fermions in graphene[J]. Nature,2005,438:197.

[7] BALANDIN A A,GHOSH S,BAO W,et al. Superior thermal conductivity of single-layer graphene[J]. Nano Letters,2008,8(3):902.

[8] DUAL J,MAZZA E,SCHILTGES G,et al. Mechanical properties of microstructures:experiments and theory[J]. Proceedings of SPIE,1997,3225: 12.

[9] NORTON P,BRAGGINS T,LEVINSTEIN H. Impurity and lattice scattering parameters as determined from Hall and mobility analysis in n-type silicon[J]. Physical Review B,1973,8:5632.

[10] GAD-EL-HAK M. MEMS Handbook[M]. New York:CRC press,2002.

[11] ABELE N,FRITSCHI R,BOUCART K,et al. Suspended-gate MOSFET: bringing new MEMS functionality into solid-state MOS transistor[J]. IEDM Technical Digest,2005:479.

[12] BUKS E,ROUKES M L. Stiction,adhesion energy,and the casimir effect in micromechanical systems[J]. Physical Review B,2001,63:033402.

[13] GARCIA-RAMIREZ M A,TSUCHIYA Y,MIZUTA H. Hybrid circuit analysis of a suspended gate silicon nanodot memory (SGSNM) cell[J]. Microelectronic Engineering,2010,87:1284.

[14] TSUCHIYA T,HIRATA M,CHIBA N,et al. Cross comparison of thin-film tensile-testing methods examined using single-crystal silicon,polysilicon,nickel, and titanium films [J]. Microelectromechanical Systems, 2005,14-5:1178.

[15] NOVOSELOV K S,GEIM A K,MOROZOV S V,et al. Electric field effect in atomically thin carbon films[J]. Science,2004,306:666.

[16] EMTSEV K V,BOSTWICK A,HORN K,et al. Towards wafer-size graphene layers by atmospheric pressure graphitization of silicon carbide [J]. Nature Materials,2009,8:203.

[17] DE HEER W A,BERGER C,WU X,et al. Epitaxial graphene[J]. Solid State Communications,2007,143:92.

[18] KIM K S,ZHAO Y,HOUK J,et al. Large-scale pattern growth of graphene films for stretchable transparent electrodes [J]. Nature, 2009, 457:706.

[19] LI X,CAI W,AN J,et al. Large-area synthesis of high-quality and uniform graphene films on copper foils[J]. Science,2009,324:1312.

[20] LI X,ZHU Y,CAI W,et al. Transfer of large-area graphene films for high-performance transparent conductive electrode [J]. Nano Letters, 2009,9(12):4359.

[21] OBRAZTSOV A N,OBRAZTSOVA E A,TYURNINA A V,et al. Chemical vapour deposition of thin graphite films of nanometer thickness[J]. Carbon,2007,45:2017.

[22] HASS J,JEFFREY C A,FENG R,et al. Highly ordered graphene for two dimensional electronics[J]. Applied Physics Letters,2006,89:143106.

[23] SATO K,YOSHIOKA T,ANDO T,et al. Tensile testing of silicon film

having different crystallographic orientations carried out on a silicon chip[J]. Sensors and Actuators A: Physical,1998,70:148.
[24] TANAKA M,HIGASHIDA K,HARAGUCHI T. Microstructure of plastic zones around crack tips in silicon revealed by HVEM and AFM[J]. Materials Science and Engineering A,2004,387-389:433.
[25] WONG E,SHEEHAN P,LIEBER C. Nanobeam mechanics: elasticity, strength,and toughness of nanorods and nanotubes[J]. Science,1997, 277:1971.
[26] YU M-F,LOURIE O,DYER M J,et al. Strength and breaking mechanism of multiwalled carbon nanotubes under tensile load[J]. Science, 2000,287:637.
[27] MARSZALEK P E,GREENLEAF W J,LI H,et al. Atomic force microscopy captures quantized plastic deformation in gold nanowires[J]. Proceedings of the National Academy of Sciences,2000,97:6282.
[28] BUNCH J S,VAN DER ZANDE A M,VERBRIDGE S S,et al. Electromechanical resonators from graphene sheets[J]. Science,2007,315: 490.
[29] QIAN D,WAGNER G J,LIU W K,et al. Mechanics of carbon nanotubes[J]. Applied Mechanics Reviews,2002,55:495.
[30] BLAKSLEE O L,PROCTORT D G,SELDIN E J,et al. Elastic constants of compression annealed pyrolytic graphite[J]. Journal of Applied Physics,1970,41:3373.
[31] LIU F,MING P,LI J. Ab initio calculation of ideal strength and phonon instability of graphene under tension[J]. Physical Review B,2007,76: 064120.
[32] JIANG J-W,WANG J-S,LI B. Young's modulus of graphene: a molecular dynamics study[J]. Physical Review B,2009,80:113405.
[33] ATALAYA J,ISACSSON A,KINARET J M. Continuum elastic modelling of graphene resonators[J]. Nano Letters,2008,8(12):4196.
[34] CADELANO E,PALLA P L,GIORDANO S,et al. Nonlinear elasticity of monolayer graphene[J]. Physics Review Letters,2009,102:235502.
[35] WEI X,FRAGNEAUD B,MARIANETTI C A,et al. Nonlinear elastic behavior of graphene: ab initio calculations to continuum description[J]. Physical Review B,2009,80:205407.

[36] KIM K, CHOI J Y, KIM T, et al. A role for graphene in silicon-based semiconductor devices[J]. Nature,2011,479:338.

[37] LEE Y, BAE S, JIANG H, et al. Wafer-scale synthesis and transfer of graphene films[J]. Nano Letters,2010,10:490.

[38] HUANG P Y, RUIZ-VARGAS C S, VAN DER ZANDE A M, et al. Grains and grain boundaries in single-layer graphene atomic patchwork quilts[J]. Nature,2011,469:389.

[39] WANG C, MORTON K J, FU Z, et al. Printing of sub-20 nm wide graphene ribbon arrays using nanoimprinted graphite stamps and electrostatic force assisted bonding[J]. Nanotechnology,2011,22:445301.

[40] LIN Y-C, LU C-C, YEH C-H, et al. Graphene annealing: how clean can it be[J]. Nano Letters,2012,12(1):414.

[41] MEYER J C, GEIM A K, KATSNELSON M I, et al. The structure of suspended graphene sheets[J]. Nature,2007,446:60.

[42] REYNTJENS S, PUERS R. A review of focused ion beam applications in microsystem technology[J]. Journal of Micromechanics and Microengineering,2001,11:287.

[43] BODEN S, MOKTADIR Z, BAGNALL D, et al. Focused helium ion beam milling and deposition[J]. Microelectronic Engineering,2010,88:2452.

[44] LEE K M, NEOGI A, PEREZ J M, et al. Focused-ion-beam-assisted selective control of graphene layers: acquisition of clean-cut ultra-thin graphitic film[J]. Nanotechnology,2010,21:205303.

[45] PRÉVEL B, BENOIT J M, BARDOTTI L, et al. Nanostructuring graphene on SiC by focused ion beam: effect of the ion fluence[J]. Applied Physics Letters,2011,99:083116.

[46] BELL D C, LEMME M C, STERN L A, et al. Precision cutting and patterning of graphene with helium ions[J]. Nanotechnology,2009,20:455301.

[47] MOKTADIR Z, BODEN S, MIZUTA H, et al. MME2010: 21st Micromechanics and Micro systems Europe Workshop[C]. Netherlands: Enschede,2010.

[48] HANG S, MOKTADIR Z, MIZUTA H. The 7th international conference on the fundamental science of graphene and applications of graphene-based devices (Graphene Week 2013)[C]. Germany: Chemnitz,2013.

[49] CHEN C, ROSENBLATT S, BOLOTIN K I, et al. Performance of monolayer graphene nanomechanical resonators with electrical readout[J]. Nature Nanotechnology, 2009, 4:861.

[50] GARCIA-SANCHEZ D, VAN DER ZANDE A M, SAN PAULO A, et al. Imaging mechanical vibrations in suspended graphene sheets[J]. Nano Letters, 2008, 8(5):1399.

[51] VAN DER ZANDE A M, BARTON R A, ALDEN J S, et al. Large-scale arrays of single-layer graphene resonators[J]. Nano Letters, 2010, 10(12):4869.

[52] BOLOTIN K I, SIKES K J, JIANGA Z, et al. Ultrahigh electron mobility in suspended graphene[J]. Solid State Communications, 2008, 146:351.

[53] WITKAMP B, POOT M, VAN DER ZANT H S J. Bending-mode vibration of a suspended nanotube resonator[J]. Nano Letters, 2006, 6:2904.

[54] XU Y, CHEN C, DESHPANDE V V, et al. Radio frequency electrical transduction of grapheme mechanical resonators[J]. Applied Physics Letters, 2010, 97:243111.

[55] RESERBAT-PLANTEY A, MARTY L, ARCIZET O, et al. A local optical probe for measuring motion and stress in a nanoelectromechanical system[J]. Nature Nanotechnology, 2012, 7:151.

[56] SAPMAZ S, BLANTER Y M, GUREVICH L, et al. Carbon nanotubes as nanoelectromechanical systems[J]. Physical Review B, 2003, 67:235414.

[57] SEOÁNEZ C, GUINEA F, CASTRO NETO A H. Dissipation in graphene and nanotube resonators[J]. Physical Review B, 2007, 76:125427.

[58] BARTON R A, PARPIA J, CRAIGHEAD H G. Fabrication and performance of graphene nanoelectromechanical systems[J]. Journal of Vacuum Science & Technology B, 2011, 29:050801.

[59] SINGH V, SENGUPTA S, S SOLANKI H, et al. Probing thermal expansion of graphene and modal dispersion at low-temperature using graphene nanoelectromechanical systems resonators[J]. Nanotechnology, 2010, 21:165204.

[60] EICHLER A, MOSER J, CHASTE J, et al. Nonlinear damping in mechanical resonators made from carbon nanotubes and graphene[J]. Na-

ture Nanotechnology,2011,6:339.

[61] OHNO Y,MAEHASHI K,YAMASHIRO Y,et al. Electrolyte-Gated graphene field-effect transistors for detecting pH and protein adsorption[J]. Nano Letters,2009,9:3318.

[62] GARAJ S,HUBBARD W,REINA A,et al. Graphene as a subnanometre trans-electrode membrane[J]. Nature,2010,467:190.

[63] VENKATESAN B M,ESTRADA D,BANERJEE S,et al. Stacked graphene-Al_2O_3 nanopore sensors for sensitive detection of DNA and DNA-Protein complexes[J]. ACS Nano,2012,6(1):441.

[64] STOKEY W. Shock and vibration handbook[M]. McGraw-Hill, New York,1988.

[65] JENSEN K,KIM K,ZETTL A. An atomic-resolution nanomechanical mass sensor[J]. Nature Nanotechnology,2008,3:533.

索　引

1.57 eV 能量下的 2D 拉曼谱带可拟合为 4 个洛伦兹峰　2D Raman measured with 1.57 eV laser energy and fitted with four lorentzian peaks / 160

200 K 下 G 带 FWHM 和 G 带位置与电子浓度关系　G band FWHM and G band position as function of electron concentration at 200K / 150

2D 带和 G 带强度比与电子空穴掺杂的关系　intensity ratio of 2D band to G band as function of electron and hole doping / 157

2D 谱带强度　2D band intensity / 158

3LG 样品 G 带和 G'带的原位拉曼电化学光谱　in situ Raman spectroelectrochemistry of G band and G' band of 3LG sample / 35

Anderson 杂质模型　Anderson impurity model / 216

Baker-Hausdorff 方程　Baker-Hausdorff formula / 300

Berry 相位　Berry phase / 206

C 1s 的芯能级　C 1s core level / 170

CVD 法制备石墨烯退火前后电阻率与栅极电压关系　resistivity vs gate voltage of CVD graphene before and after annealing / 272

D'yakonov-Perel 型松弛机制　D'yakonov-Perel type / 306

Debye-Waller 因子（温度因子）　Debye-Waller factor / 95

Dékány 模型　Dékány model / 49

Doniach-Sunjic 模型　Doniach-Sunjic profile / 169

Drude 公式　Drude formula / 181

D 带的谷内散射　intra-valley scattering of D band / 156

Elliot-Yafet 型自旋弛豫　Elliot-Yafet type spin relaxation / 306

Fizeau 干涉仪　Fizeau interferometry / 151

Foldy-Wouthuysen 转换　Foldy-Wouthuysen transformation / 299

Friedel 振荡　Friedel oscillations / 223

G 带位置和 FWHM 与电子空穴掺杂的关系　position and FWHM of G band as function of electron and hole doping / 150

G 带位置与单轴应变的关系　position of G band as function of uniaxial strain / 151

G 带线宽与栅极电压关系　line width of G band as function of gate voltage / 149

Hartree-Fock 方法（一种求解多粒子体系的近似方法）　Hartree-Fock / 230

Hartree-Fock 重正质量　Hartree-Fock renormalised mass / 225

HRTEM 形貌及不同成像条件下傅里叶变换　HRTEM image example and Fourier transforms for different imaging conditions / 99

Hüffner 曲线　Hüffner curve / 168

Hummers 方法　Hummers method / 47

IAM 方法和 DFT 方法获得电荷密度投影对比　difference in projected charge density between IAM and DFT / 106

IL 电解质和 HOPG 阳极的时间演化　time evolution of IL electrolyte and HOPG anode / 82

Klein 隧穿　Klein tunnelling / 178

Landauer-Buttiker 形式　Landauer-Buttiker formalism / 213

Landau 能级（朗道能级）　Landau level / 115

Lerf-Klinowski 模型　Lerf and Klinowski model / 49

Lippmann-Schwinger 公式　Lippmann-Schwinger equation / 217

Monkhorst-Pack 体系　Monkhorst-Pack scheme / 249

PMMA 的末端随机断裂以及共价键的形成　terminal and random scission of PMMA and formation of covalent bond / 258

RKKY 相互作用　Ruderman-Kittel-Kasuya-Yosida（RKKY）interactions / 216

SEM 获得的石墨烯晶粒的形状　examples of domain shapes imaged by SEM / 29

Shubnikov-de Haas 振荡开启场　onset field of Shubnikov-de Haas oscillations / 181

SiO_2 衬底上剥离石墨烯　exfoliated grapheme on SiO_2/ 116

Thomas-Fermi 屏蔽矢量　Thomas-Fermi screening vector

X 射线光电子能谱　X-ray photoelectron spectroscopy（XPS）/ 3

K 点附近抛物线形结构　parabolic band structure near K point / 153

埃米级间距　Angström fractions / 119

埃瓦尔德球　Ewald sphere / 93

半导体衬底上外延生长样品的形貌、完整性及电子结构　morphology, perfection and electronic structure epitaxially grown on semiconductor / 119

索　引

半峰宽　full width at half maximum（FWHM）/ 13
贝纳尔堆垛多层石墨烯　Bernal stacked multilayers / 32
贝纳尔堆垛石墨烯　Bernal stacked graphene / 7
变程跳跃　variable range hopping（VRH）/ 55
波恩-奥本海默近似　Born-Oppenheimer（BO）approximation / 149
玻耳兹曼常数　Boltzmann constant / 55
玻耳兹曼动力学方程　Boltzmann kinetic equation / 184
玻色-爱因斯坦凝聚　Bose-Einstein condensation / 179
玻碳电极　glassy carbon electrode（GCE）/ 83
泊松比　Poisson ratio / 318
不规则库仑震荡　irregular Coulomb oscillation / 281
不同 MMA-石墨烯复合物的价带结构　band structures of various MMA-graphene complexes / 251
不稳定性及对称性打破　instability and broken symmetry / 195
布里渊区　Brillouin zone / 144
采用 514 nm 和 633 nm 激光下谱图随层数的演变　evolution of spectra excited by 514 and 633 nm laser lines with number of layers / 160
采用 ARPES 垂直于布里渊区方向所获得 π 带　dispersion of π bands measured with ARPES perpendicular to direction of Brillouin zone / 6
采用 $C(v)$ 数据分析示意图　schematic illustration of data analysis using $C(v)$ / 285
插层形成　formed by intercalation / 45
常压化学气相沉积　atmospheric pressure chemical vapour deposition（APCVD）/ 24
场发射共振　field emission resonances（FERs）/ 128
场效应晶体管　field effect transistor（FET）/ 84
场效应晶体管的红外光谱　field-effect transistor infrared microspectroscopy / 209
场效应迁移率　field effect mobility / 76
超声　ultrasonification / 45
冲猾导　ballistic conduction / 195
传递积分矩阵(传递矩阵积分)　transfer integral matrix / 207
传感器　sensors / 60
传输的多体效应　many-body effects of transport properties / 221

传输性能的多体效应　many-body effects of transport properties / 221
传输性质　transport properties / 178
纯自旋流　pure spin current / 302
磁传输　magneto transport / 209
单壁碳纳米管　single-walled carbon nanotubes(SWCNT) / 167
单层石墨烯　monolayer graphene (MLG) / 4
单层石墨烯的各向异性自旋弛豫　anisotropic spin relaxation in single-layer graphene / 309
单层石墨烯内典型的 Hanle 型自旋进动信号　typical Hanle-type spin precession signals in single-layer graphene / 305
单层石墨烯谐振器温度依赖性倒数　temperature dependence of inverse for monolayer grapheme resonators / 328
单电荷传输　single-charge transport / 266
单电荷隧穿　single-charge tunnelling / 266
单晶 SiC 多晶 SiC 衬底的热分解　thermal decomposition of single crystal SiC polycrystalline SiC substrates / 1
单晶 SiC 高真空热分解　ultrahigh vacuum (UHV) thermal decomposition of single-crystal SiC / 2
弹性参数　elastic parameters / 318
弹性模量　Young's modulus / 317
弹性散射过程　elastic scattering process / 141
氮化硼　boron nitride / 36
导电性　conductivity / 205
倒易空间方法　reciprocal-space methods / 93
等离子体刻蚀技术　plasma etching technique / 320
低能电子显微术　low-energy electron microscopy (LEEM) / 7
低能电子衍射　low-energy electron diffraction (LEED) / 2
低维碳材料系统光电子发射　low-dimensional carbon systems photoemission / 167
低压化学气相沉积　low-pressure chemical vapour deposition (LP-CVD) / 29
狄拉克点　Dirac points / 16
狄拉克方程　Dirac equation / 270
狄拉克费米子　Dirac fermions / 178

狄拉克锥　Dirac cone / 10
第一性原理从头计算　ab initio calculation / 118
点缺陷的扫描隧道谱　scanning tunnelling spectroscopy (STS) of point defects / 131
电传输　electrical transport / 53
电导和互导　conductance and transconductance graphs / 283
电荷局域化　charge localization / 282
电荷压缩　charge compressibility / 230
电化学剥离石墨烯　electrochemical exfoliation graphene / 75
电化学剥离制备石墨烯示意图　diagram of synthesis of grapheme via electrolytic exfoliation / 81
电化学充电　electrochemical charging / 35
电化学顶栅法　electrochemical top gating method / 150
150 电化学还原　electrochemical reduction / 83
电化学势　electrochemical potential / 267
电解剥离　electrolytic exfoliation / 77
电介质工程　dielectric engineering / 192
电偶极子　electric dipoles / 242
电学性质　electrical properties / 178
电中性点　charge neutral point / 182
电子掺杂　electron doping / 116
电子传输　electronic transport / 178
电子-电子相互作用　electron-electron interactions / 157
电子-辐射相互作用　electron-radiation interaction / 143
电子光学畸变　electron optical aberrations / 97
电子空穴坑　electron-hole puddles / 192
电子-声子耦合　electron-photon coupling / 116
电子-声子散射　electron-phonon scattering / 167
电子-声子相互作用　electron-phonon interaction / 116
电子束光刻　electron beam lithography / 320
电子显微研究　electron microscopic studies / 102
电子性质和 C 1s 芯能级　the electronic properties and C 1s core level / 170
电子衍射分析　electron diffraction analysis / 91
冻结脉冲散射　frozen ripple scattering / 184

对比传递函数　contrast transfer function (CTF) / 97

对称和反对称声子涉及的双共振拉曼过程　double-resonance Raman processes involving symmetric and antisymmetric phonon / 159

对称性破坏　broken symmetry / 195

多重散射机制　multiple scattering mechanisms / 187

惰性衬底沉积片层形貌、完整性及电子结构　morphology, perfection and electronic structure flakes deposited on inert substrates / 119

二阶拉曼模式　second-order Raman modes / 154

二维倒易晶格　two-dimensional reciprocal lattice / 94

二维电子气长程行为的衰竭　breakdown of results on long-range behaviour in two-dimensional electron gas / 216

二元碰撞模型　binary collision model / 36

反常量子　anomalous quantum / 207

反斯托克斯过程　anti-Stokes process / 143

反铁磁有序　antiferromagnetic ordering / 216

非弹性散射过程　inelastic scattering process / 141

费米黄金定律　Fermi's golden rule / 142

费米能　Fermi energy / 3

费米能级　Fermi level / 114

费米能级上下 LDOS 的空间分布　spatial distribution of LDOS below and above Fermi level / 126

费米速度　Fermi velocity / 146

分数量子霍尔态　fractional quantum Hall states / 179

扶手椅型和锯齿型方向单轴应变的应力应变第一性原理从头计算　stress-strain ab initio calculation for uniaxial strain in armchair and zigzag directions / 319

干法刻蚀　dry etching / 320

高定向热解石墨　highly oriented pyrolytic graphite (HOPG) / 114

高频混合法　high-frequency mixing approach / 323

高迁移率的物理现象　physical phenomena in highmobility / 195

高迁移率石墨烯　graphene towards high mobility / 195

格林函数　Green's function / 217

各向异性磁阻　anisotropic MR / 298

谷内散射机理　intravalley scattering mechanism / 156

光伏　photovoltaics／58
哈密顿模型　model Hamiltonian／205
还原的氧化石墨烯的结构缺陷　structural defect in reduced graphene oxide（RGO）／49
氦离子显微镜　helium ion microscope（HIM）／321
核磁共振光谱　nuclear magnetic resonance（NMR）spectroscopy／48
虎克定律　Hook's law／318
化学传感器　chemical sensors／1
化学法制备石墨烯　chemically derived graphene／45
化学分析电子能谱　electron spectroscopy for chemical analysis（ECSA）／173
化学计量学　stoichiometry／48
化学气相沉积　chemical vapour deposition（CVD）／1
化学势　chemical potential／172
化学势和电荷压缩　chemical potential and charge compressibility／230
化学态鉴定　chemical state identification／169
环境压力条件　ambient pressure conditions／1
缓冲层衍射　buffer-layer diffraction／7
灰度再现图　greyscale rendition illustration／285
混合金属性单壁碳纳米管　metallicity-mixed SWCNT／171
获得纯自旋流及物理性质的实验　experiments for generating pure spin current and physical properties／304
霍尔迁移率　Hall mobility／14
霍尔条形状　Hall-bar geometry／13
机械剥离　mechanical exfoliation／45
机械剥离石墨烯　mechanically cleaved graphene（MCG）／49
机械强度　mechanical strength／319
计算石墨烯声子色散关系　calculated phonon dispersion relation of graphene／145
价带电子结构　valence-band electronic structure／174
键合环境检测　bonding environments inspection／173
角分辨光电子能谱　angle-resolved photoemission spectroscopy（ARPES）／166
解析势能分子动力学模拟　analytical potential molecular dynamics simula-

tion / 36
金属氧化物半导体场效应晶体管　metal oxide semiconductor field effect transistors (MOSFETs) / 298
紧束缚计算　tight-binding (TB) calculation / 277
紧束缚模型　tight-binding model / 277
近边 X 射线结构精修　near-edge x-ray and fine structure (NEXAFS) / 52
晶格无序散射　lattice disorder scattering / 186
晶界　grain boundaries / 4
精确的电导率自洽波恩近似　exact self-consistent Born approach for conductivity / 214
静电力　electrostatic force / 327
局部霍尔效应　local Hall effect / 302
局部态密度　local density of states (LDOS) / 249
具有 SW 缺陷片层在不同空间分辨率的形貌　sheet with SW defect at various spatial resolutions / 100
具有不同 C 同位素层组成 3LG 样品的拉曼图谱　Raman spectra of 3LG sample composed from layers with different C isotopes / 34
聚合物残留去除　removal of polymer residues / 188
聚甲基丙烯酸甲酯　poly(methyl methacrylate) (PMMA) / 27
聚焦离子束技术　focused-ion beam (FIB) technology / 321
聚碳酸酯　polycarbonate (PC) / 80
绝缘-半导体-半金属过渡　insulator-semiconductor-semimetal transition / 55
科恩反常　Kohn anomaly / 223
可压缩性量子点　compressible quantum dots / 283
库仑菱形结构　Coulomb diamonds / 269
库仑屏蔽　Coulomb screening / 221
库仑散射　Coulomb scattering / 183
库仑相互作用　Coulomb interactions / 301
库仑阻塞现象　Coulomb blockade / 267
快速傅里叶变换分析　fast Fourier transform analysis / 119
拉曼光谱　Raman spectroscopy / 141
拉曼散射原理　principles of Raman scattering / 141
劳厄区　Laue zone / 94
冷墙反应器　cold-wall reactor / 26

离子液体辅助电化学剥离　ionic liquid-assisted electrochemical exfoliation / 79

理化结构　physicochemical structure / 48

锂离子电池　lithium ion batteries / 29

粒子隧穿势垒示意图　schematic view of particle tunneling from potential barrier / 212

量子霍尔机制　quantum Hall regime / 282

量子霍尔平台　quantum Hall plateaus / 213

量子霍尔效应　quantum Hall effect / 195

量子霍尔效应下量子点的可压缩性　compressible quantum dots in quantum Hall regime / 283

临界点干燥　critical point drying / 194

磷酸三甲酯(TMP)基电解质　trimethylphosphate (TMP)-based electrolyte / 79

硫酸插层石墨剥离示意图　schematic depiction of exfoliation of graphite intercalated with sulfuric acid / 46

氯酸钾　potassium chlorate ($KClO_3$) / 46

锰酸钾　potassium permanganate ($KMnO_4$) / 47

密度泛函理论　density functional theory (DFT) / 52

纳米力学传感器　nanomechanical sensors

纳米谐振器　nanoresonators / 322

逆热力学密度　inverse thermodynamic density / 230

镍　nickel / 26

泡利不相容原理　Pauli exclusion principle / 267

硼氢化钠　sodium borohydride ($NaBH_4$) / 48

膨胀的氧化石墨　exfoliated graphite oxide / 45

偏置电流/电压　bias electric current/voltage / 307

偏置双层的带隙　band gap of biased bilayer / 220

偏置双层石墨烯带隙　biased bilayer grapheme band gap / 205

品质因子　quality factor / 327

平行板电容近似　parallel-plate capacitor approximation / 181

平均场理论　mean-field theory / 225

平均自由程　mean free path / 241

普朗克常数　Planck's constant

其他长程机制　other long-range mechanisms / 184
气相化学刻蚀过程　gas-phase chemistry etching process / 278
氢传感器　hydrogen sensing / 61
氢等离子体　hydrogen plasma / 48
氢氟酸缓冲溶液　buffered hydrofluoric acid（BHF）/ 193
缺失0、1和2个原子时缺陷形貌　elementary defects observed with 0, 1 and 2 missing atoms / 101
缺陷辅助的双声子过程　two-phonon defect-assisted process / 161
染料敏化太阳能电池　dye-sensitised solar cells（DSSC）/ 58
热导率　thermal conductivity / 269
热力学态密度的Hartree-Fock逆转换　Hartree-Fock inverse thermodynamic density-of-states / 225
热力学态密度的随机相位近似逆转换　random phase approximation inverse thermodynamic density of states / 206
瑞利近似　Rayleigh approximation / 329
弱相位物体近似　weak phase object approximation / 93
三角变形（三角翘曲）　trigonal warping / 209
散射源　sources of scattering / 241
扫描电子显微镜　Scanning electron microscopy（SEM）/ 27
扫描隧道显微镜　scanning tunnelling microscopy（STM）/ 105
少层石墨烯　few-layer graphene（FLG）/ 24
生物传感器　biological sensors / 329
生物传感器膜　biosensing membranes / 329
湿法刻蚀　wet etching / 320
石墨　graphite / 75
石墨箔片　graphite foils / 77
石墨和剥离石墨烯形貌及其表面粗糙度　digital image of as-received and exfoliated graphite and surface roughness parameters / 77
石墨片层及分散石墨烯的照片及示意图　schematic illustration and photograph and graphite flakes and dispersed graphene sheets / 81
石墨上的石墨烯　graphene on graphite / 114
石墨烯　graphene / 1
石墨烯边缘处碳原子分布的光谱分析　spectroscopic analysis of carbon atomic configurations at graphene edge / 105

索 引

石墨烯表面固体原子的退火温度　annealing temperature of solid atoms at graphene surface / 2
石墨烯单电子晶体管　graphene single-electron transistors
石墨烯的传输性　transport properties of graphene
石墨烯的电子结构　electronic structure of graphene / 269
石墨烯的电子衍射分析　electron diffraction analysis graphene / 91
石墨烯的恒电流 STM 形貌及零栅压下 d/dV 曲线　constant current STM topograph and d/dV spectrum of grapheme at zero gate voltage / 117
石墨烯的锯齿型边缘　graphene edge with zigzag orientation / 103
石墨烯的原子结构及其价带导带的三维形貌　atomic structure of graphene and three-dimensional image shows conduction and valence bands / 271
石墨烯的原子力显微镜形貌　atomic force microscopy image of graphene / 12
石墨烯和氧化石墨烯采用[10-10]峰归一化后强度对比　intensities for graphene vs graphene oxide normalised to the [10-10] peaks / 96
石墨烯基材料的应用　applications of graphene-based materials / 84
石墨烯基锂离子电池材料与性能　graphene-based lithium-ionbatterymaterials and properties / 60
石墨烯基质量传感器　graphene-based mass sensors / 329
石墨烯及钾掺杂石墨烯的电导率与栅极电压关系　conductivity vs gate voltage for pristine graphene and potassium-doped graphene / 252
石墨烯及吸附钛的石墨烯的电导率与栅极电压关系　conductivity vs gate voltage for pristine graphene, graphene with titanium adsorbates / 252
石墨烯-聚苯乙烯磺酸钠　graphene-polystyrene sulfonate (PSS)
石墨烯力学性能　graphene mechanical attributes / 317
石墨烯纳米带　graphene nanoribbons (GNR) / 82
石墨烯纳米带的 STM 和 STS　STM and STS on grapheme nanoribbons (GNR) / 132
石墨烯纳米电机学　graphene nanoelectromechanics (NEMS) / 316
石墨烯纳米梁谐振器温度依赖性倒数　temperature dependence of inverse of graphene resonator beam / 327
石墨烯内电子传输吸附剂效应　electronic transport adsorbents effect in graphene / 242
石墨烯内声子　phonons in graphene / 144

347

石墨烯片的初步形成及表面划痕的 SEM 形貌　SEM images of initial formation of graphene flakes and schematics of scratched surface / 29
石墨烯片和石墨的隧道谱以及 Landau 能级峰位置　tunnelling spectrum of grapheme flake vs graphite and STS and Landau level peak positions / 115
石墨烯器件　graphene devices / 266
石墨烯器件边缘通道示意图　schematic illustration of edge channels in graphene device / 285
石墨烯修饰电极　graphene-modified electrode / 84
石墨烯与晶界的环形暗场 STEM 形貌　annular dark field STEM images of graphene and grain boundaries / 102
石墨烯中的电子传输　electronic transport in graphene / 134
石墨烯转移至 Si/SiO_2 后的光学形貌（石墨烯为 Ni 上生长）　optical micrographs of grapheme transferred to Si/SiO_2 growth / 25
石墨烯自旋电子学　graphene spintronics / 297
石墨烯自旋阀的局部和非局部磁致电阻　local and non-local magnetoresistance in grapheme spin valve / 305
石墨烯自旋信号的鲁棒性　robustness of spin signals in graphene / 307
石墨柱电极　graphite column electrode (GCE) / 78
手性　chirality / 205
双层石墨烯　bilayer graphene / 9
双层石墨烯上纳米结构氦离子束图刻　helium ion patterning of nanostructures on bilayer graphene / 321
双端电流的温度依赖性　temperature dependence of two-terminal current / 55
双共振过程　double resonance process / 154
双共振机制　double-resonance mechanisms / 156
四波段连续模型响应函数　four-band continuum model response function / 221
随机相位近似　random phase approximation / 206
态密度　density of states (DOS) / 55
碳 sp^2 杂化系统　carbon sp^2 hybridised systems / 48
碳单原子层/石墨单原子层　carbon monolayer/graphite monolayer / 113
碳化硅的碳终端面　c-terminated SiC / 119
碳纳米管　carbon nanotubes / 48

碳源料　carbon feedstocks / 27
提高载流子迁移率的方法　approaches to increase carrier mobility / 188
铁磁电极　ferromagnetic electrode / 307
铁磁有序　ferromagnetic ordering / 216
同位素标记　isotope labeling / 32
铜箔上 N 掺杂石墨烯的 STM 形貌以及氮原子的 d/dV 曲线　STM image of n-doped grapheme on copper foil and d/dV graphs taken on N atom / 132
铜衬底上大尺寸石墨烯片　graphene with large domain sizes on copper / 27
铜催化制备样品(石墨烯片)大小、形状和层数　examples of Cu-catalysed CVD domain sizes, shapes and layer numbers / 28
铜单晶　copper single crystals / 30
透明导体　transparent conductors / 27
透射电子显微镜　transmission electron microscopy(TEM) / 91
退火　annealing / 2
退火参数　annealing parameters / 254
退火后含有 PMMA 残留的 CVD 制备石墨烯的 TEM 形貌　TEM image of CVD grapheme with PMMA residue after annealing / 255
退火后拉曼 2D 带发生蓝移　blueshift of Raman 2D band after annealing / 250
退火温度　annealing temperature / 254
外延生长多层石墨烯的两种 moiré 结构的 STM 形貌　STM topographs of two moiré patterns on multilayer epitaxial graphene / 11
外延生长石墨烯 Landau 能级的隧穿谱　tunnelling spectroscopy of Landau levels of epitaxial graphene / 115
外延石墨烯　epitaxial graphene / 1
弯曲声子　flexural phonons / 187
网格分析仪实验设定　measurement setup using network analyser / 323
微波辐射　microwave radiation / 79
微扰理论　perturbation theory / 142
污染吸附　contaminative adsorptions / 99
无标记核酸传感器　label-free nucleic acids sensors / 63
无缺陷双声子过程　two-phonon defect-free process / 144
吸附对场效应晶体管的影响　influence of adsorbates on field effect transistors / 248

吸附剂作用　adsorbents effect / 31
吸附相互作用　interaction of adsorbates / 242
线性响应函数　linear response function / 222
相关函数　correlation function / 285
相同位置处 d/dV 曲线和狄拉克点的能量位置　d/dV spectra taken at same point and energy position of Dirac point / 118
像差校正 TEM 和 STEM 所获得缺陷　defects observed by aberrationcorrected TEM and STEM / 97
校正和未校正情况下 TEM 对比传递函数　contrast transfer function of TEM for corrected and uncorrected cases / 98
谐波振荡器　harmonic oscillator / 329
谐振频率　resonant frequency / 323
信噪比　signal-to-noise ratio / 246
悬浮石墨烯　suspended graphene / 96
悬浮石墨烯层的粗糙度及层法线的变化　roughness of suspended sheets and variation in surface normal and patterns / 96
衍射分析　diffraction analysis / 91
赝自旋相关部分　pseudospin-dependent part of Hartree-Fock / 226
阳极电化学插层　anodic electrochemical intercalation / 78
阳极氧化　anodic oxidation / 81
氧化石墨烯　graphene oxide (GO) / 48
氧化石墨烯和还原氧化石墨烯的应用　applications of graphene oxide (GO) and reduced graphene oxide (RGO) / 57
氧化铟锡　indium tin oxide (ITO) / 84
一阶拉曼　first-order Raman modes / 144
以费米能级的态密度为单位的双层石墨烯静态响应　bilayer graphene static response in units of density of states at Fermi energy / 224
阴极电化学插层　cathodic electrochemical intercalation / 78
用于测量谐振频率的高频混合实验设置　high-frequency mixing experimental setup used to measure resonant frequency / 323
有机光伏器件　organic photovoltaic devices (OPV) / 58
有效质量重正　effective mass renormalisation / 225
有效质量重正化　effective mass renormalisation / 225

与金属性有关的信号及谱线概述　main panel overview of signal and spectrum corresponding to metallicity / 171
原子分辨率图像　atomic resolution imaging / 106
原子晶格间距　atomic lattice spacing / 119
原子力显微镜　atomic force microscopy（AFM）/ 12
远程界面声子散射机制　remote interfacial phonon（RIP）scattering mechanism / 187
约束控制升华法　confinement controlled sublimation(CCS) / 14
载流子　charge carriers / 13
载流子迁移率　carrier mobility / 13
栅极依赖性及灰度再现　gate dependence and grayscale rendition / 285
栅极影响因子　gate influence factor / 267
栅压　gate voltage / 116
整数量子霍尔效应　integer quantum Hall effect（IQHE）/ 195
置换掺杂　substitutional doping
中间能隙态共振散射　midgap states resonant scattering / 184
周期性堆叠的多层膜　periodically stacked multilayers / 31
转移过程导致的金属和分子吸附　transfer-induced metal and molecule adsorptions / 244
转移过程引起的金属和分子吸附　transfer-induced metal and molecule adsorptions / 244
准粒子干涉图样　quasiparticle interference patterns / 132
紫外光电子能谱　ultraviolet photoemission spectroscopy(UPS) / 170
自洽能　self-consistent energy / 227
自旋相干　spin coherence / 297
自旋注入技术　spin injection technique / 302
总基态能量　total ground-state energy / 230
纵向光学(LO)边界声子　longitudinal optical（LO）zoneboundary phonons / 145
纵向声子　longitudinal acoustic phonons / 144
最稳定配置结构的 MMA 的俯视和侧视图　top and side views of most stable confi gurations for MMA / 250